WILLIAM F. MAAG LIBRARY
YOUNGSTOWN STATE UNIVERSITY

Advances in

Heterocyclic Chemistry

Volume 26

Editorial Advisory Board

R. A. Abramovitch
A. Albert
A. T. Balaban
S. Gronowitz
T. Kametani
C. W. Re
Yu. N. Shein
H. A. S
M. Ti

Advances in

HETEROCYCLIC CHEMISTRY

Edited by

A. R. KATRITZKY

A. J. BOULTON

School of Chemical Sciences
University of East Anglia
Norwich, England

1980

Volume 26

ACADEMIC PRESS

A Subsidiary of Harcourt Brace Jovanovich, Publishers
New York London Toronto Sydney San Francisco

COPYRIGHT © 1980, BY ACADEMIC PRESS, INC.
ALL RIGHTS RESERVED.
NO PART OF THIS PUBLICATION MAY BE REPRODUCED OR
TRANSMITTED IN ANY FORM OR BY ANY MEANS, ELECTRONIC
OR MECHANICAL, INCLUDING PHOTOCOPY, RECORDING, OR ANY
INFORMATION STORAGE AND RETRIEVAL SYSTEM, WITHOUT
PERMISSION IN WRITING FROM THE PUBLISHER.

ACADEMIC PRESS, INC.
111 Fifth Avenue, New York, New York 10003

United Kingdom Edition published by
ACADEMIC PRESS, INC. (LONDON) LTD.
24/28 Oval Road, London NW1 7DX

LIBRARY OF CONGRESS CATALOG CARD NUMBER: 62–13037

ISBN 0–12–020626–9

PRINTED IN THE UNITED STATES OF AMERICA

80 81 82 83   9 8 7 6 5 4 3 2 1

Contents

CONTRIBUTORS vii

PREFACE ix

Heterocyclic Betaine Derivatives of Alternant Hydrocarbons
CHRISTOPHER A. RAMSDEN

I. Introduction 3
II. Heterocyclic Betaines Isoelectronic with Odd Alternant
Hydrocarbon Anions 4
III. The Chemistry of Heterocyclic Mesomeric Betaines 10
IV. The Electronic Structure of Mesomeric Betaines 74
V. Pericyclic Reactions of Mesomeric Betaines 89
VI. Heterocyclic Betaines Isoelectronic with Even Alternant
Hydrocarbon Dianions 97
VII. The Chemistry of Heterocyclic Betaines Derived from Even Alternant
Hydrocarbon Dianions 100
VIII. Conclusion 104
IX. Appendix Added in Proof 105
References Added in Proof 113

Thiocoumarins
O. METH-COHN AND B. TARNOWSKI

I. Introduction 115
II. Synthetic Methods 116
III. Applications of Thiocoumarins 122
IV. Spectroscopic and Physical Properties of Thiocoumarins 123
V. Reactions of Thiocoumarins 124
VI. Dithiocoumarins 131
VII. 3,4-Dihydrothiocoumarins 132

Benzo[c]furans
WILLY FRIEDRICHSEN

I. Introduction 135
II. Theoretical Aspects 137
III. Benzo[c]furan and Its Alkyl- and Monoaryl-Substituted Derivatives 142
IV. 1,3-Diarylbenzo[c]furans 161

V. Spectroscopic Properties	215
VI. Cyclobuta[c]furans and Cyclobutabenzo[c]furans	218
VII. Benz-Annelated and Hetero-Substituted Benzo[c]furans and Larger Ring[c]-Fused Furans	219
VIII. Benzo[c]furan-4,7-diones	234
CUMULATIVE INDEX OF TITLES	243

Contributors

Numbers in parentheses indicate the pages on which the authors' contributions begin.

WILLY FRIEDRICHSEN, *Institut für Organische Chemie der Universität Kiel, D-2300 Kiel, West Germany* (135)

O. METH-COHN, *Department of Chemistry and Applied Chemistry, University of Salford, Salford M5 4WT, England* (115)

CHRISTOPHER A. RAMSDEN, *The Research Laboratories, May and Baker Ltd., Dagenham, Essex RM10 7XS, England* (1)

B. TARNOWSKI, *Department of Chemistry and Applied Chemistry, University of Salford, Salford M5 4WT, England* (115)

Preface

Volume 26 contains three chapters. C. A. Ramsden reviews heterocyclic betaines derived from six-membered heteroaromatic rings. These compounds have been intensively studied over the past 10 years and have not previously been comprehensively reviewed. O. Meth-Cohn and B. Tarnowski summarize the chemistry of the thiocoumarins, a somewhat neglected group of heterocycles. Finally, W. Friedrichsen has reviewed the chemistry of the benzo[c]furans which, unlike the aza and thia analogs, have not previously been considered in this series. Here again we have a group of compounds the chemistry of which has advanced considerably over the last decade.

A. R. Katritzky
A. J. Boulton

Heterocyclic Betaine Derivatives of Alternant Hydrocarbons

CHRISTOPHER A. RAMSDEN

*The Research Laboratories, May and Baker Ltd.,
Dagenham, Essex, England*

I. Introduction 3
II. Heterocyclic Betaines Isoelectronic with Odd Alternant Hydrocarbon Anions . . 4
   A. Reducible and Irreducible Odd Alternant Hydrocarbon Anions 4
   B. Heterocyclic Betaines Isoelectronic with Irreducible Anions
      (Mesomeric Betaines). 7
   C. Heterocyclic Betaines Isoelectronic with Reducible Anions
      (Cross-Conjugated Betaines) 9
III. The Chemistry of Heterocyclic Mesomeric Betaines. 10
   A. Systems with Seven Conjugated Atoms. 11
     1. Pyrylium-3-olates 11
     2. Pyridinium-3-olates 16
     3. Pyridinium-3-aminides 24
     4. Thiopyrylium-3-olates 25
     5. Pyridazinium-3-olates 25
     6. Pyridazinium-5-olates 26
   B. Systems with Eleven Conjugated Atoms 27
     1. 2-Benzopyrylium-4-olates 27
     2. Isoquinolinium-4-olates 30
     3. Isoquinolinium-4-aminides 32
     4. Quinolizinium-1-olates. 33
     5. Quinolinium-8-olates 34
     6. 2-Benzothiopyrylium-4-olates 34
     7. Phthalazinium-1-olates (Pseudophthalazones) 35
     8. Cinnolinium-4-olates 42
     9. Cinnolinium-4-thiolates 45
     10. Cinnolinium-8-olates 46
     11. 1,7-Naphthyridinium-4-olates 46
     12. 1,9-Naphthyridinium-4-olates 47
     13. 1,2,3-Benzotriazinium-4-olates 47
     14. 1,2,3-Benzotriazinium-4-aminides 49
     15. 1,2,3-Benzotriazinium-4-thiolates 49
     16. Pyrido[2,1-*f*][1,2,4]triazinium-1-olates 50
     17. 2,1,3-Benzothiadiazinium-4-olates 50
     18. Quinolinium-3-olates 51
     19. Cinnolinium-3-olates 51

20. Quinolizinium-3-olates. 51
21. Quinolinium-6-olates 52
C. Systems with Thirteen Conjugated Atoms. 52
1. 2-(Pyridinium)phenolates 52
2. 6-(Aryl)-1,2,4-triazinium-5-olates 53
3. 4-(Pyridinium)phenolates 53
4. 4-(Pyridinium)thiophenolates 54
5. Benz[de]isoquinolines (Naphtho[1,8-cd]pyridines) 54
6. Naphtho[1,8-cd]thiopyrans 56
7. Naphtho[1,8-de]triazines 58
8. Quinolino[8,1-ef][1,2,4]triazines 60
9. Naphtho[1,8-cd][1,2,6]thiadiazines 60
10. Naphtho[1,8-cd][1,2,6]selenadiazines 61
D. Systems with Fifteen Conjugated Atoms 61
1. Pyrido[1,2-b]cinnolinium-11-olates 61
2. Pyrido[1,2-b]cinnolinium-11-aminides 63
3. Pyrido[1,2-b]cinnolinium-11-methylates 63
4. Pyrido[1,2-b]cinnolinium-11-thiolates 64
5. Acridinium-4-olates 64
6. Acridinium-2-olates 64
7. 5,8,10-Triazabenzo[a]quinolizinium-11-olates 64
8. Phenanthridinium-2-olates 65
E. Systems with Seventeen Conjugated Atoms 65
1. Neooxyberberines. 65
2. 3-Arylcinnolinium-4-olates. 69
F. Systems with Nineteen Conjugated Atoms 70
1. Dibenzo[a,g]quinolizinium-13-olates 70
2. 8-Azadibenzo[a,g]quinolizinium-13-olates 71
3. Phenanthro[4,10-bc]azinium-6,11-diolates 71
IV. The Electronic Structure of Mesomeric Betaines 74
A. A Perturbation Molecular Orbital (PMO) Model 74
1. Ionization Potentials 74
2. Absorption Spectra 75
3. Frontier Orbitals 83
4. Sulfur d-Orbitals 84
5. Thermodynamic Stability 85
6. pK_a Values 86
B. Molecular Orbital Calculations 88
V. Pericyclic Reactions of Mesomeric Betaines 89
A. Valence Tautomerism 90
B. Dimerization 92
C. Cycloadditions. 94
1. 2π-Electron Addends. 94
2. 4π-Electron Addends. 96
VI. Heterocyclic Betaines Isoelectronic with Even Alternant Hydrocarbon Dianions . 97
A. Kekulé and Non-Kekulé Alternant Hydrocarbons 97
B. Heterocyclic Betaines Isoelectronic with Non-Kekulé Dianions 98
VII. The Chemistry of Heterocyclic Betaines Derived from Even Alternant
Hydrocarbon Dianions 100
A. Mesomeric Betaines 100
B. Cross-Conjugated Betaines. 102

VIII. Conclusion	104
IX. Appendix Added in Proof	105
References Added in Proof	113

I. Introduction

In recent years increasing attention has been directed toward those heterocyclic molecules that participate in 1,3-dipolar cycloaddition reactions, and these studies have led to the discovery of a variety of novel and valuable transformations. Previous articles have described the structure and chemistry of two classes of these heterocyclic 1,3-dipoles, i.e., mesoionic compounds (**1**)[1] and mesomeric betaine derivatives of heteropentalenes (**2**).[2] In this review, which is the final chapter of a trilogy, a third major group of heterocyclic mesomeric betaines (those which are isoelectronic with odd alternant hydrocarbon anions) is recognized and the chemistry of this family is discussed.

Remarkably little is known about the chemistry of this large class of heterocyclic betaines. One of the earliest encounters with these molecules was made by F. M. Rowe at the University of Leeds who, in a series of studies in the 1930s, prepared N-aryl derivatives of the phthalazones (**3**) and demonstrated their acid-catalyzed thermal rearrangement to the isomers (**4**).[3] However, general interest in this area did not begin to awaken until the 1960s when a number of novel mesomeric betaines were fully characterized: These included the pyrylium-3-olates (**5**),[4] the cinnolinium-4-olates (**6**)[5] and the triazaphenalenes (**7**).[6] Subsequently, a variety of other betaines have been prepared and their participation in cycloaddition reactions demonstrated. Furthermore, a degree of respectability has been bestowed upon these heterocycles by the inclusion among their ranks of members of the chemical aristocracy (i.e., natural products). Thus, the oxoaporphine alkaloids corunnine (**8**; $R^1 = R^2 = Me$)[7] and nandazurine (**8**; $R^1R^2 = CH_2$)[8] have been isolated and characterized. Finally, this brief preamble would not be complete without mention of the elegant work of Katritzky and co-workers who, in a series of studies from 1970 onwards, have uncovered a wealth of cycloadditions, dimerizations, and rearrangements of the pyridinium-3-olates (**9**).[9]

---

[1] W. D. Ollis and C. A. Ramsden, *Adv. Heterocycl. Chem.* **19**, 1 (1976).
[2] C. A. Ramsden, *Tetrahedron* **33**, 3203 (1977).
[3] F. M. Rowe, D. A. W. Adams, A. T. Peters, and A. E. Gillam, *J. Chem. Soc.* 90 (1937).
[4] G. Suld and C. C. Price, *J. Am. Chem. Soc.* **83**, 1770 (1961).
[5] D. E. Ames and H. Z. Kucharska, *J. Chem. Soc.*, 4924 (1963).
[6] M. J. Perkins, *J. Chem. Soc.*, 3005 (1964).
[7] I. Ribas, J. Sueiras, and L. Castedo, *Tetrahedron Lett.*, 3093 (1971).
[8] J. Kunitomo, M. Ju-ichi, Y. Yoshikawa, and H. Chikamatsu, *Experientia* **29**, 518 (1973).
[9] N. Dennis, A. R. Katritzky, and Y. Takeuchi, *Angew. Chem., Int. Ed. Engl.* **15**, 1 (1976).

The aims of the present review are threefold: (i) to provide an account of general aspects of the structure, bonding, and reactivity of this class of betaine; (ii) to provide a comprehensive survey of betaine chemistry; and (iii) to direct attention to new areas worthy of exploration.

The literature available to the author up to July, 1978 has been surveyed. An appendix (Section IX) extends the coverage to October, 1979. The subject has not previously been reviewed.

II. Heterocyclic Betaines Isoelectronic with Odd Alternant Hydrocarbon Anions

A. Reducible and Irreducible Odd Alternant Hydrocarbon Anions

Alternant hydrocarbons (AH) are defined as conjugated systems in which the atoms can be divided into two sets, starred and unstarred, in such a way that no atoms of like parity are directly linked. Benzene is typical.

Since only atoms of opposite parity are covalently bonded, it follows that double bonds in these species must necessarily link starred and unstarred atoms. This fundamental feature of double bonding in alternant systems is an important factor governing the representation of mesomeric betaines. The relationship between mesomeric betaines and AHs will be developed later in the section.

The bonding of AHs is well understood and accounts of their electronic structure are numerous.[10,11] Because the π-electron charge density at each carbon atom of an AH is unity (or very close to unity), the Hückel molecular orbital (HMO) method provides a particularly good yet simple model for their study. Two types of neutral AH are recognized: (a) even AHs associated with m starred atoms and m unstarred atoms and (b) odd AHs associated with m starred atoms and $m - 1$ unstarred atoms. The latter species are necessarily radicals (e.g., the allyl radical) and have the special property that the highest occupied molecular orbital (HOMO) is a nonbonding molecular orbital (NBMO) which, within the framework of the HMO method, vanishes on the unstarred atoms (inactive positions). A rule for calculating the coefficients of these NBMOs has been described by Longuet-Higgins.[12] Introduction of a second electron into the NBMO gives an odd AH anion.

This review is primarily concerned with heterocyclic betaines that are isoelectronic with odd AH anions. For the purpose of recognizing and classifying these species, it is convenient to make a further subdivision by distinguishing between two types of odd AH anion.

Many odd AHs contain even AH fragments on which the NBMO completely vanishes, i.e., the coefficients are zero on both the starred and unstarred sets of atoms. These regions to which the NBMO does not extend are referred to as inactive segments. For the purpose of this review we will describe those odd AHs that contain one or more inactive segments as *reducible odd alternant hydrocarbons* (ROAH). Examples of ROAH anions are the 3-phenylbenzyl anion (**10**), which contains an inactive phenyl group (**11**), and the 2,4(1,8-naphthalenediyl)pentadienyl anion (**12**), which contains an inactive napthalenediyl group (**13**).

We shall refer to odd AHs which do not contain inactive segments as *irreducible odd alternant hydrocarbons* (IOAH). Thus the allyl anion (**14**), the benzyl anion (**16**), and the perinaphthenyl anion (**18**) are examples of IOAH anions. Within the framework of HMO theory, the NBMO of an IOAH

---

[10] M. J. S. Dewar, "The Molecular Orbital Theory of Organic Chemistry." McGraw-Hill, New York, 1969.

[11] M. J. S. Dewar and R. C. Dougherty, "The PMO Theory of Organic Chemistry." Plenum, New York, 1975.

[12] H. C. Longuet-Higgins, *J. Chem. Phys.* **18**, 265, 275, 283 (1950).

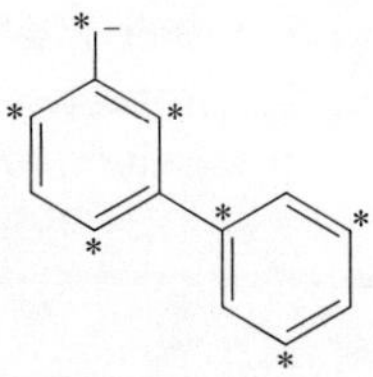

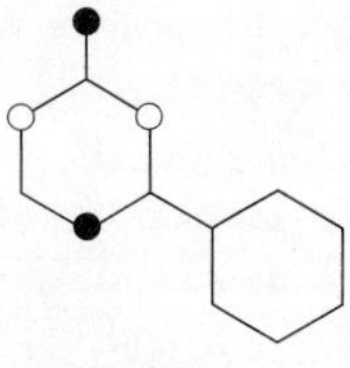

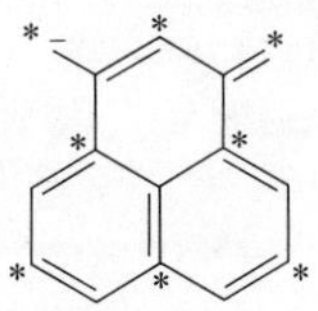

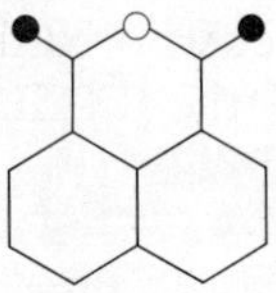

(10) **(11)** **(12)** **(13)**

anion extends over the whole π-system, having a finite value at starred positions and vanishing at unstarred positions (e.g., **15**, **17** and **19**). Occasionally the NBMO vanishes also at a starred position—for example, the central atom in the perinaphthenyl system (**19**).

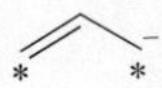

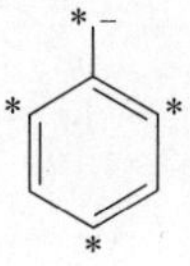

(14) **(15)** **(16)**

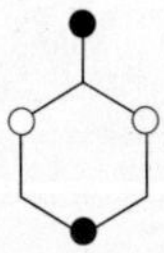

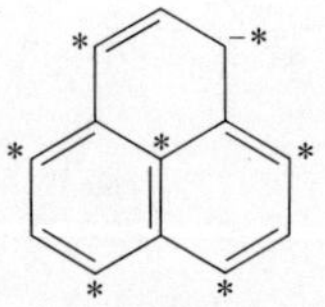

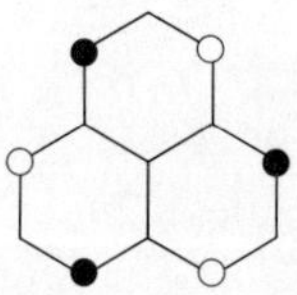

(17) **(18)** **(19)**

The relationship between IOAHs and ROAHs is well-defined. ROAHs are formed by union (←u→) between even AHs and inactive positions of IOAHs (e.g., **20 → 21**). ROAH anions, therefore, can be conveniently regarded as derivatives of IOAH anions. The even AH substituents, which have little or no influence on the nature of the anion, can be treated independently. These even and odd fragments are described as cross-conjugated.[10]

This division of odd AH anions into an IOAH anion fragment and cross-conjugated even AH substituents is a convenient simplification that will be employed in the following sections to discuss isoelectronic heterocycles. However, it must not be forgotten when considering heteroderivatives that

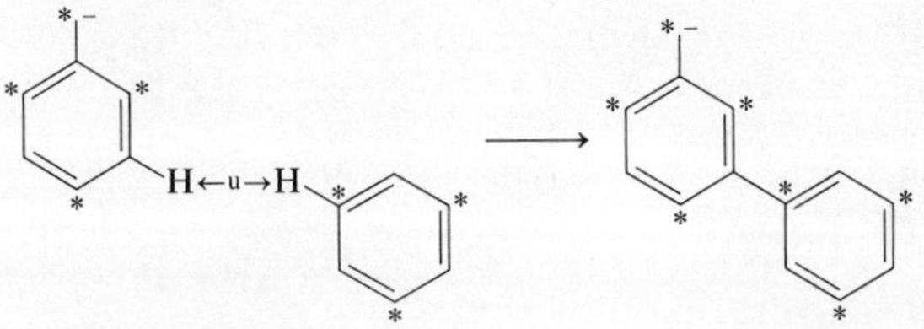

(20) **(21)**

it is only the NBMO that is localized on the IOAH segment. Other molecular orbitals extend over the whole ROAH network, and the properties of the IOAH fragment will be influenced accordingly, e.g., by σ-inductive effects.

B. Heterocyclic Betaines Isoelectronic with Irreducible Anions (Mesomeric Betaines)

Irreducible odd AH anions are associated with m starred atoms and $m - 1$ unstarred atoms, and isoconjugate heterosystems are characterized by a lone pair of electrons which originates on a heteroatom. In this context we regard the point of origin of a lone pair as that heteroatom upon which the lone pair is localized if the molecule is divided into its component atoms (i.e., it is that heteroatom which donates two electrons to the π-electron system). Let us consider those heterosystems isoconjugate with IOAH anions in which the lone pair originates on one of the m starred atoms. A typical example is N-methylpyrid-2-one (23) which is isoconjugate with the benzyl anion (24). If the lone pair is depicted as being localized at its origin (starred), the remainder of the conjugated system is composed of equal numbers of atoms of opposite parity (i.e., $m - 1$ starred and $m - 1$ unstarred), and these $2m - 2$ atoms constitute an even AH network. It follows that each of the remaining $m - 1$ starred atoms can be linked by a double bond with one of the $m - 1$ unstarred atoms (i.e., 22 → 23). These heterocycles, therefore, are satisfactorily represented by covalent structures in which the lone pair is associated with a starred heteroatom and the remainder of the conjugated system is depicted by conventional single and double bonds (e.g., 23).

Now consider neutral heteroderivatives of IOAH anions in which the lone pair is donated by a heteroatom occupying one of the $m - 1$ unstarred positions. When the lone pair is localized at its origin, the remainder of the conjugated system is, in this case, an even AH which is composed of m starred and $m - 2$ unstarred atoms. Since the numbers of atoms of opposite parity are unequal, it follows that it is impossible to draw conventional uncharged structures for these species.[13] This situation is illustrated by N-methylpyridinium-3-olate (27) which like N-methylpyrid-2-one (23) is isoconjugate with the benzyl anion (24). Species of this type are not diradicals (e.g., 25 → 26) and can only be represented satisfactorily by dipolar structures (e.g., 27). Further examples of systems of this type are molecules 5–8. Since they can only be depicted by dipolar structures and since several dipolar canonical forms are possible, they are aptly described as mesomeric betaines. Many more examples of betaines having this relationship to IOAH anions can now be envisaged. The structural features that give rise to this family of heterocyclic molecules can be summarized by the following rule:

Neutral heterosystems isoelectronic with IOAH anions cannot be represented by nonpolar structures if the lone pair originates at an unstarred position.

These mesomeric betaines are not of interest simply because they can only be represented by dipolar structures. Their special relationship to odd AH anions endows them with a particularly interesting electronic structure.[13] Since IOAH anions are characterized by a NBMO which vanishes at unstarred positions, the substitution of heteroatoms at these positions leaves the NBMO unperturbed. It follows that mesomeric betaines derived in this way will be characterized by a highest occupied molecular orbital (HOMO) that is high in energy and has the topological features of a NBMO.[13] This property is largely responsible for the ready participation of these heterocycles in 1,3-dipolar cycloaddition reactions with electron-deficient dipolarophiles. It would be premature to discuss general aspects of the reactivity of mesomeric betaines in detail before their chemistry has been reviewed. In Section III the chemistry of these mesomeric betaines is discussed systematically, and later sections deal with general aspects of their structure and reactivity.

It is important to emphasize that systems in which the lone pair originates at a starred position can be represented by dipolar canonical forms as well as by purely covalent structures and that sometimes these dipolar structures are a better representation of the molecule. Examples of this are the antibiotic pyocyanine (28 ↔ 29)[14] and the berberine derivative 30 ↔ 31.[15] In

---

[13] C. A. Ramsden, *J. C. S., Chem. Commun.*, 109 (1977).
[14] J. C. MacDonald, in "Antibiotics" (D. Gottlieb and P. D. Shaw, eds.), Vol. II, p. 52. Springer-Verlag, Berlin and New York, 1967.
[15] P. W. Jeffs and J. D. Scharver, *J. Am. Chem. Soc.* **98**, 4301 (1976).

spite of being represented by dipolar structures, these molecules are quite distinct from the family of mesomeric betaines that we have just described.

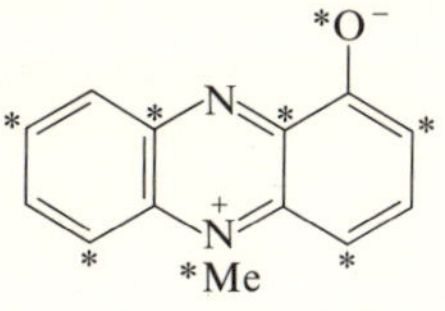

(28)

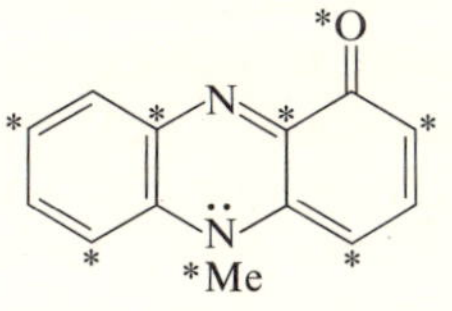

(29)

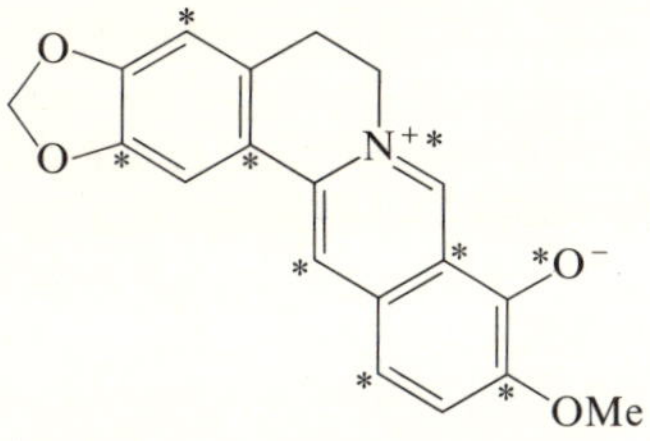

(30)

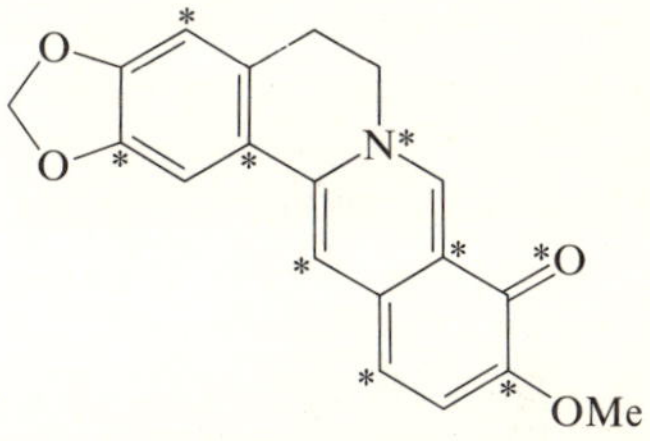

(31)

C. Heterocyclic Betaines Isoelectronic with Reducible Anions (Cross-Conjugated Betaines)

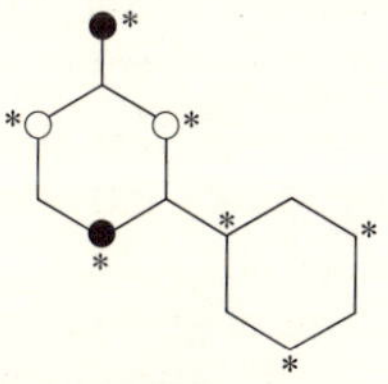

(32)

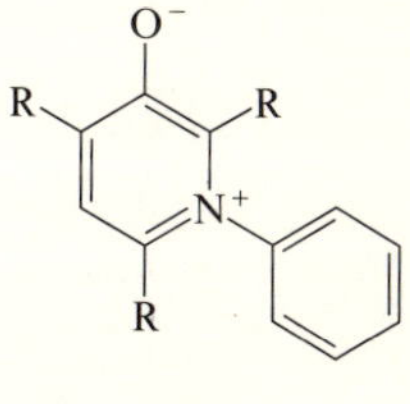

(33)

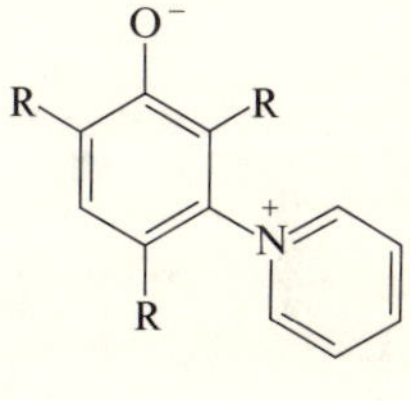

(34)

Heterosystems isoconjugate with ROAH anions have properties analogous to molecules derived from IOAH anions (Section II,B) if the lone pair "originates" in the IOAH fragment of the skeleton. For example, the N-phenylpyridinium-3-olates (33) are typical mesomeric betaines. In contrast, if the lone pair originates in an inactive fragment of the ROAH skeleton then a fundamentally different dipolar species is obtained. The negative charge is necessarily restricted to the IOAH fragment whereas the positive charge is localized on a cross-conjugated even AH fragment, and the charged segments of the molecule can be treated independently. The pyridinium phenolate 34 (R = NO_2)[16] is an example of these *cross-conjugated dipoles*

---

[16] K. Okon, G. Adamska, and W. Waclawek, *Rocz. Chem.* **43**, 1219 (1969).

or inner salts. They are quite distinct from the *conjugated dipoles* or mesomeric betaines **33** and they have quite different chemical properties. Another example is the natural product trigonelline (**35**) which occurs in the seeds of many plants.[17]

(**35**)

III. The Chemistry of Heterocyclic Mesomeric Betaines

The heterocycles discussed in this section are classified by reference to the size and structure of their isoconjugate IOAH anions. Heterocyclic mesomeric betaines isoconjugate with fourteen different IOAH anions are known to the author. These AH skeletons (**37–50**) are shown in Table I. Table II gives examples of corresponding mesomeric betaines (**51–64**).

Among the structural types that are within the scope of this review are many heterocyclic *N*-oxides, *N*-imides and *N*-ylides (e.g., **36**). The chemistry of these systems, which is extensive, has already been the subject of comprehensive reviews[18–20] and a discussion of their chemistry is not included here. Similarly the structure and chemistry of prototropic tautomers has also been reviewed[21] recently and will not be duplicated here.

(**36**) X = O, NR, CR$_2$

The following nomenclature has been used in this review: The ring system is named as the appropriate heterocation and the exocyclic group carrying the negative charge is named as the appropriate anion (i.e., olate, thiolate, aminide, or methylate). Thus, compounds of type **9** are named as

---

[17] E. Späth and G. Bobenberger, *Ber. Dtsch. Chem. Ges. B* **77**, 362 (1944).
[18] E. Ochiai, "Aromatic Amine Oxides." Elsevier, Amsterdam, 1967.
[19] A. R. Katritzky and J. M. Lagowski, "The Chemistry of Heterocyclic *N*-Oxides." Academic Press, New York, 1971.
[20] I. Zugravescu and M. Petrovanu, "*N*-Ylid Chemistry." McGraw-Hill, New York, 1976.
[21] J. Elguero, C. Marzin, A. R. Katritzky, and P. Linda, "The Tautomerism of Heterocycles." Academic Press, New York, 1976.

pyridinium-3-olates. This method has the appeal of being consistent with the method of naming mesoionic compounds which is now advocated by The Chemical Society. Mesomeric betaines which are not associated with an exocyclic fragment are named in the conventional manner. For example, compounds of type **7** are named as naphtho[1,8-*cd*]triazines.

A. Systems with Seven Conjugated Atoms

1. *Pyrylium-3-olates* (**65**)

(65) (66) (67)

Substituted derivatives of the pyrylium-3-olates (**65**) have been isolated. These brightly colored solids are transformed to pale yellow or colorless valence tautomers (**66**) by ultraviolet (UV) radiation.[22,23] The process (**65** → **66**) is reversed by visible light and such photochromic behavior is of potential value in information storage technology.[23]

Photochemical ring opening of 4,5-epoxy-2,3,4,5-tetraphenyl-2-cyclopentenone (**66**; $R^1 = R^2 = R^3 = R^4 = Ph$) (prepared by oxidation of tetraphenylcyclopentadienone or by base treatment of its chlorohydrin) provides a route to the tetraphenyl derivative (**65**; $R^1 = R^2 = R^3 = R^4 = Ph$).[24] Rapid formation of the mesomeric betaine is accompanied by a slower rearrangement of the epoxide (**66**; $R^1 = R^2 = R^3 = R^4 = Ph$) to tetraphenyl-2-pyrone (**67**). 2,4,6-Triphenylpyrylium-3-olate (**65**; $R^1 = R^2 = R^4 = Ph, R^3 = H$) has been obtained by quite different methods.[4,25] Treatment of 2,4,6-triphenylthiopyrylium perchlorate (**68**) with phenyllithium gives a deep purple intermediate which may be the thiabenzene **69**. Oxidation of this compound (**69**) by oxygen followed by treatment with hydrogen chloride gives the betaine **65** ($R^1 = R^2 = R^4 = Ph, R^3 = H$) as purple needles.[25] Alternatively, 3-acetoxy-2,4,6-triphenylpyrylium perchlorate (**70**),

---

[22] E. F. Ullman, *J. Am. Chem. Soc.* **85**, 3529 (1963).
[23] E. F. Ullman (American Cyanamid Co.), U.S. Patent 3,329,502 (1967).
[24] J. M. Dunston and P. Yates, *Tetrahedron Lett.*, 505 (1964).
[25] G. Suld and C. C. Price, *J. Am. Chem. Soc.* **84**, 2094 (1962).

TABLE I
ODD ALTERNANT SKELETONS OF KNOWN HETEROCYCLIC MESOMERIC BETAINES

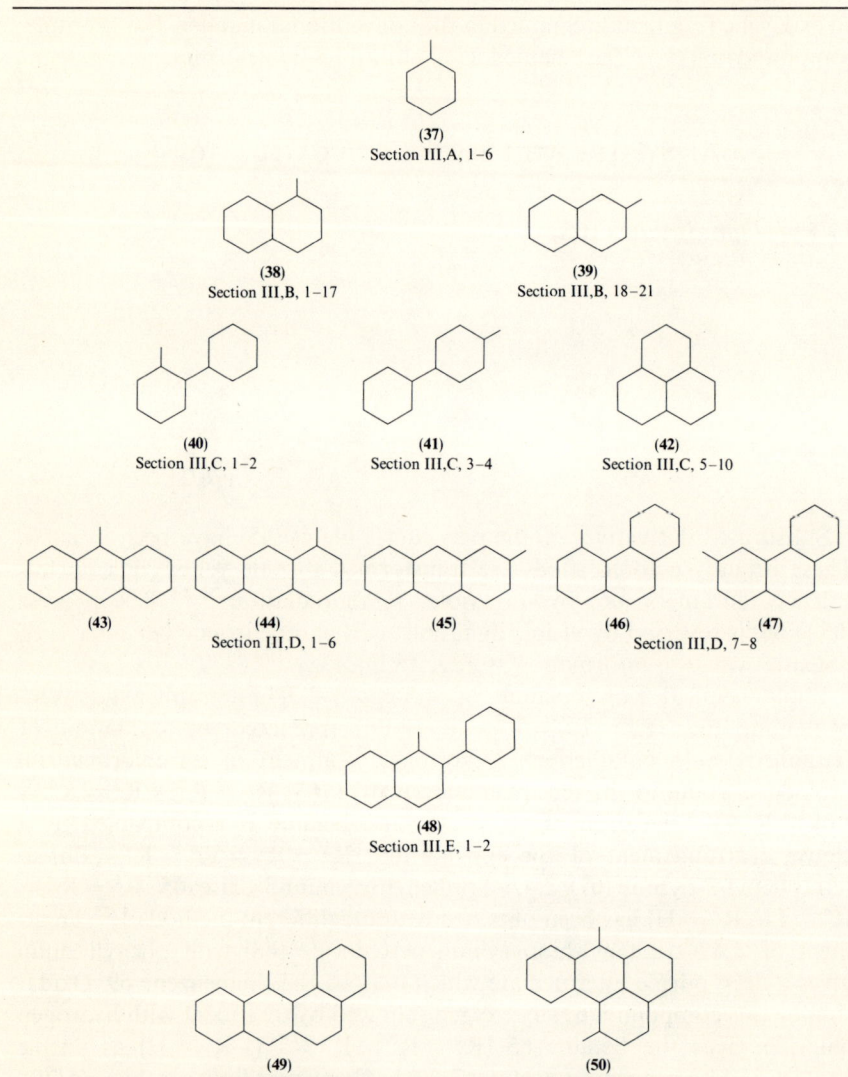

TABLE II
EXAMPLES OF KNOWN HETEROCYCLIC MESOMERIC BETAINES DERIVED FROM THE
FOURTEEN STRUCTURAL TYPES SHOWN IN TABLE I

prepared in the normal manner, gives compound **65** ($R^1 = R^2 = R^4 = Ph$, $R^3 = H$) when reacted with aqueous alkali[25] or pyridine–water.[26]

The pyrylium-3-olates (**65**) form salts under acidic conditions. The triphenyl compound (**65**; $R^1 = R^2 = R^4 = Ph$, $R^3 = H$) is oxidized by air to the butenolide **73**, which is also formed when a solution of the same precursor is photolyzed ($\lambda = 3660$ Å) in the presence of a benzophenone sensitizer. This oxidative rearrangement (**65**; $R^1 = R^2 = R^4 = Ph$, $R^3 = H \rightarrow$ **73**) has been rationalized in terms of the triketone intermediate **71** cyclizing to the zwitterion **72** followed by rearrangement.[27]

An example of 2,4,6-triphenylpyrylium-3-olate (**65**; $R^1 = R^2 = R^4 = Ph$, $R^3 = H$) reacting as a 1,3-dipole was first provided by Suld and Price who obtained a maleic anhydride adduct ($C_{25}H_{18}O_5$).[25] Subsequently, an extensive study of the cycloadditions of this species has been published by Potts, Elliott, and Sorm.[26] With acetylenic dipolarophiles, compound **65** ($R^1 = R^2 = R^4 = Ph$, $R^3 = H$) gives 1:1 adducts that have the general structure **74** and that isomerize to 6-benzoyl-2,4-cyclohexadienones (**76**) upon thermolysis. This thermal rearrangement (**74** → **76**) has been interpreted in terms of an intermediate ketene **75**. The 2,3-double bond of adduct **74** (R = Ph) is reduced by catalytic hydrogenation. Potential synthetic value of these cycloadducts (**74**) is demonstrated by the conversion of compound **74** (R = Ph) to 1,2,3,4,6-pentaphenylcyclohepta-1,3,5-triene (**79**; R = Ph) via the alcohol **78** (Scheme 1).[26]

It is interesting to note that adducts having the 8-oxabicyclo[3.2.1]octane skeleton (**74**) have been encountered via a different route.[28] Thus the 3,4-

---

[26] K. T. Potts, A. J. Elliott, and M. Sorm, *J. Org. Chem.* **37**, 3838 (1972).
[27] H. H. Wasserman and D. L. Pavia, *Chem. Commun.*, 1459 (1970).
[28] J. P. Freeman and M. J. Hoare, *J. Org. Chem.* **36**, 19 (1971).

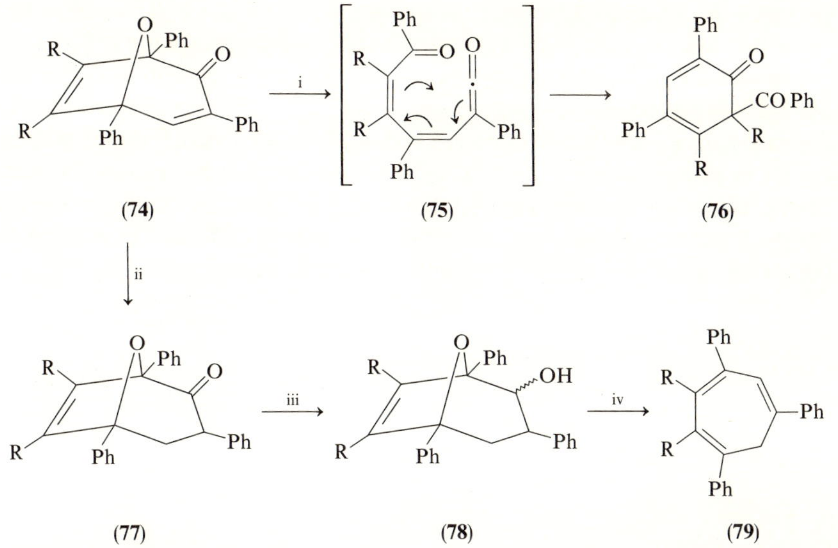

Sec. III.A]   HETEROCYCLIC BETAINES   15

SCHEME 1. Reagents: i, ΔT; ii, H_2/PtO_2; iii, $LiAlH_4$; iv, $p\text{-Me-}C_6H_4\text{-SO}_3H$.

diazacyclopentadienone N,N'-dioxide (80) with dimethyl acetylenedicarboxylate at 80°C gives compound 82, whose structure is supported by the products obtained on oxidation and reduction. The formation of compound 82 may be interpreted in terms of addition of dimethyl acetylenedicarboxylate to an intermediate pyrylium-3-olate (81), generated as shown in Scheme 2.[28]

Olefinic dipolarophiles, such as dimethyl fumarate, fumaronitrile, ethyl vinyl ether, N-phenylmaleimide, norbornene, and norbornadiene, react with

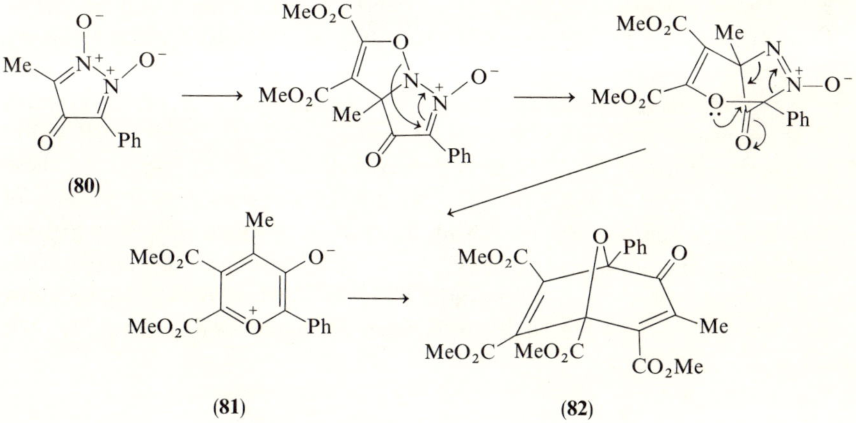

SCHEME 2

triphenylpyrylium-3-olate (**65**; $R^1 = R^2 = R^4 = Ph$, $R^3 = H$) in a manner similar to acetylenes, giving adducts **83**.[26] The stereochemistry of the adduct can vary according to the temperature of the reaction: At room temperature fumaronitrile gives the kinetically preferred cycloadduct (**84**) that is completely transformed to the thermodynamically more stable adduct (**85**) at 127°C. The cycloaddition is therefore reversible at elevated temperatures.[26] Cycloadducts are also formed with heterocumulenes.[26] For the purpose of 1,3-dipolar cycloaddition reactions, the pyrylium-3-olate (**65**) need not be isolated but can be generated *in situ* from the perchlorate salt and triethylamine.[26]

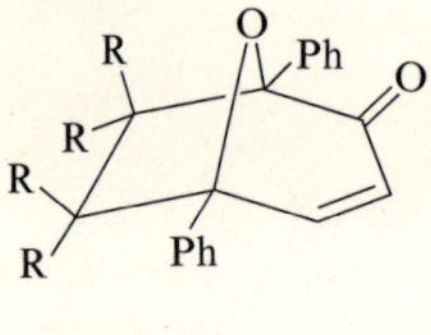

(**83**) 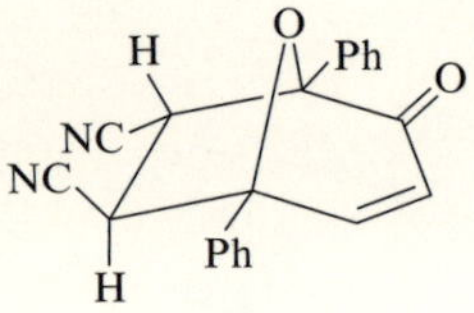　　(**84**) 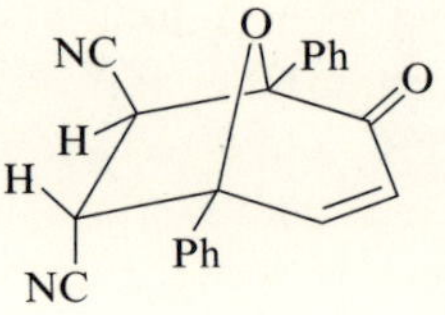　　(**85**)

2. *Pyridinium-3-olates* (**86**)

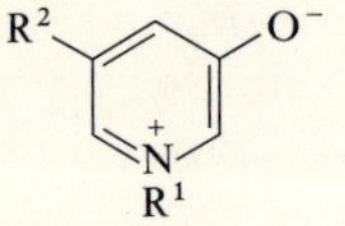

(**86**) 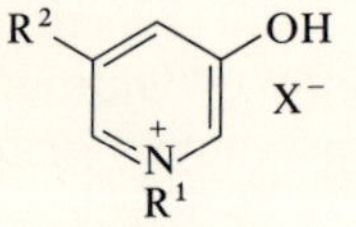　　(**87**) 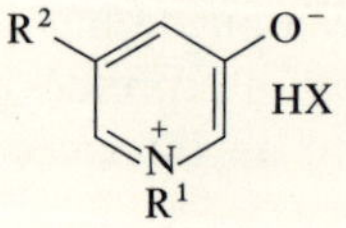　　(**88**)

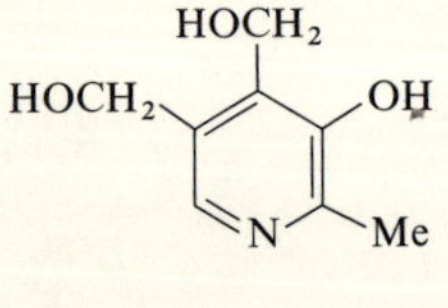

(**89**) 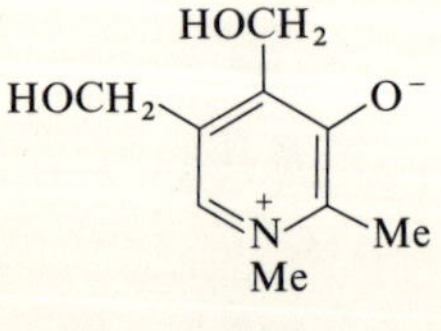　　(**90**) 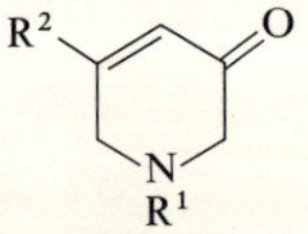　　(**91**)

Pyridinium-3-olates (**86**) have been extensively studied by Katritzky and co-workers who in particular have investigated the relationship between 1,3-dipolar reactivity and the nature of the N-substituent. Comprehensive accounts of this work have been published[9,29] and so only general features will be described here.

[29] A. R. Katritzky and N. Dennis, in "Studies in Organic Chemistry" (R. B. Mitra, ed.), Vol. 3, p. 290. Elsevier, Amsterdam, 1979.

The earliest reference to a pyridinium-3-olate (**86**) is the description of a viscous oil obtained from *N*-methyl-3-hydroxypyridinium iodide (**87**; $R^1 =$ Me, $R^2 = H$, $X = I$) and aqueous silver oxide.[30] The iodide **87** ($R^1 = $ Me, $R^2 = H$, $X = I$) ($C_6H_7NO \cdot HI$) with triethylamine gives a complex salt $[(C_6H_7NO)_2 \cdot HI]$[31] which also yields the betaine **86** ($R^1 = $ Me, $R^2 = H$) (C_6H_7NO) with silver oxide.[32] A similar procedure using vitamin B_6 (**89**) gives betaine **90**[33] which is also made from compound **89** and diazomethane.[33,34] Other betaine derivatives of vitamin B_6 have been encountered.[35–37]

A number of salts of the pyridinium-3-olates (**86**) have been prepared from 3-hydroxypyridine and alkyl halides.[38] Whether the products are 3-hydroxypyridinium salts (**87**)[9,39–41] or the betaine complexes (**88**)[38] is uncertain. Preparation of free pyridinium-3-olates (**86**) by dehydrohalogenation of 3-hydroxypyridinium salts (**87** or **88**; $X = $ Cl, Br, I) is best achieved by using an ion-exchange resin or by treatment with triethylamine.[9,29] Aqueous sodium carbonate generates the phenacyl derivative **86** ($R^1 = CH_2COPh$, $R^2 = H$) although the product is too unstable to be fully characterized.[42] A useful alternative synthesis involves oxidation of the enones **91** using metachloroperbenzoic acid[43] or of their salts using pyridinium bromide perbromide.[44] Another oxidative route which gives 2,4,6-triarylpyridinium-3-olates involves treatment of 2-(pyridinium)phenolates in ethanol with hydrogen peroxide[45] (see Section III,C,1). Acid-catalyzed rearrangement of the diazepinone salts **92** (R = H or Me) leads to stable

---

[30] R. R. Williams, *J. Ind. Eng. Chem.* **13**, 1107 (1921).
[31] K. Mecklenborg and M. Orchin, *J. Org. Chem.* **23**, 1591 (1958).
[32] L. Paoloni, M. L. Tosato, and M. Cignitti, *Theor. Chim. Acta* **14**, 221 (1969).
[33] S. A. Harris, T. J. Webb, and K. Folkers, *J. Am. Chem. Soc.* **62**, 3198 (1940).
[34] A. Itiba and K. Miti, *Sci. Pap. Inst. Phys. Chem. Res. (Jpn.)* **35**, 73 (1938) [*CA* **33**, 3430 (1939)].
[35] S. A. Harris, *J. Am. Chem. Soc.* **63**, 3363 (1941).
[36] R. Hüttenrauch and R. Tümmler, *Pharmazie* **22**, 561 (1967).
[37] E. E. Harris, J. L. Zabriskie, E. M. Chamberlin, J. P. Crane, E. R. Peterson, and W. Reuter, *J. Org. Chem.* **34**, 1993 (1969); A. Cohen, J. W. Haworth, and E. G. Hughes, *J. Chem. Soc.*, 4374 (1952).
[38] S. L. Shapiro, K. Weinberg, and L. Freedman, *J. Am. Chem. Soc.* **81**, 5140 (1959).
[39] H. Furst and H. T. Dietz, *J. Prakt. Chem.* **4**, 147 (1956).
[40] D. A. Prins, *Recl. Trav. Chim. Pays-Bas* **76**, 58 (1957).
[41] K. Undheim, P. O. Tveita, L. Borka, and V. Nordal, *Acta Chem. Scand.* **23**, 2065 (1969).
[42] N. Dennis, A. R. Katritzky, and S. K. Parton, *Chem. Pharm. Bull.* **23**, 2904 (1975).
[43] Y. Tamura, T. Saito, H. Kiyokawa, L.-C. Chen, and H. Ishibashi, *Tetrahedron Lett.*, 4075 (1977); Y. Tamura, M. Akita, H. Kiyokawa, L.-C. Chen, and H. Ishibashi, *ibid.*, 1751 (1978).
[44] N. Dennis, A. R. Katritzky, and R. Rittner, *J. C. S., Perkin 1*, 2329 (1976).
[45] A. R. Katritzky, C. A. Ramsden, Z. Zakaria, R. L. Harlow, and S. H. Simonsen, *J. C. S., Chem. Commun.*, 363 (1979).

N-aminopyridinium-3-olates **93** (R = H or Me). Acetylation of the *N*-amino compound **93** (R = H) gives the *N*-imide **94** (R = COMe), and not the tautomer **93** (R = COMe), whereas the *N*-methylamino derivative gives the pyridinium-3-olate **95**.[46,47]

(92) (93) (94)

(95) (96)

Stability of the compounds **86** varies dramatically with the nitrogen substituent. Methyl[48] and phenyl[49] derivatives (**86**; R^1 = Me or Ph) are moderately stable crystalline solids—often hydrated (**87** or **88**; X = OH). The 2,4-dinitrophenyl derivative [**86**; $R^1 = C_6H_3(NO_2)_2$, $R^2 = H$] is less stable but can be isolated as orange prisms (mp 112°C)[50]: It readily rearranges to the diaryl ether **96** (mp 128°C) with which it was originally confused.[50-53] The monomeric 5-nitro-2-pyridyl derivative is similarly unstable and in solution dimerizes giving adduct **97a**.[54-55a] The 4,6-dimethylpyrimidin-2-yl derivative can only be isolated as the dimers **97b** and

---

[46] J. A. Moore, *J. Am. Chem. Soc.* **77**, 3417 (1955).
[47] J. A. Moore and J. Binkert, *J. Am. Chem. Soc.* **81**, 6045 (1959).
[48] A. R. Katritzky and Y. Takeuchi, *J. Chem. Soc. C*, 874 (1971).
[49] N. Dennis, A. R. Katritzky, T. Matsuo, S. K. Parton, and Y. Takeuchi, *J. C. S., Perkin 1*, 746 (1974).
[50] N. Dennis, B. Ibrahim, A. R. Katritzky, I. G. Taulov, and Y. Takeuchi, *J. C. S., Perkin 1*, 1883 (1974).
[51] A. F. Vompe and N. F. Turitsyna, *Zh. Obshch. Khim.* **27**, 3282 (1957).
[52] N. Dennis, A. R. Katritzky, S. K. Parton, and Y. Takeuchi, *J. C. S., Chem. Commun.*, 707 (1972).
[53] N. Dennis, B. Ibrahim, and A. R. Katritzky, *Org. Mass. Spectrom.* **11**, 814 (1976).
[54] N. Dennis, B. Ibrahim, and A. R. Katritzky, *J. C. S., Chem. Commun.*, 500 (1974).
[55] N. Dennis, B. Ibrahim, and A. R. Katritzky, *J. C. S., Perkin 1*, 2296 (1976); A. R. Katritzky, N. Dennis, and H. A. Dowlatshahi, *J. C. S., Chem. Commun.*, 316 (1978).
[55a] A. R. Katritzky, N. Dennis, M. Chaillet, C. Larrieu, and M. E. Mouhtadi, *J. C. S., Perkin 1*, 408 (1979).

98b which in chloroform solution are in dynamic equilibrium (1:4)—isomerization occurs by weak dissociation to the monomer (**99b** ≡ **100b**).[55-56] Compound **97b** is the kinetically favored dimer, being the only product obtained by immediate work-up of the basified salt **87** (R^1 = 4,6-dimethylpyrimidin-2-yl, R^2 = H, X = Cl). 1-(4,6-Diphenyl)- and 1-(4,6-dimethoxy-s-triazin-2-yl)pyridinium-3-olates also dimerize spontaneously when generated from their salts, but in this case the products are exclusively the syn dimers (**97c** and **97d**).[57] The effect of other nitrogen substituents is under investigation.[29,58]

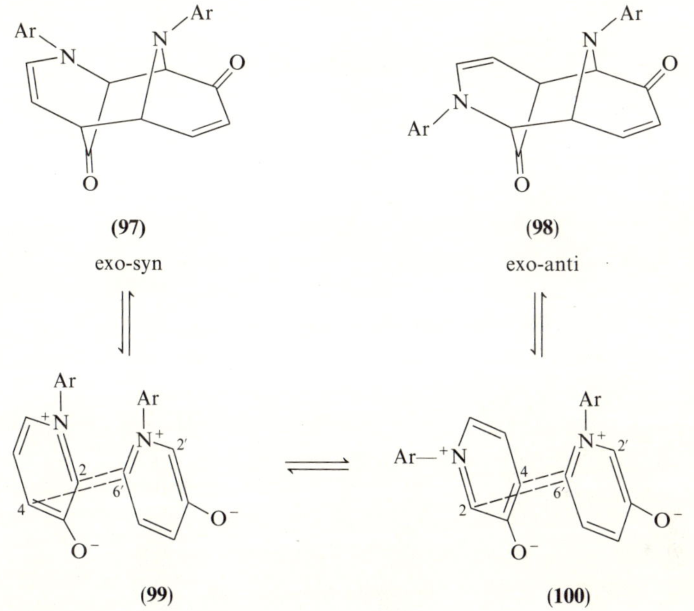

(97) exo-syn

(98) exo-anti

(99)

(100)

SCHEME 3. In structures **97-100**: a, Ar = 5-nitropyrid-2-yl; b, Ar = 4,6-dimethylpyrimidin-2-yl; c, Ar = 4,6-diphenyl-s-triazin-2-yl; d, Ar = 4,6-dimethoxy-s-triazin-2-yl.

Ease of dimerization (Scheme 3) is related to HOMO–LUMO splitting.[56] Within the framework of the frontier molecular orbital (FMO) approximation,[59] reducing the frontier orbital gap favors symmetry allowed cycloadditions (e.g., **99** → **97**) by stabilizing the transition state. Since the

---

[56] N. Dennis, B. Ibrahim, and A. R. Katritzky, *J. C. S., Chem. Commun.*, 425 (1975).
[57] N. Dennis, A. R. Katritzky, G. J. Sabounji, and L. Turker, *J. C. S., Perkin 1*, 1930 (1977).
[58] A. R. Katritzky, J. Banerji, A. Boonyarakvanich, A. T. Cutler, N. Dennis, S. Q. Abbas Rizvi, G. J. Sabongi, and H. Wilde, *J. C. S., Perkin 1*, 399 (1979); A. R. Katritzky, M. Abdallah, S. Bayyuk, A. M. A. Bolouri, N. Dennis, and G. J. Sabongi, *Polish J. Chem.* **53**, 57 (1979).
[59] I. Fleming, "Frontier Orbitals and Organic Chemical Reactions." Wiley, New York, 1976.

N-substituents of the betaines **86** are cross-conjugated (Section II,A), modifications of this group have little effect on the HOMO energy. The energy of the LUMO, however, is appreciably lowered by extended conjugation or by introduction of electronegative substituents at this position, and the frontier orbital gap is reduced accordingly. Thus the s-triazin-2-yl betaines (**99c** and **99d**) show great affinity for dimerization (Scheme 3), the N-methyl and N-phenyl derivatives (**86**; R^1 = Me or Ph, R^2 = H) do not dimerize and the 5-nitro-2-pyridyl derivative (**99a**) shows intermediate behavior. FMO methods have also been used to discuss the regioselectivity [2,2'–4,6' (**99**) or 2,6'–4,2' (**100**)] of the cycloadditions (Scheme 3).[9,55,55a] The magnitudes of the HOMO and LUMO coefficients appear to determine the structure of the kinetically preferred dimer (2,2'–4,6'). If dimerization is sufficiently reversible, the thermodynamically more stable product (2,6'–4,2') slowly forms (e.g., **97b** → **99b** ≡ **100b** → **98b**; Scheme 3). No endo dimers are observed presumably due to unfavorable second-order interactions in the transition state.[55,55a]

Dimerization of pyridinium-3-olates (**86**) is also induced photochemically. In accord with theory the structure of the photodimers differs from that of the thermal dimers. N-Phenylpyridinium-3-olate (**101**) upon irradiation gives dimer **103** together with the bicyclic valence tautomer **102** (Scheme 4). This valence tautomerism (**101** → **102**) is analogous to that of pyrylium-3-

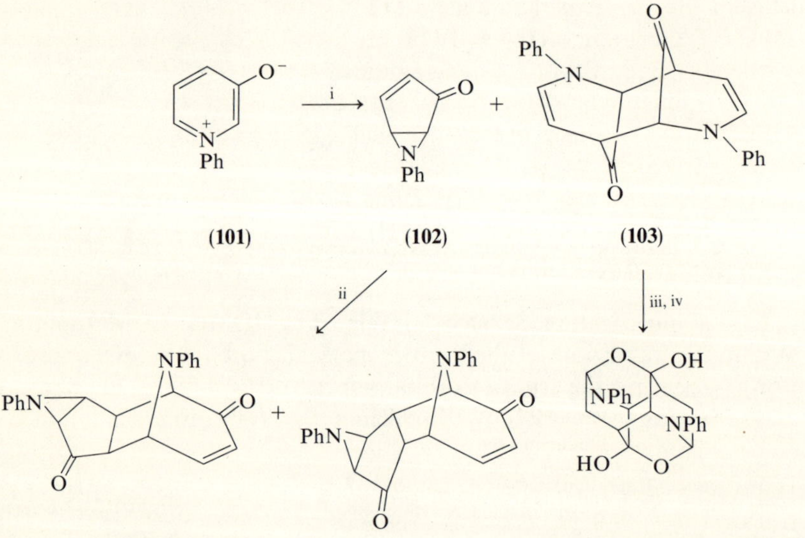

SCHEME 4. Reagents: i, hv (3500 Å)–EtOAc; ii, N-phenylpyridinium-3-olate (**101**); iii, conc HCl; iv, H_2O.

olates (**65** → **66**) (Section III,A,1). A portion of the valence tautomer **102** undergoes further reaction with the monomeric betaine **101** giving the 1,3-dipolar adducts **104** and **105** (Scheme 4).[60,61] Dimer **103** with concentrated hydrochloric acid and hydrolysis of an intermediate salt gives the dioxadiazadiamantone **106**.[62] Photochemical rearrangement of pyridinium-3-olates to 2-pyridones has also been reported.[63]

Among photooxidation products of the betaines **86** (R^1 = Me, PhCH$_2$, Ph, R^2 = H) are N-substituted maleimides (**107**).[62,64] Evidence for the intermediate **108** in these reactions is provided by the observation that photooxidation of 1-methyl-5-phenylpyridinium-3-olate (**86**; R^1 = Me, R^2 = Ph) gives the peroxy dimer **109** (11%).[62]

(**107**) (**108**) (**109**)

Numerous reactions of the betaines **86** with olefinic 1,3-dipolarophiles have been reported. N-Methylpyridinium-3-olate (**111**) with electron-deficient olefins gives the adducts **112**.[65,66] Alkynes give similar adducts (**110**).[43] These adducts (**110** and **112**) are useful intermediates for tropolone synthesis. Quaternization to methiodides (**113**, **115**) and treatment with base gives dimethylaminotropones (**114**) which are hydrolyzed to tropolones (**116**).[66] Good syntheses of stipitatic acid (**116**; R^1 = CO$_2$H, R^2 = OH) and hinokitiol (**116**; R^1 = H, R^2 = CHMe$_2$) have been achieved by this route (Scheme 5).[43] Similar transformations have been achieved using the N-phenyl betaine **86** (R^1 = Ph, R^2 = H) which is more reactive but whose adducts are difficult to quaternize.[67] A difference is observed when benzyne is

---

[60] A. R. Katritzky and H. Wilde, J. C. S., Chem. Commun., 770 (1975).
[61] N. Dennis, A. R. Katritzky, and H. Wilde, J. C. S., Perkin 1, 2338 (1976).
[62] J. Banerji, N. Dennis, A. R. Katritzky, R. L. Harlow, and S. H. Simonsen, J. Chem. Res. (S), 38 (1977); J. Chem. Res. (M), 517 (1977).
[63] T. Lærum and K. Undheim, Acta Chem. Scand., Ser. B 32, 68 (1978); P.-O. Ranger, G. A. Ulsaker, and K. Undheim, ibid., 70.
[64] A. Mori, S. Ohta, and H. Takeshita, Heterocycles 2, 243 (1974); H. Takeshita, A. Mori, and S. Ohta, Bull. Chem. Soc. Jpn. 47, 2437 (1974).
[65] A. R. Katritzky and Y. Takeuchi, J. Am. Chem. Soc. 92, 4134 (1970).
[66] A. R. Katritzky and Y. Takeuchi, J. Chem. Soc. C, 878 (1971).
[67] N. Dennis, A. R. Katritzky, S. K. Parton, Y. Nomura, Y. Takahashi, and Y. Takeuchi, J. C. S., Perkin 1, 2289 (1976).

the 1,3-dipolarophile: The phenyl derivative gives compound **117** (R = Ph)[52]; the methyl betaine gives compound **118**.[68] *N*-Alkyl adducts (**117**; R = alkyl) can be made using oxanorbornadienes as dipolarophile followed by deoxygenation: Some of these derivatives show anti-inflammatory activity.[69]

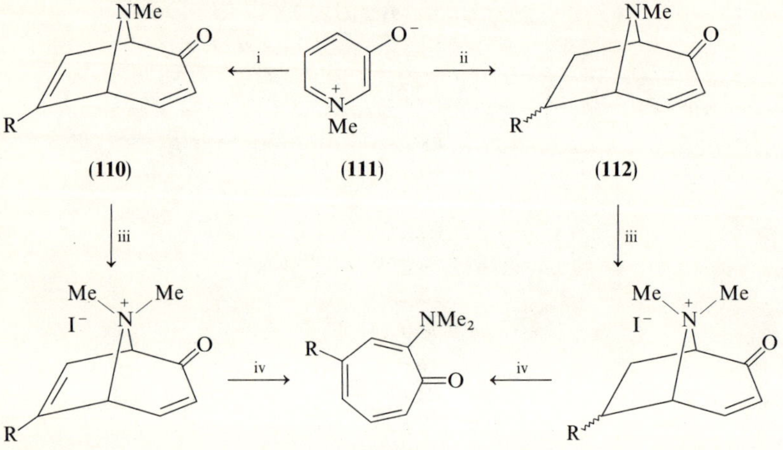

SCHEME 5. Reagents: i, RC≡CH; ii, RCH=CH$_2$; iii, MeI; iv, Ag$_2$O or NaHCO$_3$.

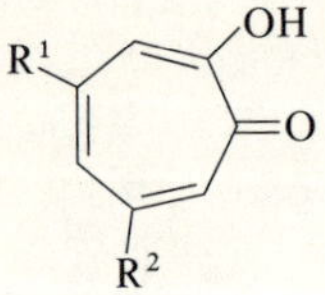

Other transformations of value are (i) cycloaddition of chloroketenes (RCCl=C=O) across the exocyclic oxygen atom and the 4-position of the betaines **86** (R^2 = H) giving, after spontaneous dehydrohalogenation, 2-oxofuro[2,3-*c*]pyridines (**119**; R^2 = H or Cl)[70] and (ii) intramolecular 1,3-dipolar cycloaddition of the *N*-pent-5-enyl derivative **120** giving the tricyclic adduct **121**.[71] Intramolecular reaction (**120** → **121**) is slow: In the

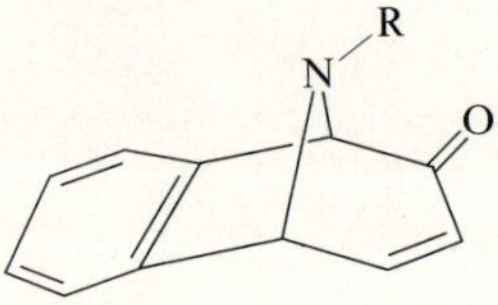

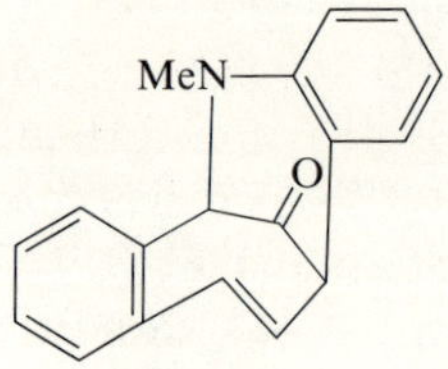

[68] N. Dennis, A. R. Katritzky, and S. K. Parton, *J. C. S., Perkin 1*, 2285 (1976).
[69] T. Sasaki (Grelan Pharm. Co., Ltd.), Japanese Patent 74 45, 098 [*CA* **81**, 105308 (1974)];
T. Sasaki, K. Kanematsu, K. Hayakawa, and M. Uchide, *J. C. S., Perkin 1*, 2750 (1972).
[70] N. Dennis, A. R. Katritzky, and G. J. Sabounji, *Tetrahedron Lett.*, 2959 (1976).
[71] P. G. Sammes and R. A. Watt, *J. C. S., Chem. Commun.*, 367 (1976).

presence of *N*-phenylmaleimide, compound **120**—like other *N*-alkyl derivatives[65,72]—gives exclusively the intermolecular exo adduct (**122**; R = pent-5-enyl).[71]

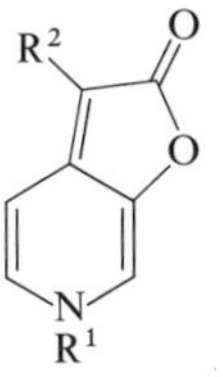

(119) 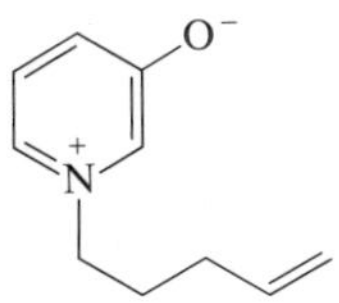   (120) 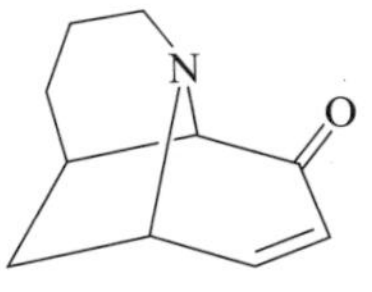   (121) 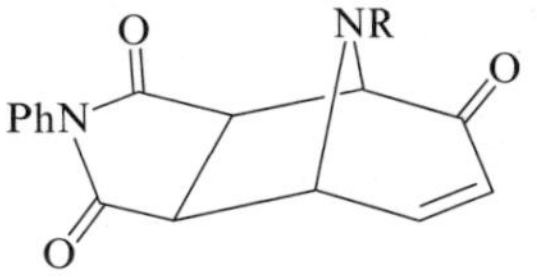   (122)

In most of these cycloaddition reactions the energy of the HOMO is the controlling factor. With the pyrid-2-yl, pyrimidin-2-yl, and *s*-triazin-2-yl derivatives (**99a–d**; Scheme 3), in which the LUMO is considerably lower in energy, LUMO interactions are also important controlling factors and these betaines (**99a–d**) are more versatile 1,3-dipoles.[9] They react with both electron-rich olefins (high HOMO) and conjugated olefins as well as showing increased activity toward electron-deficient olefins (low LUMO).[56] This substituent effect is illustrated by the following reactions of the nitropyridyl betaine **123** with 2π-, 4π-, and 6π-electron addends.[73] Solutions of the thermal dimers (**97a** and **98a**) contain a small equilibrium concentration of monomer (**123**), and this provides a convenient *in situ* source of betaine.[73]

With monosubstituted electron-deficient ethylenes, compound **123** gives a mixture of the exo and endo adducts (e.g., **124** and **125**; R = CO$_2$Me, CN, COMe). With styrenes the endo isomer (**126**) predominates. These adducts (**126**) are transformed into isomers **127** by trifluoromethanesulfonic acid.[74] Ethyl vinyl ether gives exclusively the endo adduct (**124**; R = OEt). Similar

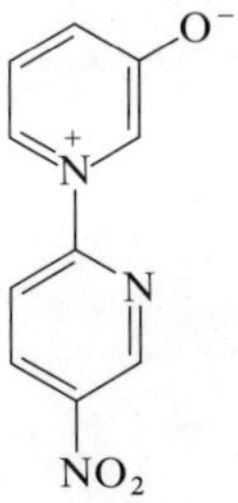

(123) 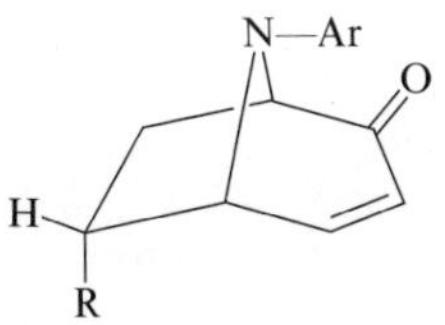   (124) 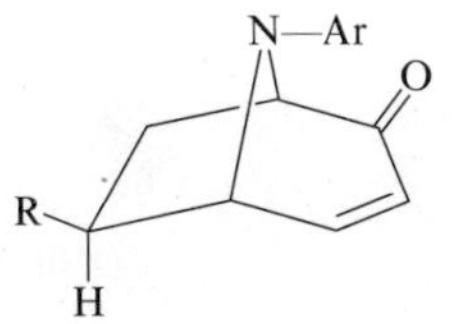   (125)

---

[72] J. Banerji, N. Dennis, J. Frank, A. R. Katritzky, and T. Matsuo, *J. C. S., Perkin 1*, 2334 (1976).
[73] N. Dennis, B. Ibrahim, and A. R. Katritzky, *J. C. S., Perkin 1*, 2307 (1976).
[74] N. Dennis, B. Ibrahim, and A. R. Katritzky, *Synthesis*, 105 (1976).

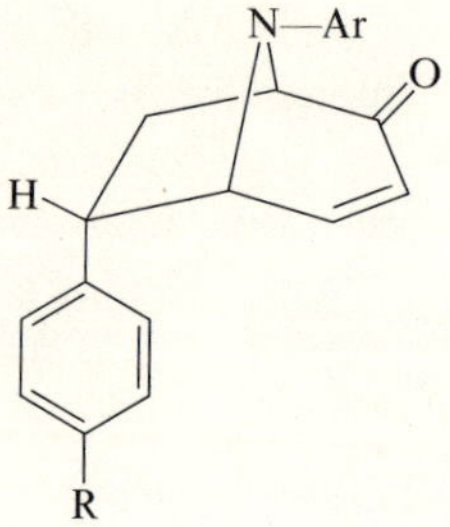

(126) 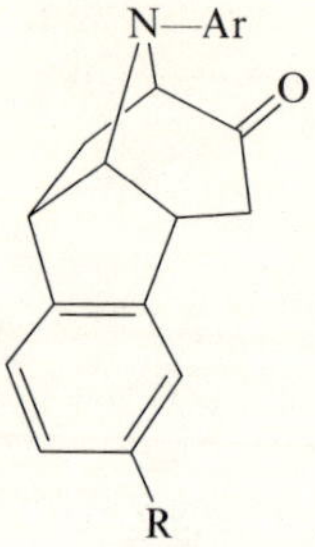   (127)

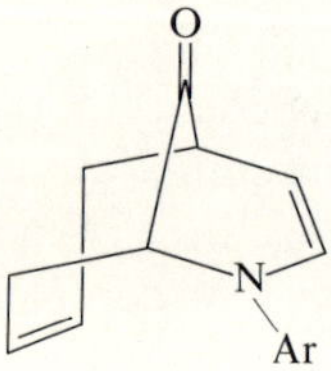

(128) 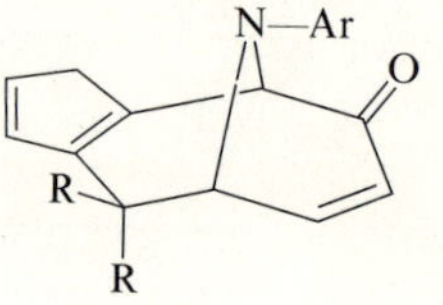   (129)

behavior is observed using disubstituted ethylenes.[73] The different orbital symmetry characteristics of 4π-electron 1,3-dipolarophiles results in addition across the betaine 2,4-positions. Compound **123** with 2,3-dimethylbuta-1,3-diene gives the 2,4-adduct **128**. Cyclopentadiene gives a mixture of products including a 2,4-adduct. 6π-Electron addends are like olefins in that 2,6-adducts are formed. Thus 6,6-dimethylfulvene gives the adduct **129** and similar compounds are obtained with other fulvenes.[73]

3. *Pyridinium-3-aminides* (**130**)

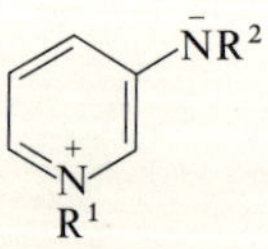

(130) 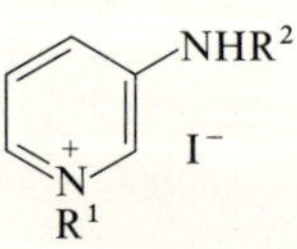   (131)

Two derivatives of this system (**130**; R^1 = Me, R^2 = SO_2Me and 2,4,6-trinitrophenyl) have been obtained by base treatment of the corresponding iodides (**131**). The crystal structure of compound **130** (R^1 = Me, R^2 = SO_2Me) has been determined. 1,3-Dipolar reactivity of these betaines (**130**) has not been observed.[75]

[75] N. Dennis, A. R. Katritzky, H. Wilde, E. Gavuzzo, and A. Vaciago, *J. C. S. Perkin 2*, 1304 (1977).

4. Thiopyrylium-3-olates (132)

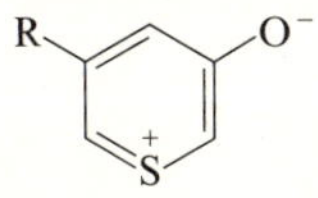

(132) 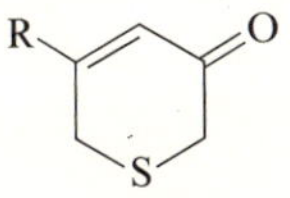   (133) 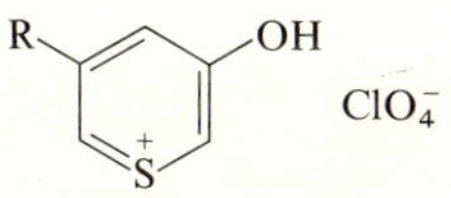   (134)

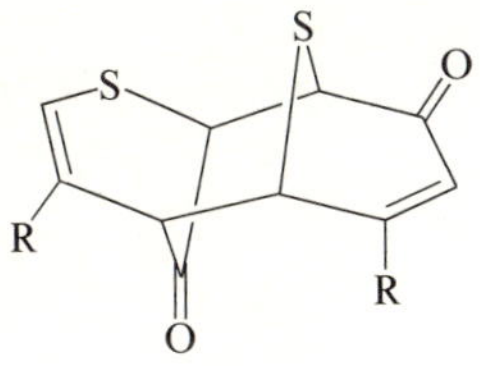

(135) exo-syn

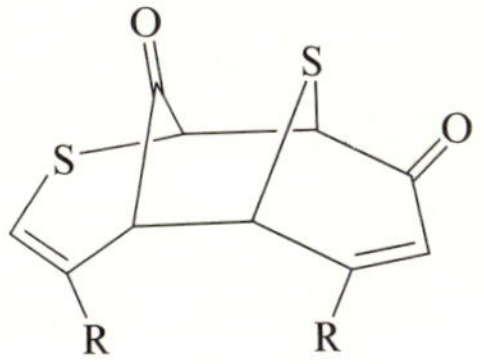

(136) endo-syn

Thiopyran-3(6H)-ones (133; R = H or Me) are converted into 3-hydroxythiopyrylium perchlorates (134; R = H or Me) by triphenylmethyl perchlorate. Deprotonation of these salts (134) by tertiary base did not liberate the thiopyrylium-3-olates (132) but instead gave dimeric products (135 and 136) in high yield. Evidence for the transient formation of the mesomeric betaines (132) is provided by the appearance of a greenish-yellow coloration which rapidly fades. The parent betaine (132; R = H) gave exclusively the endo dimer (136; R = H) (70%). In addition to the endo dimer (136; R = Me) (75%), the 5-methyl derivative (132; R = Me) also gave a small amount of the exo dimer (135; R = Me) (6%). The dimerization (132 → 136) appears to be rapid and irreversible: Attempts to trap the betaines (132) using 1,3-dipolarophiles *in situ* were not successful. It is notable that whereas the isoconjugate pyridinium-3-olates (86) (Section III,A,2) form exo dimers (Scheme 3), the thiopyrylium-3-olates (132) favor endo dimers. In both cases (2,2'–4,6') dimers (i.e., exo-syn and endo-syn) appear to be kinetically favored, a preference which is probably controlled by the magnitude of the HOMO and LUMO coefficients (see Section V).[75a]

5. Pyridazinium-3-olates (137)

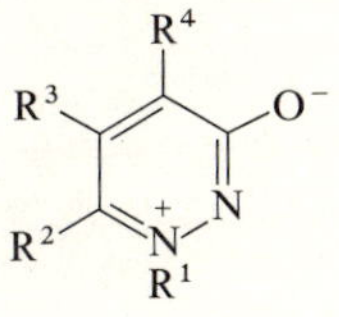

(137) 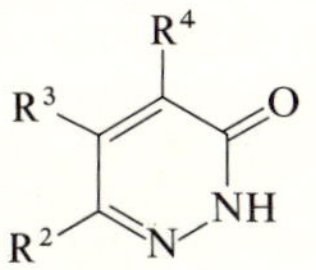   (138) 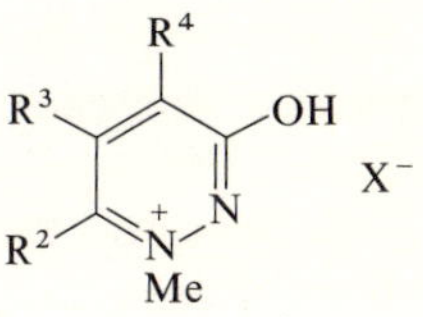   (139)

[75a] S. Baklien, P. Groth, and K. Undheim, *ibid.* 1, 2099 (1975).

(140)   (141)

Methylation (Me$_2$SO$_4$ or MeOTs) of pyridazin-3-ones (138) followed by deprotonation of the salts (139) gives good yields of the crystalline pyridazinium-3-olates (137; R^1 = Me) ($v_{C=O}$ 1550 cm^{-1}).[76–78] These betaines (137; R^1 = Me) appear to be unreactive toward dipolarophiles.[78]

Photolysis of 1,6-dimethylpyridazinium-3-olate (137; R^1 = R^2 = Me, R^3 = R^4 = H) in water gives 2,6-dimethylpyridazin-3-one (140), whereas irradiation in acetonitrile solution gives the fused diaziridine 141 which is converted into compound 140 by water. This transformation (141 → 140) probably involves initial hydrolysis of the amide bond (see Section III,B,7).[79]

6. *Pyridazinium-5-olates* (142)

(142)   (143)   (144)

Microbial oxidation of D-glucose gives calcium 2,5-diketo-D-gluconate (145) whose hydrazones (146) are converted to the betaines 148 by acid-catalyzed decarboxylation and cyclodehydration (Scheme 6).[80] Derivatives of these stable crystalline compounds have also been prepared by (i) methylation (Me$_2$SO$_4$) of 4-hydroxypyridazines (143)[81] and (ii) thermal elimination of methyl iodide from 5-methoxy-1-methylpyridazinium iodides (144).[82] Irradiation of the mesomeric betaines (148) in ethanol (Hg arc lamp)

---

[76] T. Nakagome, A. Misaki, and A. Murano, *Chem. Pharm. Bull.* **14**, 1090 (1966).
[77] F. Reicheneder and R. Kropp (Badische Anilin-und-Soda-Fabrik A.-G.), Ger. Offen. 2,003,461 (1971).
[78] N. Dennis, A. R. Katritzky, and M. Ramaiah, *J. C. S. Perkin 1*, 1506 (1975).
[79] Y. Maki, M. Kawamura, H. Okamoto, M. Suzuki, and K. Kaji, *Chem. Lett.*, 1005 (1977).
[80] K. Imada, *J. C. S., Chem. Commun.*, 796 (1973); K. Imada and K. Asano, (Daiichi Seiyaku Co., Ltd.), Japanese Patent 73 03,629 [*CA* **78**, 147982 (1973)]; K. Imada, *Chem. Pharm. Bull.* **22**, 1732 (1974).
[81] K. Eichenberger, R. Rometsch, and J. Druey, *Helv. Chim. Acta* **39**, 1755 (1956).
[82] G. B. Barlin and P. Lakshminarayana, *J. C. S., Perkin 1*, 1038 (1977).

gives good yields of the isomeric 6-hydroxymethyl-4(3*H*)-pyrimidinones **149** (Scheme 6).[83,84] The mechanism of this photoisomerization is not clear; similar rearrangements of mesoionic rings have been observed[1] (see also Section III,D,1).

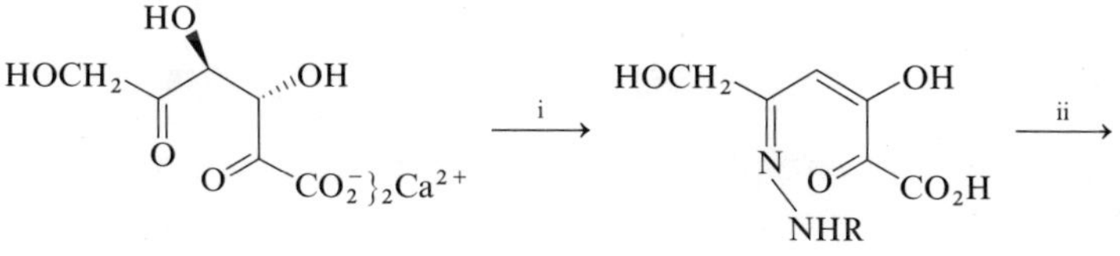

(145) **(146)**

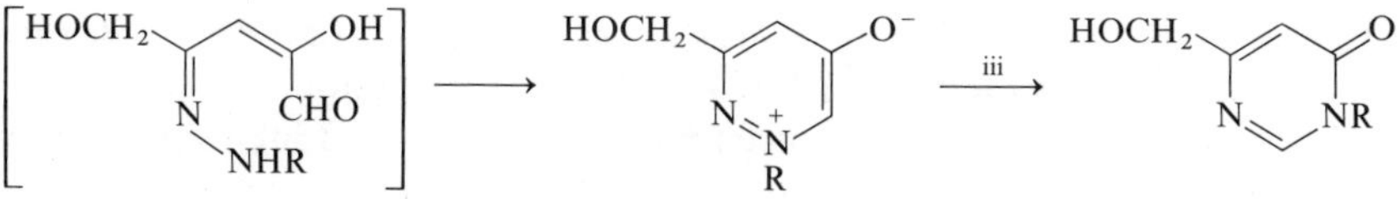

(147) **(148)** **(149)**

SCHEME 6. Reagents: i, $RNHNH_2$; ii, ΔT, H^+/H_2O; iii, *hν*.

B. SYSTEMS WITH ELEVEN CONJUGATED ATOMS

1. *2-Benzopyrylium-4-olates* (**150**)

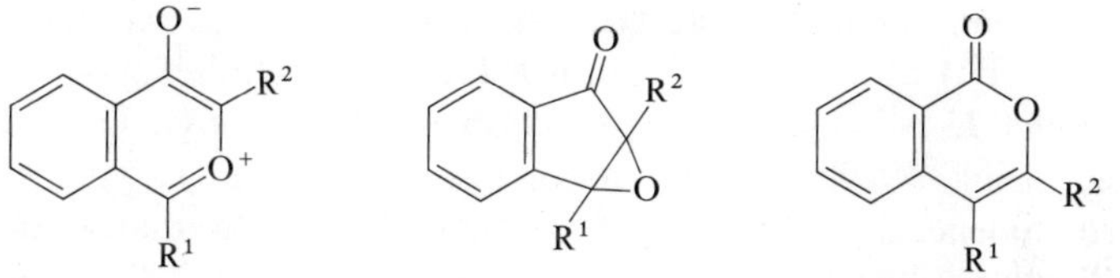

(150) **(151)** **(152)**

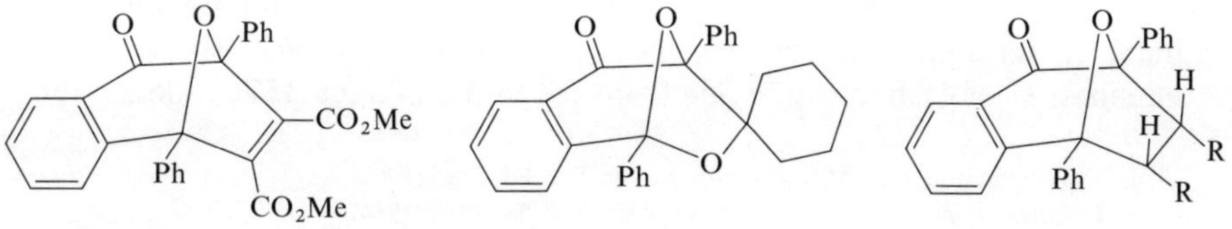

(153) **(154)** **(155)**

[83] Y. Maki, M. Suzuki, T. Furuta, T. Hiramitsu, and M. Kuzuya, *Tetrahedron Lett.*, 4107 (1974).

[84] Y. Maki and K. Imada (Daiichi Seiyaku Co., Ltd.), Japanese Patent 75 101,369 [*CA* **84**, 59532 (1976)].

(156)   (157)   (158)

Representatives of the benzopyrylium-4-olates (**150**) can be generated in solution by thermal or photolytic isomerization of indenone oxides (**151**).[85–89] Irradiation of solutions of compound **151** ($R^1 = R^2 = Ph$) with UV light gives a red coloration attributable to the mesomeric betaine (**150**; $R^1 = R^2 = Ph$). Blue-black crystals of this betaine (mp 94°–97°C) are deposited during irradiation (2537 Å) of hexane solutions of compound **151** ($R^1 = R^2 = Ph$).[89] The valence tautomerism (**151** → **150**) is reversed by visible light.[86] Dimethyl acetylenedicarboxylate traps the betaine giving the expected adduct **153**.[85,86] Prolonged irradiation gives 3,4-diphenylisocoumarin (**152**; $R^1 = R^2 = Ph$) (29%), 3-phenylphthalide (5%), and a dimer (26%), formulated as compound **160**.[88,89] Photolysis of 2-methyl-3-phenylindenone oxide (**151**; $R^1 = Ph$, $R^2 = Me$) gives comparable results but an alternative structure has been assigned to the dimeric product.[90] Irradiation of indenone oxide (**151**; $R^1 = R^2 = H$) gives only isocoumarin (**152**; $R^1 = R^2 = H$).[91]

Thermolysis of indenone oxides (**151**) is an equally useful route to the betaines (**150**). Dimethyl acetylenedicarboxylate and compound **151** ($R^1 = R^2 = Ph$) at 175°C give adduct **153**, and cyclohexanone at 150°C gives adduct **154**.[85,86] Similarly, 1,3-dipolar adducts (e.g., **155**) have been obtained using a wide variety of olefins—including *cis*- and *trans*-1,2-dichloroethylene, dimethyl maleate, dimethyl fumarate, maleic anhydride, *cis*- and *trans*-stilbene, *trans*-dibenzoylethylene, *trans*-1,2-dicyanoethylene, *N*-phenylmaleimide, vinylene carbonate, acenaphthylene, and norbornadiene.[86,92] With cis olefins the endo adduct (**155**) is usually the predominant isomer. Diphenylcyclopropenone gives compound **156** by spontaneous elimination of carbon monoxide from the initial adduct (**157**). Adduct **156**

---

[85] E. F. Ullman and J. E. Milks, *J. Am. Chem. Soc.* **84**, 1315 (1962).
[86] E. F. Ullman and J. E. Milks, *J. Am. Chem. Soc.* **86**, 3814 (1964).
[87] E. F. Ullman, *J. Am. Chem. Soc.* **86**, 5357 (1964).
[88] E. F. Ullman and W. A. Henderson, *J. Am. Chem. Soc.* **86**, 5050 (1964).
[89] E. F. Ullman and W. A. Henderson, *J. Am. Chem. Soc* **88**, 4942 (1966).
[90] H. E. Zimmerman and R. D. Simkin, *Tetrahedron Lett.*, 1847 (1964).
[91] K. Undheim and B. P. Nilsen, *Acta Chem. Scand.*, *Ser. B* **29**, 503 (1975).
[92] J. W. Lown and K. Matsumoto, *Can. J. Chem.* **49**, 3443 (1971).

Sec. III.B] HETEROCYCLIC BETAINES 29

is also formed from the betaine **150** ($R^1 = R^2 = Ph$) and diphenylacetylene. With benzyne, compound **158** is formed.[92]

Prolonged thermolysis of compound **151** ($R^1 = R^2 = Ph$) results in the formation of a pair of dimeric products which have been formulated as compounds **160** (47%) and **161** (4%) and which can be regarded as being formed by addition of the indenone carbonyl group to the betaine. Precedent for this type of addition is formation of the cyclohexanone adduct **154**.[86] Evidence for structure **160** is provided by the observation that acid hydrolysis gives a monohydrate which is formulated as compound **162**—this product **162** "displayed ... *ultraviolent* absorption"![86] Isomer **161** does not form a monohydrate.

It is notable that only dimer **160** is produced photochemically, and it is possible that this product (**160**) is kinetically preferred whereas the thermodynamically preferred isomer (**161**) is formed only at elevated temperatures.

Nilsen and Undheim have prepared the unsubstituted betaine **166** by oxidation of 4-acetoxyisochromene (**164**). Thus, compound **164** with tetrachloro-1,2-benzoquinone (TBQ) or 2,3-dichloro-5,6-dicyano-1,4-benzoquinone (DDQ) forms adducts **163** and **165** which in trifluoroacetic acid give 2-benzopyrylium-4-olate (**166**). Compound **166** cannot be isolated because dimerization readily occurs. Surprisingly, the TBQ adduct (**165**) gives the endo dimer (**168**) (19%) together with a small amount of the exo dimer (**167**) (1%) whereas the DDQ adduct (**163**) gives exclusively the exo dimer (**167**) (27%).[93]

It is an interesting facet of the chemistry of 2-benzopyrylium-4-olates (**150**) that they appear to form two types of dimer (cf. **160** and **167**) depending upon the nature of the ring substituents.

(**163**)   (**164**)   (**165**)

(**166**)   (**167**) exo-anti   (**168**) endo-anti

2. *Isoquinolinium-4-olates* (**170**)

The preceding sections demonstrate two important general routes to six-membered heterocyclic mesomeric betaines. These are (i) deprotonation of appropriate quaternary salts and (ii) valence tautomerism of bicyclic isomers. Both approaches have been used to prepare isoquinolinium-4-olates (**170**) (Scheme 7).

Indano[1,2-*b*]aziridin-6-ones (**171**), which have been prepared by three different routes,[94–96] are converted to isoquinolinium-4-olates (**170**) by either photolysis[96] or thermolysis.[94,96] The betaines (**170**) have not been

---

[93] B. P. Nilsen and K. Undheim, *Acta Chem. Scand., Ser. B* **30**, 619 (1976).
[94] D. L. Garling and N. H. Cromwell, *J. Org. Chem.* **38**, 654 (1973).
[95] K. Undheim and P. E. Hansen, *Chem. Scr.* **3**, 113 (1973).
[96] P. E. Hansen and K. Undheim, *J. C. S., Perkin 1*, 305 (1975).

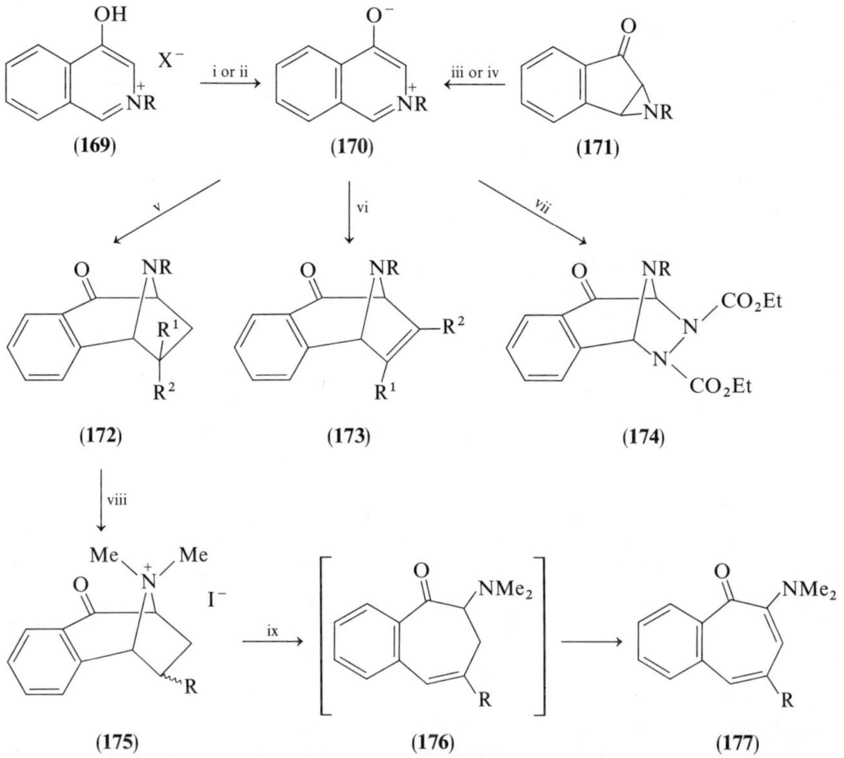

SCHEME 7. Reagents: i, ion-exchange resin; ii, Et₃N; iii, hv; iv, ΔT; v, $R^1R^2C{=}CH_2$; vi, $R^1C{\equiv}CR^2$; vii, $EtO_2CN{=}NCO_2Et$; viii, MeI; ix, Ag_2O or $NaHCO_3$.

isolated by this method but they can be trapped either as olefin adducts or as quaternary salts.[96] Alternatively, the *N*-methyl derivative (**170**; R = Me) has been prepared (as a hydrate) by treatment of the iodide (**169**; R = Me, X = I) with an ion-exchange resin.[97] A similar deprotonation of the iodide **169** (R = 2,4-dinitrophenyl, X = I) using triethylamine generates the betaine **170** (R = 2,4-dinitrophenyl) *in situ*.[98]

The betaines **170** (R = Me, 2,4-dinitrophenyl) are reactive 1,3-dipolar-species.[94,97,98] With electron-deficient olefins they give tricyclic adducts (**172**). Acetylenes and diethyl azodicarboxylate give similar adducts (**173** and **174**) (Scheme 7). Menschutkin methylation of **172** gives the methiodides (**175**) which undergo Hofmann elimination to 2-dimethylamino-6,7-benzotropones (**177**) (Scheme 7). The benzocycloheptadienones (**176**) are not

---

[97] N. Dennis, A. R. Katritzky, and Y. Takeuchi, *J. C. S., Perkins 1*, 2054 (1972).
[98] N. Dennis, A. R. Katritzky, and S. K. Parton, *Chem. Pharm. Bull.* **23**, 2899 (1975).

isolated but aromatize spontaneously. The sequence **170** → **172** → **175** → **177** (Scheme 7), which is analogous to the preparation of tropones from pyridinium-3-olates (Section III,A,2), provides a valuable synthetic route to benzotropones (**177**).[97]

3. *Isoquinolinium-4-aminides* (**178**)

Thermolysis of 1-cyclohexyl-6-(cyclohexylimino)-1a-phenylindano[1,2-*b*]aziridine (**179**; R^1 = Ph, R^2 = R^4 = C_6H_{11}, R^3 = H) in toluene or xylene at 135°C gives a deep purple solution of the isoquinolinium betaine **178** (R^1 = Ph, R^2 = R^4 = C_6H_{11}, R^3 = H). This valence tautomerism (**179** → **178**) is reversed by cooling or sunlight, but prolonged heating results in slow elimination of cyclohexene and formation of 1-phenyl-4-cyclohexylamino-isoquinoline. In the presence of 1,3-dipolarophiles, the betaine is trapped giving adducts of the type **181**. Typically, acrylonitrile gives compound **181** (R = C_6H_{11}, R^1 = R^3 = R^4 = H, R^2 = CN). Other olefinic 1,3-dipolar-ophiles that have been used are *N*-phenylmaleimide, dimethyl fumarate, cyclohexene, norbornene, and norbornadiene. The exclusive formation of endo adducts (**181**; R^1 = R^3 = H) may be due to the bulky N-substituent. In a similar manner acetylenes give 1,3-dipolar cycloadducts: dimethyl acetylenedicarboxylate gives **182** (R = C_6H_{11}, R^1 = R^2 = CO_2Me); methyl propiolate gives mainly **182** (R = C_6H_{11}, R^1 = H, R^2 = CO_2Me); methyl phenylpropiolate gives a mixture of stereoisomeric adducts (**182**; R = C_6H_{11}, R^1 = CO_2Me or Ph, R^2 = Ph or CO_2Me); and benzyne gives compound **183**. All these cycloaddition reactions closely parallel the reactions of the 2-benzopyrylium-4-olates (**150**) (Section III,B,1) and isoquinolinium-4-olates (**170**) (Section III,B,2). Spontaneous elimination of carbon monoxide from the diphenylcyclopropenone adduct (**184**) gives, in a manner similar to adduct **157**, the diphenylacetylene adduct **182** (23%), but in this case the ring-opened product **185** (42%) is also formed. If betaine **178** (R^1 = Ph, R^2 = R^4 = C_6H_{11}, R^3 = H) is generated in the presence of ammonium halides, it is then trapped as the isoquinolinium salt **180**.[99,100]

(**178**)   (**179**)   (**180**)

[99] J. W. Lown and K. Matsumoto, *Chem. Commun.*, 692 (1970).
[100] J. W. Lown and K. Matsumoto, *J. Org. Chem.* **36**, 1405 (1971).

Photolysis of compound **179** ($R^1 = Ph$, $R^2 = R^4 = C_6H_{11}$, $R^3 = H$) also generates betaine **178** ($R^1 = Ph$, $R^2 = R^4 = C_6H_{11}$, $R^3 = H$) which can be trapped as the chloride (**180**; $R = C_6H_{11}$, $X = Cl$), but attempts to intercept the photoproduct (**178**) with dipolarophiles were less successful.[99,100]

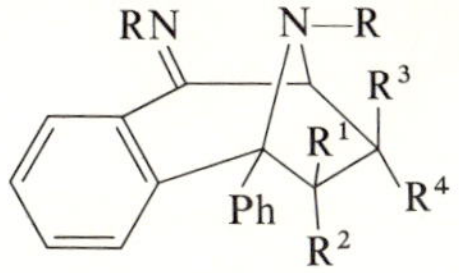

(181) 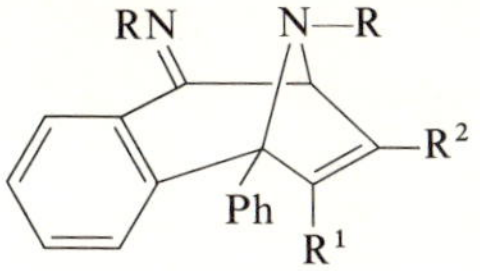   (182) 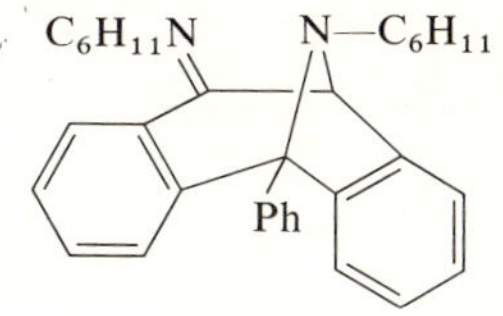   (183)

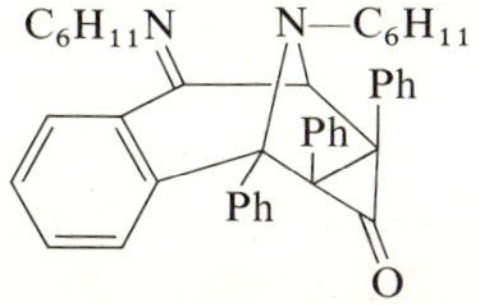

(184) 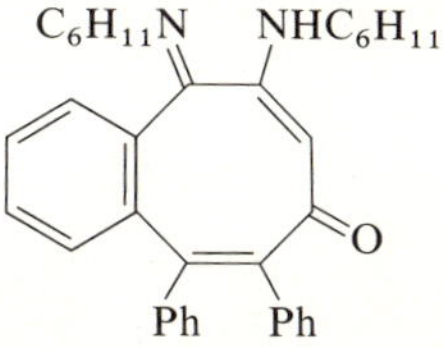   (185)

4. Quinolizinium-1-olates (186)

The conversion of 1-hydroxyquinolizinium bromide (**187**; $R^1 = R^2 = H$, $X = Br$) into three quinolizinium-1-olate derivatives (**186**) has been achieved as follows: (i) Treatment with aqueous sodium carbonate gives the hydrated betaine (**186**; $R^1 = R^2 = H$) as a yellow solid; (ii) brief reaction with hot dilute nitric acid gives red needles of the 4-bromo-2-nitro derivative (**186**; $R^1 = NO_2$, $R^2 = Br$); and (iii) prolonged reaction with hot dilute nitric acid gives the dinitro derivative (**186**; $R^1 = R^2 = NO_2$). 1-Hydroxyquinolizinium nitrate (**187**; $R^1 = R^2 = H$, $X = NO_2$) and dilute nitric acid give the 2-nitro derivative (**186**; $R^1 = NO_2$, $R^2 = H$).[101,102] Betaine **186** ($R^1 = Me$, $R^2 = H$) has been detected in aqueous solutions of 1-hydroxy-2-methylquinolizinium bromide (**187**; $R^1 = Me$, $R^2 = H$).[103]

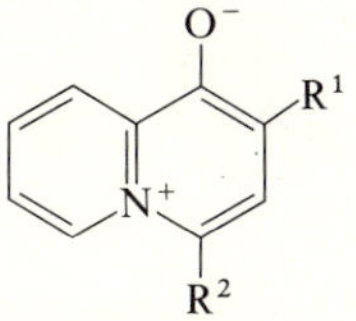

(186) 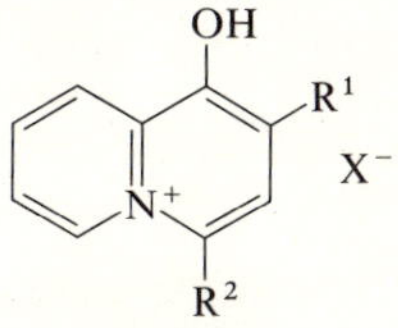   (187) 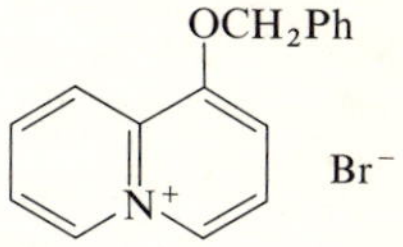   (188)

[101] P. A. Duke, A. Fozard, and G. Jones, *J. Org. Chem.* **30**, 526 (1965).
[102] A. Fozard and G. Jones, *J. Chem. Soc.*, 3030 (1964).
[103] J. Adamson and E. E. Glover, *J. Chem. Soc. C*, 861 (1971).

Little is known about the chemistry of this class of mesomeric betaines. Catalytic reduction of the 2-nitro betaine (**186**; $R^1 = NO_2$, $R^2 = H$) gives 2-hydroxy-3-aminoquinolizinium bromide.[102] Compound **186** ($R^1 = R^2 = H$) and benzyl bromide give 1-benzyloxyquinolizinium bromide (**188**).[101]

5. *Quinolinium-8-olates* (**189**)

(**189**)   (**190**)

Diazoxine, a red product which accompanies the formation of 8-methoxyquinoline from 8-hydroxyquinoline and diazomethane, was first encountered by Caronna and Sansone in 1939.[104] Later, it was suggested that the properties of this product were consistent with structure **189** ($R^1 = Me$, $R^2 = R^3 = H$).[105] Subsequently, this product has been prepared by treatment of the iodide **190** with potassium carbonate, and the betaine structure has been confirmed by spectroscopic and chemical studies. Compound **189** ($R^1 = Me$, $R^2 = R^3 = H$) is isolated as hydrated violet-red needles. The UV and visible spectra are strongly dependent on the nature of the solvent; the colors of solutions vary from yellow to blue.[106,107] Bromination gives the 5,7-dibromo derivative (**189**; $R^1 = Me$, $R^2 = R^3 = Br$), which is also obtained from 5,7-dibromo-8-hydroxyquinoline and diazomethane. Treatment with hydrochloric acid gives a hydrochloride.[107]

6. *2-Benzothiopyrylium-4-olates* (**191**)

(**191**)   (**192**)   (**193**)

---

[104] G. Caronna and B. Sansone, *Gazz. Chim. Ital.* **69**, 24 (1939).
[105] H. Schenkel-Rudin and M. Schenkel-Rudin, *Helv. Chim. Acta* **27**, 1456 (1944).
[106] J. P. Saxena, W. H. Stafford, and W. L. Stafford, *J. Chem. Soc.*, 1579 (1959).
[107] J. P. Phillips and R. W. Keown, *J. Am. Chem. Soc.* **73**, 5483 (1951).

(194)
exo-anti

(195)
endo-anti

Derivatives of this system (**191**) have not been isolated. The parent molecule (**191**; R = H) and a methyl derivative (**191**; R = Me) have been generated in solution but dimerization occurs rapidly. Thus, cyclodehydration of (benzylthio)acetic acid (PhCH$_2$SCH$_2$CO$_2$H) gives isothiochroman-4-one (**192**; R = H) which with triphenylmethyl perchlorate gives 4-hydroxy-2-benzothiopyrylium perchlorate (**193**; R = H). Deprotonation of this salt with triethylamine gives a transient deep yellow coloration that is attributable to the betaine (**191**; R = H). The color rapidly fades and the final products are the betaine dimers **195** (74%) and **194** (10%). A similar sequence employing the 1-methyl-2-benzothiopyrylium perchlorate **193** (R = Me) gives only the exo dimer (**194**; R = Me) (70%). This affinity for dimerization is analogous to that of 2-benzopyrylium-4-olate (**166**) (Section III,B,1). Attempts to trap the short-lived mesomeric betaines (**191**; R = H or Me) by 1,3-dipolarophiles were unsuccessful.[108–110]

7. *Phthalazinium-1-olates* (*Pseudophthalazones*) (**196**)

(196)   **(197)**   **(198)**

Examples of phthalazinium-1-olates (**196**) were first prepared in 1926 by Rowe[111] and co-workers using the sequence shown in Scheme 8. Subsequent

---

[108] K. Undheim and S. Baklien, *J. C. S., Perkin 1*, 1366 (1975).
[109] S. Baklien, P. Groth, and K. Undheim, *Acta Chem. Scand., Ser. B* **30**, 24 (1976).
[110] P. Groth, *Acta Chem. Scand., Ser. A* **29**, 298 (1975).
[111] E. H. Rodd, *J. Chem. Soc.*, 2323 (1948).

studies by this group have described the chemistry of a large number of 3-aryl derivatives (**196**; $R^1 = Ar$).[112–137] Detailed accounts of this work have been published[138–140] and only general features are included in this review.

Coupling of 2-naphthol-1-sulfonic acid (**199**) with diazonium salts leads in several well-defined steps (Scheme 8) to 1-hydroxy-3,4-dihydrophthalazine-4-acetic acids (**204**).[112,115,124] Treatment of these intermediates (**204**) with hot aqueous sulfuric acid results in elimination of acetic acid giving 4-unsubstituted 3-arylphthalazinium-1-olates (**205**) whereas oxidation with cold acidic potassium dichromate solution gives 4-methyl derivatives (**206**).[113–123] This route is particularly successful for (but not exclusive to) derivatives in which the 3-aryl substituent carries a nitro group. The *N*-phenyl derivatives (e.g., **205**; Ar = Ph) (mp 208°C), which are unusual in that they are water soluble, can be made in good yield.[137]

[112] F. M. Rowe, E. Levin, A. C. Burns, J. S. H. Davies, and W. Tepper, *J. Chem. Soc.*, 690 (1926).
[113] F. M. Rowe and E. Levin, *J. Chem. Soc.*, 2550 (1928).
[114] F. M. Rowe, M. A. Himmat, and E. Levin, *J. Chem. Soc.*, 2556 (1928).
[115] F. M. Rowe and A. T. Peters, *J. Chem. Soc.*, 1065 (1931).
[116] F. M. Rowe, E. Levin, and A. T. Peters, *J. Chem. Soc.*, 1067 (1931).
[117] F. M. Rowe, C. Dunbar, and N. H. Williams, *J. Chem. Soc.*, 1073 (1931).
[118] F. M. Rowe and A. T. Peters, *J. Chem. Soc.*, 1918 (1931).
[119] F. M. Rowe and C. Dunbar, *J. Chem. Soc.*, 11 (1932).
[120] F. M. Rowe and F. J. Siddle, *J. Chem. Soc.*, 473 (1932).
[121] F. M. Rowe and F. S. Tomlinson, *J. Chem. Soc.*, 1118 (1932).
[122] F. M. Rowe and A. T. Peters, *J. Chem. Soc.*, 1067 (1933).
[123] F. M. Rowe, G. B. Jambuserwala, and H. W. Partridge, *J. Chem. Soc.*, 1134 (1935).
[124] F. M. Rowe, W. C. Dovey, B. Garforth, E. Levin, J. D. Pask, and A. T. Peters, *J. Chem. Soc.*, 1796 (1935).
[125] F. M. Rowe, J. G. Gillan, and A. T. Peters, *J. Chem. Soc.*, 1808 (1935).
[126] F. M. Rowe, A. S. Haigh, and A. T. Peters, *J. Chem. Soc.*, 1098 (1936).
[127] F. M. Rowe and H. J. Twitchett, *J. Chem. Soc.*, 1704 (1936).
[128] F. M. Rowe, D. A. W. Adams, A. T. Peters, and A. E. Gillam, *J. Chem. Soc.*, 90 (1937).
[129] F. M. Rowe, M. A. Lécutier, and A. T. Peters, *J. Chem. Soc.*, 1079 (1938).
[130] F. M. Rowe and E. J. Cross, *J. Chem. Soc.*, 461 (1947).
[131] F. M. Rowe, W. T. McFadyen, and A. T. Peters, *J. Chem. Soc.*, 468 (1947).
[132] F. M. Rowe and W. Osborn, *J. Chem. Soc.*, 829 (1947).
[133] F. M. Rowe, A. T. Peters, and Y. I. Rangwala, *J. Chem. Soc.*, 206 (1948).
[134] F. M. Rowe, R. L. Desai, and A. T. Peters, *J. Chem. Soc.*, 281 (1948).
[135] A. T. Peters, G. T. Pringle, and F. M. Rowe, *J. Chem. Soc.*, 597 (1948).
[136] C. I. Brodrick, A. T. Peters, and F. M. Rowe, *J. Chem. Soc.*, 1026 (1948).
[137] A. T. Peters, F. M. Rowe, and C. I. Brodrick, *J. Chem. Soc.*, 1249 (1948).
[138] J. C. E. Simpson, in "The Chemistry of Heterocyclic Compounds" (A. Weissberger, ed.), Vol. 5, p. 119. Wiley (Interscience), New York, 1953.
[139] N. R. Patel, in "The Chemistry of Heterocyclic Compounds" (R. N. Castle, ed.), Vol. 27, p. 652. Wiley (Interscience), New York, 1973.
[140] R. C. Elderfield and S. L. Wythe, in "Heterocyclic Compounds" (R. C. Elderfield, ed.), Vol. 6, p. 203. Wiley, New York, 1957.

An alternative route that also gives good yields of 3-arylphthalazinium-1-olates (e.g., **205**; Ar = Ph) has been described by Lund and involves thermolysis (100–120°C) of *N*-arylamino-3-hydroxyphthalimidines (**197**).[141,142] This transformation (**197** → **196**) probably proceeds by formation of a hydrazide (**198**) prior to cyclodehydration (**198** → **196**). Compound **205** (Ar = Ph) is obtained in low yields by lithium aluminum hydride reduction of 4-hydroxy-2-phenyl-1(2*H*)-phthalazinone **207** (R^1 = OH, R^2 = Ph).[143]

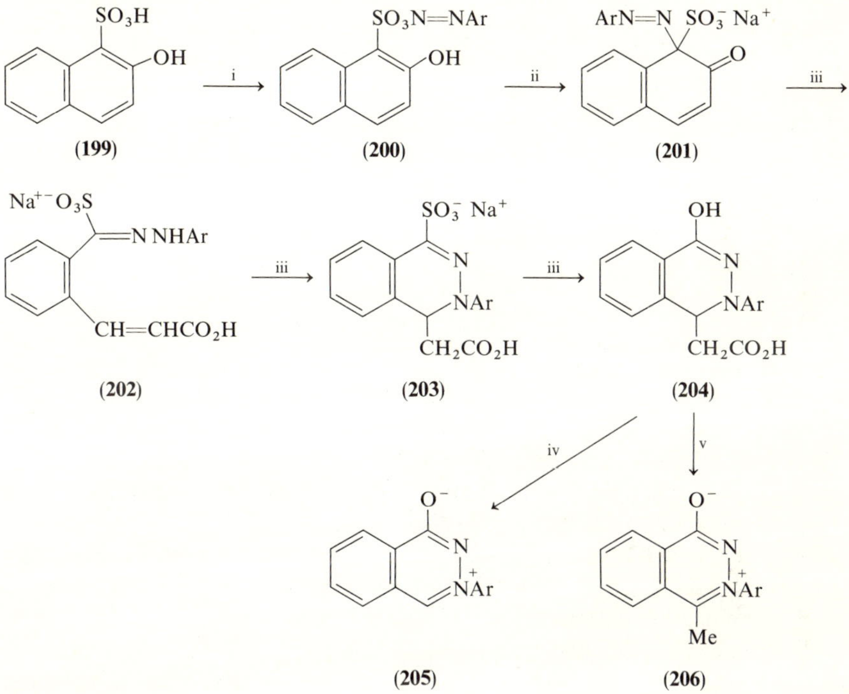

SCHEME 8. Reagents: i, $ArN_2^+X^-$; ii, aq Na_2CO_3; iii, aq NaOH; iv, aq H_2SO_4 at 140°C; v, aq H_2SO_4/K_2CrO_4.

3-Methylphthalazinium-1-olate (**196**; R^1 = Me, R^2 = H) has been prepared in several ways. Phthalazin-1-one (**207**; $R^1 = R^2$ = H) with methyl tosylate gives **208** which at pH 6 liberates the free betaine.[144] The same transformation [**207** ($R^1 = R^2$ = H) → **196** (R^1 = Me, R^2 = H)] is also achieved by treatment of compound **207** ($R^1 = R^2$ = H) with methyl iodide

[141] H. Lund, *Tetrahedron Lett.*, 3973 (1965).
[142] H. Lund, *Collect. Czech. Chem. Commun.* **30**, 4237 (1965).
[143] B. K. Diep and B. Cauvin, *C. R. Hebd. Seances Acad. Sci.*, Ser. C **262**, 1010 (1966).
[144] N. Dennis, A. R. Katritzky, and M. Ramaiah, *J. C. S., Perkin 1*, 2281 (1976).

and silver oxide.[145] Methylation of 4-carboxyphthalazin-1-one (207; R^1 = CO_2H, R^2 = H) with methyl iodide and alkali is accompanied by decarboxylation giving compound 196 (R^1 = Me, R^2 = H).[146] Alternatively, N-methyl and N-benzyl derivatives (196; R^1 = Me, $PhCH_2$) have been obtained by reaction of 1-alkoxy-3-alkylphthalazinium iodides (209) with silver oxide.[145]

(207)   (208)   (209)

(210)   (211)   (212)

Phthalazinium-1-olates (196) are stable, yellow or colorless crystalline compounds that undergo an interesting variety of reactions. Reduction of aryl derivatives with zinc and dilute hydrochloric acid gives phthalimidines (210) whereas reduction with alkaline sodium hyposulfite gives 3-aryl-3,4-dihydro-1(2H)-phthalazinones (211).[113,114,116,118–120,137] 3-Methylphthalazinium-1-olate (196; R^1 = Me, R^2 = H) is converted to 3-methyl-3,4-dihydro-1(2H)-phthalazinone (211; R^1 = Me, R^2 = H) by sodium borohydride[146]; electrochemical reduction of the 3-phenyl derivative (205; Ar = Ph) similarly gives 3-phenyl-3,4-dihydro-1(2H)-phthalazinone (211; R^1 = Ph, R^2 = H).[142] Methylation of 4-methyl derivatives with dimethyl sulfate yields 1-methoxyphthalazinium salts that are readily deprotonated giving 4-methylene compounds (212).[116,125,127,130,136] The betaines (196) similarly form salts with dilute acids. It is notable that with hydrogen iodide, 3-methyphthalazinium-1-olate (196; R^1 = Me, R^2 = H) forms a complex salt [$(C_9H_8N_2O)_2$ · HI] that is analogous to a salt [$(C_6H_7NO)_2$ · HI] formed by N-methylpyridinium-3-olate (Section III,A,2).[145] Heating 3-arylphthalazinium-1-olates (213) with 1.2 N hydrochloric acid at 180°C (sealed tube) results in rearrangement to the isomeric phthalazones 215 (Scheme 9).[128,134,136] This reaction has become known as the Rowe rearrangement and its mechanism

---

[145] T. Ikeda, S. Kanahara, and K. Aoki, *Yakugaku Zasshi* **88**, 521 (1968) [*CA* **69**, 86934 (1968)].
[146] A. N. Kost, K. V. Grabliauskas, V. G. Vinokurov, and A. M. Zjakun, *J. Prakt. Chem.* **312**, 542 (1970).

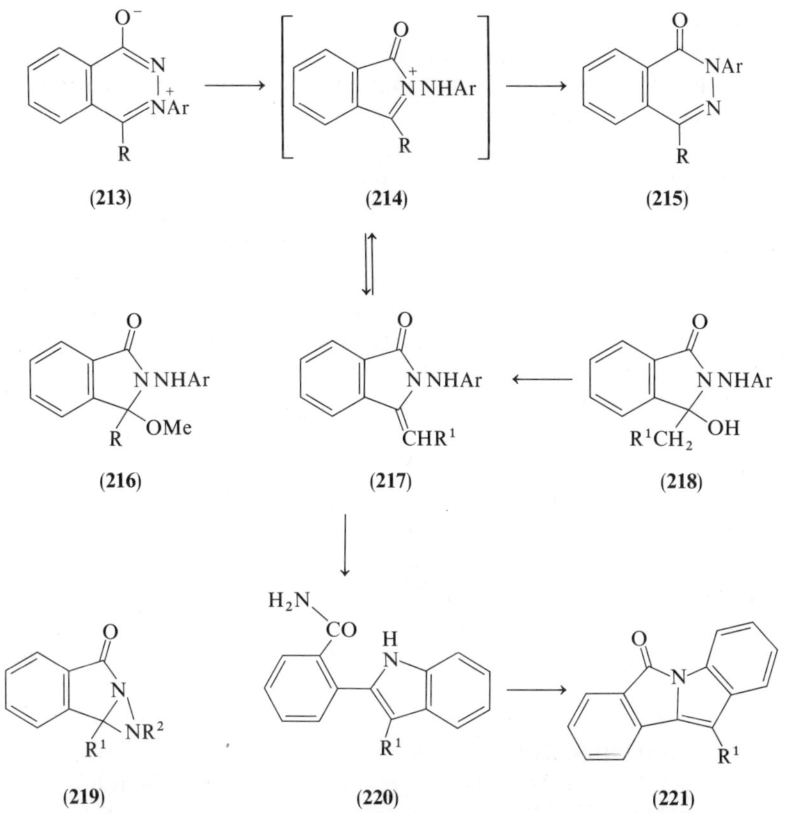

SCHEME 9

is of some interest.[147,148] The isomerization is particularly favored by 2'- and 4'-nitrophenyl substituents and, although the nitro group is not essential, most of the known examples are of this type. ^{15}N-Labeling has shown that aryl group migration does not occur,[148] and a further mechanistic clue is provided by the observation that by using 0.1–0.2 N hydrochloric acid as reaction medium, 4-methyl derivatives (213; R = Me) are transformed into isolatable 2-arylamino-3-methyleneisoindolinones (217; R^1 = H). Concentrated sulfuric acid or 1.2 N hydrochloric acid at 180° converts these isoindolinones (217) into the phthalazinones (215). Vaughan has proposed that the Rowe rearrangement proceeds by initial formation of cations of the type 214 which then undergo ring enlargement (214 → 215).[147,148] Alternatively, in more dilute acid the methyl cations (214; R = Me) may be deprotonated

---

[147] W. R. Vaughan, *Chem. Rev.* **43**, 447 (1948).
[148] W. R. Vaughan, D. I. McCane, and G. J. Sloan, *J. Am. Chem. Soc.* **73**, 2298 (1951).

giving the observed 3-methyleneisoindolinones (**217**), which regenerate the cation in strong acid. In view of the inclination of mesomeric betaines to undergo valence tautomerism, it is tempting to extend the speculation by suggesting that the cations (**214**) arise by initial formation of a diaziridine (**219**) followed by protonation, and that they then ring-open. However, the actual mechanism may well be more mundane. 4-Benzyl-3-phenylphthalazinium-1-olate (**213**; R = PhCH$_2$, Ar = Ph) does not undergo a Rowe rearrangement but instead gives 11-phenylisoindolo[2,1-*a*]indol-6-one (**221**).[149] This product (**221**) together with its precursor, 2-(*o*-carbamoylphenyl)-3-phenylindole (**220**), is also formed when compound **218** (R^1 = Ar = Ph) is heated in a mixture of acetic and concentrated hydrobromic acids, and it is reasonable to suppose that cation **214** (R = PhCH$_2$, Ar = Ph) is a common intermediate in both routes to compound **221**.[149] Presumably, in both cases the enehydrazine **217** (R^1 = Ar = Ph) follows the course of the Fischer indole synthesis.[150]

Photochemical generation of the diaziridine **219** (R^1 = PhCH$_2$, R^2 = Ph) may account for the formation of 2-anilino-3-benzyl-3-methoxyphthalimidine **216** (R = PhCH$_2$, Ar = Ph) when betaine **213** (R = PhCH$_2$, Ar = Ph) is photolyzed in methanol solution.[149] An almost quantitative yield of 2-methylphthalazinone (**207**; R^1 = H, R^2 = Me) is obtained by irradiation of the *N*-methyl betaine **222** in water. If the photolysis is carried out in dry acetonitrile, the product is the valence tautomer **219** (R^1 = H, R^2 = Me) which is converted to 2-methylphthalazinone by water. It has been suggested that this transformation **219** → **207** (R^1 = H, R^2 = Me) involves intermediates **223** (R = OH) and **224**, and the observation that methanolysis and aminolysis of **219** (R^1 = H, R^2 = Me) gives the ester **223** (R = OMe), and the amide **223** (R = NHEt), supports this mechanism.[79]

(**222**)   (**223**)   (**224**)

Like isoconjugate mesomeric betaines, which have been discussed in preceding sections, phthalazinium-1-olates (**196**) give 1,3-dipolar adducts with acetylenes. Thermally induced isomerization reactions of these adducts (**255**) make them of special interest. The *N*-methyl derivative and diphenyl-

---

[149] V. Scartoni, I. Morelli, A. Marsili, and S. Catalano, *J. C. S., Perkin 1*, 2332 (1977).
[150] R. B. Van Order and H. G. Lindwall, *Chem. Rev.* **30**, 78 (1942); P. L. Julian, E. W. Meyer, and H. C. Printy, in "Heterocyclic Compounds" (R. C. Elderfield, ed.), Vol. 3, p. 8. Wiley, New York, 1952.

acetylene give compound **225** ($R^1 = R^2 = Ph$, $R^3 = Me$), but no reaction is observed with phenylacetylene, acrylonitrile, N-phenylmaleimide, or tetracyanoethylene. The N-phenyl betaine **196** ($R^1 = Ph$, $R^2 = H$) is more reactive. Diphenyl acetylene gives the expected adduct **225** ($R^1 = R^2 = R^3 = Ph$), but with dimethyl acetylenedicarboxylate the structure of the product is determined by the solvent. In hot xylene the primary adduct **225** ($R^1 = R^2 = CO_2Me$, $R^3 = Ph$) is formed whereas in hot chloroform the isomeric diazocine **226** ($R^1 = R^2 = CO_2Me$, $R^3 = Ph$) is obtained. Both products are transformed into a third isomer having structure **227** ($R^1 = R^2 = CO_2Me$, $R^3 = Ph$) by heating at 180–190°C. Similar products are obtained using phenylacetylene as 1,3-dipolarophile.[144,151] The precise mechanisms of these rearrangements (**225** or **226** → **227**) are not known. A closely related

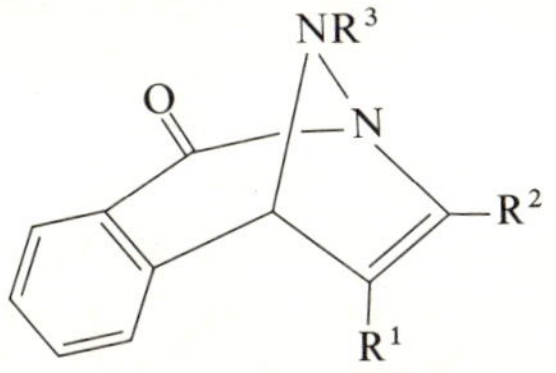

(225) 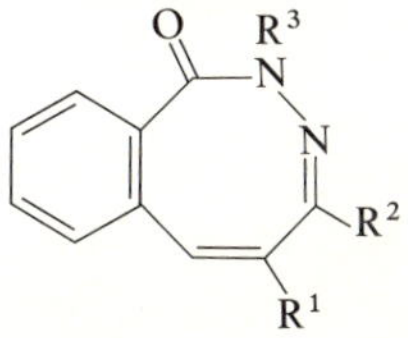   (226) 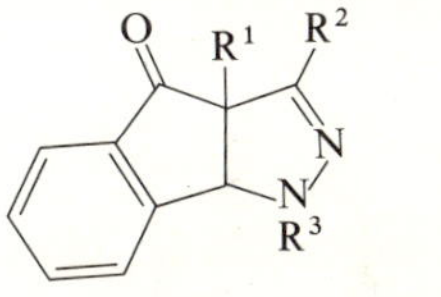   (227)

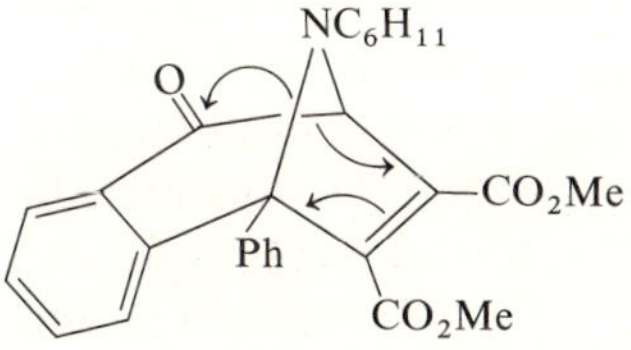

(228) 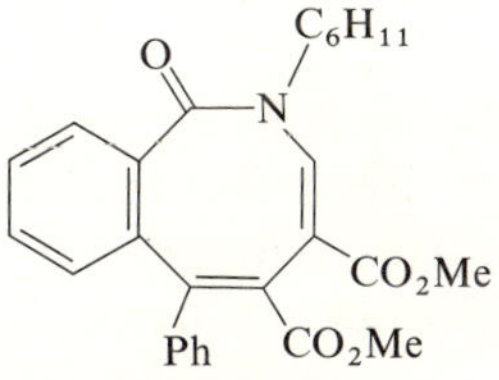   (229) 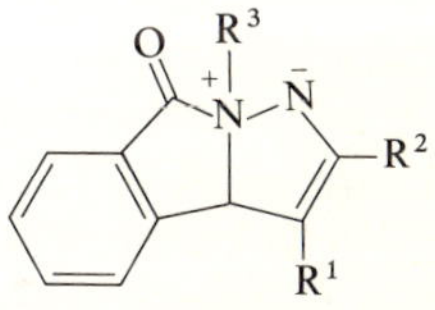   (230)

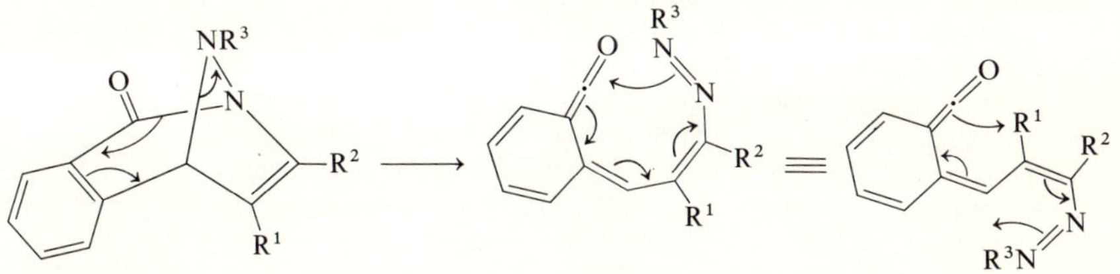

(225)   (231)   (232)

(226)   (227)

[151] N. Dennis, A. R. Katritzky, E. Lunt, M. Ramaiah, R. L. Harlow, and S. H. Simonsen, *Tetrahedron Lett.*, 1569 (1976).

reaction is the thermal transformation of the bridged ketone **228** into the benzazocine **229**.[152] A synchronous mechanism (**228**; arrows) has been suggested for this reaction (**228** → **229**), and a similar mechanism may account for the formation of the benzodiazocines (**226**). Alternatively, the formation of an intermediate zwitterion (**230**) has been suggested: Ring opening leads to the benzodiazocine (**225** → **230** → **226**), and a 1,3-acyl shift (**226** → **230** → **227**) gives compound **229**.[144,151] The author prefers a mechanism in which ring opening of adduct **225** gives the ketene derivative **231** which then undergoes reversible cyclization to the benzodiazocine (**231** → **226**) or which alternatively may undergo "criss-cross" cyclization to the bicyclic product (**232** → **227**). This sequence has the appeal of having precedent in other thermal transformations of 1,3-dipolar cycloadducts (Sections III,A,1 and III,D,1).

8. *Cinnolinium-4-olates* (**233**)

N-Alkyl derivatives of this class of heterocycle have been fully characterized but *N*-aryl derivatives are unknown.

Methylation of 4-hydroxycinnoline with dimethyl sulfate gives a colorless crystalline product, mp 165–166.5°C, which was originally assumed to be 1-methylcinnolin-4-one (**234**; $R^1 = Me$, $R^2 = R^3 = R^4 = R^5 = R^6 = H$)[153] but later reformulated as 2-methylcinnolinium-4-olate (**233**; $R^1 = Me$, $R^2 = R^3 = R^4 = R^5 = R^6 = H$) on the basis that reduction with lithium aluminum hydride gave 1,2,3,4-tetrahydro-2-methylcinnoline (**235**).[154] Authentic 1-methylcinnolin-4-one (**234**; $R^1 = Me$, $R^2 = R^3 = R^4 = R^5 = R^6 = H$), mp 114–116°C, is in fact a minor product of the same reaction.[154] A series of detailed studies by Ames and co-workers[154–166] have demonstrated that alkylation

---

[152] A. Padwa, P. Sackman, E. Shefter, and E. Vega, *J. C. S., Chem. Commun.*, 680 (1972).
[153] K. Schofield and J. C. E. Simpson, *J. Chem. Soc.*, 512 (1945).
[154] D. E. Ames and H. Z. Kucharska, *J. Chem. Soc.*, 4924 (1963).
[155] D. E. Ames and H. Z. Kucharska, *J. Chem. Soc.*, 283 (1964).
[156] D. E. Ames, *J. Chem. Soc.*, 1763 (1964).
[157] D. E. Ames, R. F. Chapman, and H. Z. Kucharska, *J. Chem. Soc.*, 5659 (1964).
[158] D. E. Ames, R. F. Chapman, H. Z. Kucharska, and D. Waite, *J. Chem. Soc.*, 5391 (1965).
[159] D. E. Ames and A. C. Lovesey, *J. Chem. Soc.*, 6036 (1965).
[160] D. E. Ames, R. F. Chapman, and D. Waite, *J. Chem. Soc. C*, 470 (1966).
[161] D. E. Ames and R. F. Chapman, *J. Chem. Soc. C*, 40 (1967).
[162] D. E. Ames, B. Novitt, D. Waite, and H. Lund, *J. Chem. Soc. C*, 796 (1969).
[163] D. E. Ames and B. Novitt, *J. Chem. Soc. C*, 2355 (1969).
[164] D. E. Ames, H. R. Ansari, A. D. G. France, A. C. Lovesey, B. Novitt, and R. Simpson, *J. Chem. Soc. C*, 3088 (1971).
[165] D. E. Ames, S. Chandrasekhar, and R. Simpson, *J. C. S., Perkin 1*, 2035 (1975).
[166] G. M. Singerman, in "The Chemistry of Heterocyclic Compounds" (R. N. Castle, ed.), Vol. 27, p. 87. Wiley (Interscience), New York, 1973.

of 4-hydroxycinnolines in alkaline media is a general route to the *N*-alkyl betaines **233**. Alkylation usually occurs predominantly at N-2, presumably due to a less sterically hindered approach.[154,155,158] Evidence for this steric effect is provided by observation that, when using bulky alkyl groups, or with 8-substituted-4-hydroxycinnolines, almost exclusive formation of the mesomeric betaine (**233**) occurs.[158,160,161] In contrast, the introduction of 3-substituents (e.g., methyl,[160] bromo,[158] phenyl,[162,167] alkylcarboxy,[157,168] CH_2CO_2R[159]) leads to predominance of the 1-alkylcinnolin-4-one (**234**) which is sometimes the only product. Interestingly, methylation of 3-bromo-4-hydroxy-8-nitrocinnoline gives exclusively the betaine (**233**; $R^1 = Me$, $R^2 = Br$, $R^3 = R^4 = R^5 = H$, $R^6 = NO_2$).[161] Methylation of 4-hydroxy-6-bromo-3-cinnolinecarboxylic acid at room temperature (Me_2SO_4/aq KOH) gives equal amounts of the inner salt **236** and the 1-methyl isomer (**234**; $R^1 = Me$, $R^2 = CO_2H$, $R^3 = R^5 = R^6 = H$, $R^4 = Br$): Compound **236** is decarboxylated in hot dimethylformamide giving the mesomeric betaine (**233**; $R^1 = Me$, $R^2 = R^3 = R^5 = R^6 = H$, $R^4 = Br$).[168]

Several other sequences based on the alkylation of cinnoline derivatives have been reported to furnish betaines (**233**). Treatment of 4-methoxy- and 4-phenoxycinnolines with methyl iodide gives 2-methylcinnolinium iodides (**237**; R = Me or Ph) which are converted to 2-methylcinnolinium-4-olate (**243**) by hot hydrobromic acid.[156,157] In a closely related approach 4-amino- or 4-methylaminocinnolines have been methylated and the 2-methyl-4-aminocinnolinium salts (**238**; R = H or Me) subsequently hydrolyzed to the betaine (**243**) by hot aqueous alkali.[156,169]

[167] H. S. Lowrie, *J. Med. Chem.* **9**, 784 (1966).
[168] R. P. Brundage and G. Y. Lesher, *J. Heterocycl. Chem.* **13**, 1085 (1976).
[169] H. J. Barber and E. Lunt, *J. Chem. Soc.*, 1468 (1965).

The 2-alkylcinnolinium-4-olates (**233**) are stable, colorless compounds. In the ^{1}H NMR the *N*-methyl resonance is observed at lower field than for nonpolar isomers (i.e., **234**; R^1 = Me and **246**); this trend is useful for diagnosing the position of alkylation of 4-hydroxycinnolines.[170,171]

The chemical reactions of the cinnolinium-4-olates (**233**) are conveniently illustrated by considering specific reactions of the 2-methyl derivative (**243**) (Scheme 10). Compound **243** readily forms salts: Ethanolic hydrogen chloride gives 2-methyl-4-hydroxycinnolinium chloride (**239**)[154] and triethyloxonium tetrafluoroborate gives 2-methyl-4-ethoxycinnolinium tetra-

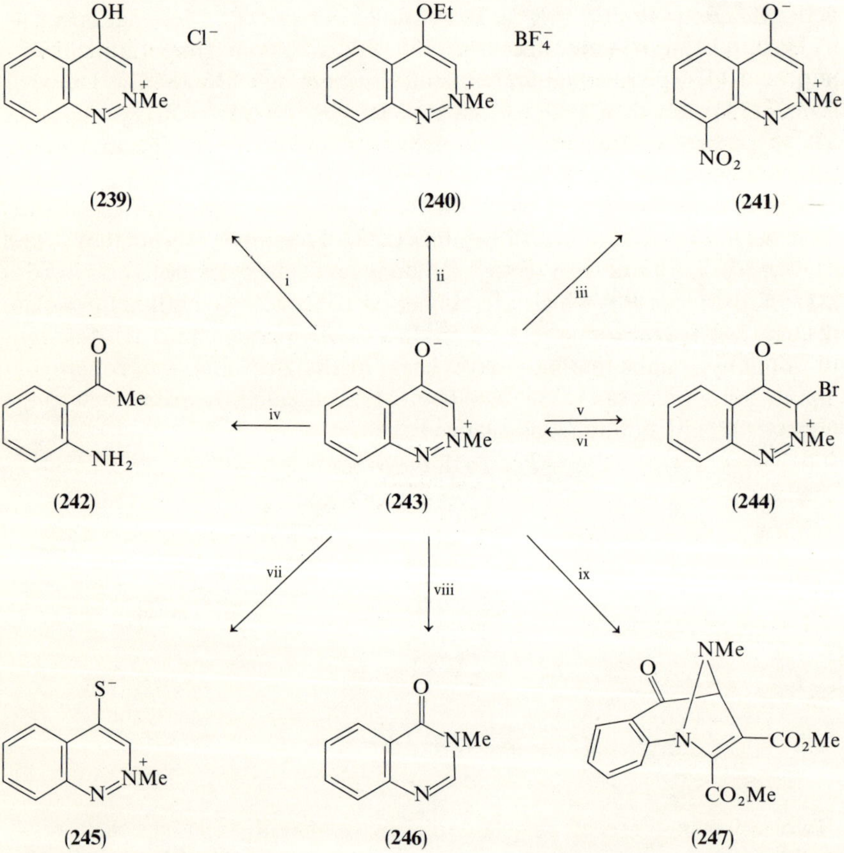

SCHEME 10. Reagents: i, HCl/EtOH; ii, $Et_3O^+BF_4^-$; iii, HNO_3/H_2SO_4; iv, Zn/NH_4OH; v, $Br_2/AcOH$; vi, KI/AcOH; vii, P_4S_{10}, viii, hv; ix, $MeO_2C-C\equiv C-CO_2Me$.

[170] A. W. Ellis and A. C. Lovesey, *J. Chem. Soc. B*, 1285 (1967).
[171] A. W. Ellis and A. C. Lovesey, *J. Chem. Soc. B*, 1393 (1968).

fluoroborate (**240**).[163] Nitration gives 3-methyl-8-nitrocinnolinium-4-olate (**241**) which is also obtained by methylation of 4-hydroxy-8-nitrocinnoline.[160] In contrast, bromination occurs in the 3-position, giving 2-methyl-3-bromocinnolinium-4-olate (**244**) which is isomeric with the product obtained by methylation of 3-bromo-4-hydroxycinnoline. The 3-bromo derivative (**244**) is reduced back to the parent molecule by potassium iodide in hot acetic acid.[158] Whereas reduction of compound **243** with lithium aluminum hydride gives 1,2,3,4-tetrahydro-2-methylcinnoline (**235**), reduction by zinc dust in aqueous ammonia gives 2-aminoacetophenone (**242**).[154] 3-Methylcinnolinium-4-thiolate (**245**) (Section III,B,9) is made by reaction of compound **243** with phosphorus pentasulfide.[172]

A reaction of particular interest is the photochemical rearrangement of compound **243** to 3-methyl-4(3H)-quinazolinone (**246**). This transformation (**243** → **246**) (Scheme 10), which takes place in high yield in ethanol, has been applied to several other derivatives and is clearly analogous to the photorearrangement of pyridazinium-5-olates to 4-pyrimidones (Section III,A,6). The mechanism of the reaction is not clear.[165]

Several 1,3-dipolar cycloaddition reactions of cinnolinium-4-olates have been reported. The methyl derivative (**243**) with dimethyl acetylenedicarboxylate gives adduct **247**.[163] Previously, the formation of a 1:1 adduct between compound **243** and phenylacetylene had been reported by Lunt and Threlfall.[173] The 6-chloro derivative (**233**; R^1 = Me, R^2 = R^3 = R^5 = R^6 = H, R^4 = Cl) forms similar adducts with dimethyl acetylenedicarboxylate and diphenylacetylene and a single regioisomer with phenylacetylene.[78] No reaction with olefinic dipolarophiles was observed.

9. *Cinnolinium-4-thiolates* (**248**)

(**248**)   (**249**)   (**250**)

Two representatives of this system (**248**; R = H, OMe) have been prepared by reaction of the corresponding cinnolinium-4-olates (**249**; R = H, OMe) with phosphorus pentasulfide.[158,164,172] Menschutkin methylation gives the cinnolinium iodide **250** (R = H).[161]

---

[172] G. B. Barlin, *J. Chem. Soc.*, 2260 (1965).
[173] E. Lunt and T. L. Threlfall, *Chem. Ind. (London)*, 1805 (1964).

10. Cinnolinium-8-olates (251)

(251)

(252)

8-Hydroxycinnoline (252) slowly dissolves in ethereal diazomethane "producing a brilliant emerald green colour which changes to deep blue." Comparison with the analogous reaction of 8-hydroxyquinoline (Section III,B,5) has led to the suggestion that this blue product, mp ~120°C, is 1-methylcinnolinium-8-olate (251; R = Me).[174]

11. 1,7-Naphthyridinium-4-olates (253)

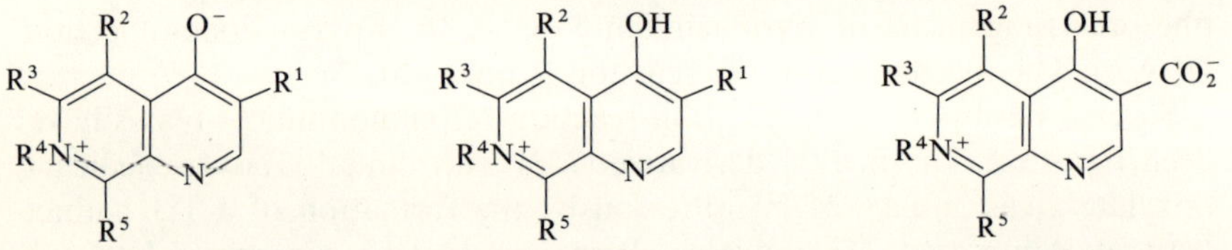

(253)

(254)

(255)

Alkylation of 4-hydroxy-1,7-naphthyridines gives 4-hydroxy-1,7-naphthyridinium iodides (254) which on treatment with aqueous alkali yield 1,7-naphthyridinium-4-olates (253). A number of derivatives (253; R^1 = H, CN, CO_2H) have been prepared in this manner. On the basis of IR spectroscopy, the acid derivatives (253; R^1 = CO_2H) have been formulated as mesomeric betaines rather than the inner salts 255.[175]

Treatment of 8-methyl derivatives (256) with potassium carbonate and alkyl halide results in C-alkylation of the 8-methyl group (i.e., 256 → 257) (Scheme 11). Reaction of the betaines (257) with hot aqueous potassium

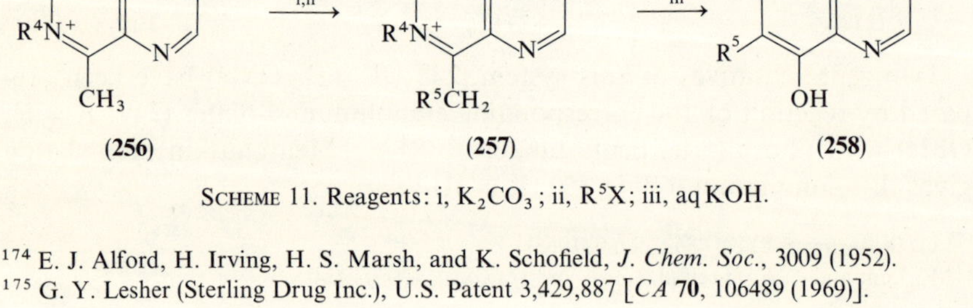

(256)

(257)

(258)

SCHEME 11. Reagents: i, K_2CO_3; ii, R^5X; iii, aq KOH.

[174] E. J. Alford, H. Irving, H. S. Marsh, and K. Schofield, J. Chem. Soc., 3009 (1952).
[175] G. Y. Lesher (Sterling Drug Inc.), U.S. Patent 3,429,887 [CA 70, 106489 (1969)].

hydroxide gives moderate yields of 4,8-dihydroxyquinolines (**258**) (Scheme 11).[175]

12. *1,9-Naphthyridinium-4-olates* (**259**)

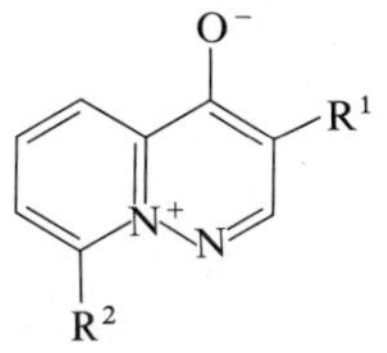

(**259**) 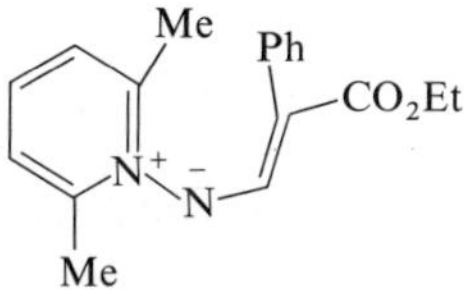 (**260**)

Photolysis of the lutidine *N*-imide **260** gives a crystalline compound, mp 226–228°C, which appears to be betaine **259** (R^1 = Ph, R^2 = Me) (15%).[176] (See also Section III,B,16.)

13. *1,2,3-Benzotriazinium-4-olates* (**261**)

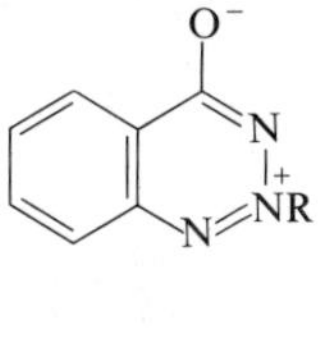

(**261**) 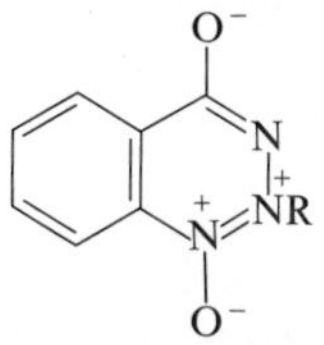 (**262**) 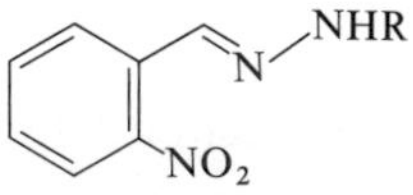 (**263**)

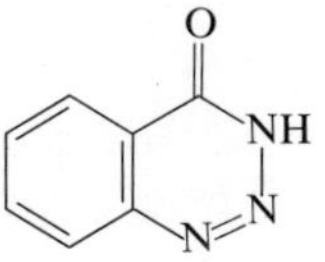

(**264**) 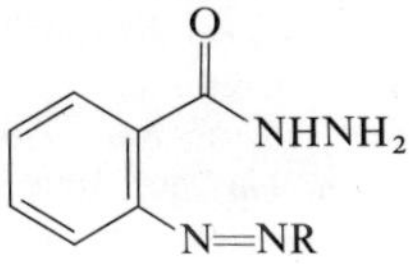 (**265**) 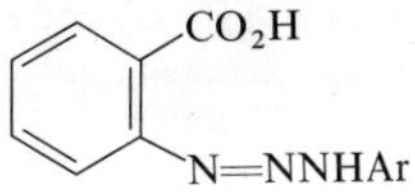 (**266**)

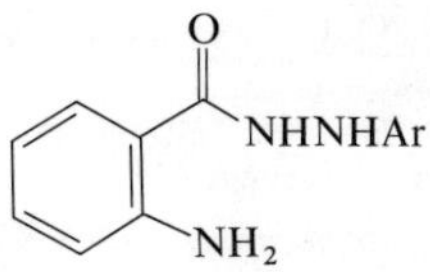

(**267**) 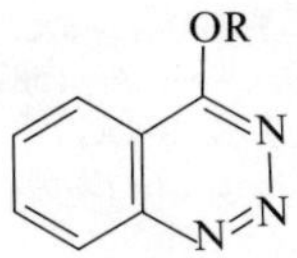 (**268**)

Two methods for preparing *N*-aryl or *N*-alkyl 1,2,3-benzotriazinium-4-olates (**261**) have been reported. Oxidation of *o*-nitrobenzaldehyde

[176] A. Kakehi, S. Ito, T. Funahashi, and Y. Ota, *J. Org. Chem.* **41**, 1570 (1976).

hydrazones (**263**) gives the explosive *N*-oxides **262** which are reduced to betaines **261** in good yield by stannous chloride.[177-189] Alternatively, these compounds (**261**) can be prepared, also in good yield, by N-alkylation (R_2SO_4)[189-191] or N-arylation $(Ar_2I^+Cl^-)$[189] of 1,2,3-benzotriazin-4(3*H*)-one (**264**). Since they were first prepared by Chattaway and co-workers,[177-183] various structures have been assigned to these betaines (**261**) and their precursors (**262**).[185-189] Structures **261** and **262** were first recognized by Kerber[188] in 1972 and confirmed in 1974 by McKillop and Kobylecki[189] who have also summarized the conclusions of earlier structural investigations.[192] The *p*-(*N*,*N*-dimethylamino)phenyl derivative (**261**; R = *p*-Me$_2$NC$_6$H$_4$) is reported to be formed by (i) condensation of indazolinone and 4-nitroso-*N*,*N*-dimethylaniline[188,193,194] and (ii) nitrosation of 2(4′-dimethylaminophenylazo)benzhydrazide (**265**)[195] but a recent review casts some doubt over these results.[192]

Betaines **261** are stable crystalline compounds. Knowledge of their chemical reactions is still limited. Alkaline hydrolysis of aryl derivatives (**261**; R = Ar) gives the 2-azobenzoic acids **266** but the mechanism of this rearrangement is unknown.[178] Reduction by tin and hydrochloric acid gives the hydrazides **267**.[178,188] Thermolysis of the *p*-tolyl compound (**261**; R = *p*-MeC$_6$H$_4$) (120°C at 0.1 mm Hg) gives the isomeric triazine (**268**; R = *p*-MeC$_6$H$_4$).[189] Phosphorus pentasulfide converts the 2-methyl derivative (**261**; R = Me) into 2-methyl 1,2,3-benzotriazinium-4-thiolate (**272**; R = Me) (Section III,B,15).

[177] F. D. Chattaway and A. J. Walker, *J. Chem. Soc.*, 2407 (1925).
[178] F. D. Chattaway and A. J. Walker, *J. Chem. Soc.*, 323 (1927).
[179] F. D. Chattaway and A. B. Adamson, *J. Chem. Soc.*, 157 (1930).
[180] F. D. Chattaway and A. B. Adamson, *J. Chem. Soc.*, 843 (1930).
[181] F. D. Chattaway and A. B. Adamson, *J. Chem. Soc.*, 2787 (1931).
[182] F. D. Chattaway and A. B. Adamson, *J. Chem. Soc.*, 2792 (1931).
[183] F. D. Chattaway and G. D. Parkes, *J. Chem. Soc.*, 1005 (1935).
[184] G. D. Parkes and E. D'A. Burney, *J. Chem. Soc.*, 1619 (1935).
[185] M. S. Gibson, *Nature (London)* **193**, 474 (1962).
[186] M. S. Gibson, *Tetrahedron* **18** 1377 (1962).
[187] W. A. F. Gladstone, J. B. Aylward, and R. O. C. Norman, *J. Chem. Soc. C*, 2587 (1969).
[188] R. C. Kerber, *J. Org. Chem.* **37**, 1587 (1972).
[189] A. McKillop and R. J. Kobylecki, *J. Org. Chem.* **39**, 2710 (1974).
[190] G. Wagner and H. Gentzsch, *Pharmazie* **23**, 629 (1968).
[191] G. Wagner and H. Gentzsch, *Arch. Pharm. (Weinheim, Ger.)* **301**, 923 (1968) [*CA* **70**, 106825 (1969)].
[192] R. J. Kobylecki and A. McKillop, *Adv. Heterocycl. Chem.* **19**, 215 (1976).
[193] J. J. Jennen, *Ind. Chim. Belge* **16**, 472 (1951).
[194] J. J. Jennen, *Meded. Vlaam. Chem. Ver.* **18**, 43 (1956). [*CA* **51**, 5094 (1957)].
[195] R. C. Kerber and P. J. Heffron, *J. Org. Chem.* **37**, 1592 (1972).

14. 1,2,3-Benzotriazinium-4-aminides (269)

(269)   (270)   (271)

Menschutkin alkylation of 4-arylamino-1,2,3-benzotriazines (270) in ethanol occurs at the 2-position. The benzotriazinium iodides (271) so formed are readily deprotonated by base giving the betaines (269; R^1 = alkyl, R^2 = aryl). In the presence of sodium ethoxide, alkylation of benzotriazines 270 occurs at positions 2 and 3.

The bright red betaines 269 form salts with dilute hydrochloric acid and are dealkylated by hot ethanolic potassium hydroxide. The mesomeric betaine structure (269) is fully supported by spectroscopic properties.[196-198]

15. 1,2,3-Benzotriazinium-4-thiolates (272)

(272)   (273)   (274)

Betaines 272 have been made by (i) treatment of 1,2,3-benzotriazinium-4-olate (261) (Section III,B,13) with phosphorus pentasulfide and (ii) glycosidation of the sodium salt of 4-mercapto-1,2,3-benzotriazine (273; R = H) using 2,3,4,5-tetra-O-acetyl-α-D-glucopyranosyl bromide. In the second method, betaine 272 [R = Gluc(Ac)$_4$(β)] is accompanied by S- and N(3)-substituted isomers 273 and 274 [R = Gluc(Ac)$_4$(β)]. When the thioether 273 [R = Gluc(Ac)$_4$(β)] is heated with mercuric bromide, rearrangement occurs giving betaine 272 [R = Gluc(Ac)$_4$(β)] (55%) and a smaller amount of isomer 274 [R = Gluc(Ac)$_4$(β)] (18%).[190,191]

---

[196] H. N. E. Stevens and M. F. G. Stevens, *J. Chem. Soc. C*, 2289 (1970).
[197] R. A. W. Johnstone, D. W. Payling, P. N. Preston, H. N. E. Stevens, and M. F. G. Stevens, *J. Chem. Soc. C*, 1238 (1970).
[198] M. F. G. Stevens, *Prog. Med. Chem.* 13, 205 (1976).

16. Pyrido[2,1-f][1,2,4]triazinium-1-olates (275)

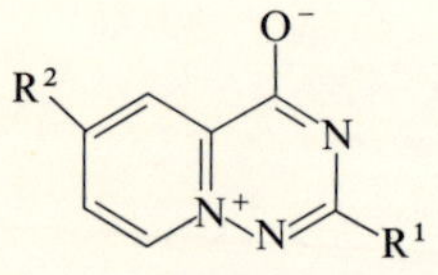

(275) 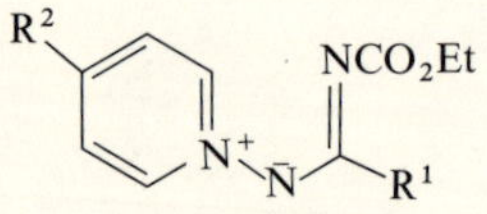 (276) 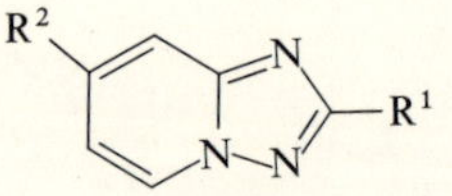 (277)

Thermolysis of the N-imidoyliminopyridinium ylides **276** (R^1 = Ph, R^2 = H, Me) gives the yellow, crystalline betaines **275** (R^1 = Ph, R^2 = H, Me) ($\nu_{C=O}$ 1600–1610 cm^{-1}) in good yield (50–60%), together with triazolopyridines **277** (R^1 = Ph, R^2 = H, Me) (~40%). A similar transformation yields the naphthyridinium betaine **259** (R^1 = Ph, R^2 = Me) (Section III,B, 12) and a cycloelimination mechanism has been proposed to account for the formation of these products (**259** and **275**).[199,200]

17. 2,1,3-Benzothiadiazinium-4-olates (278)

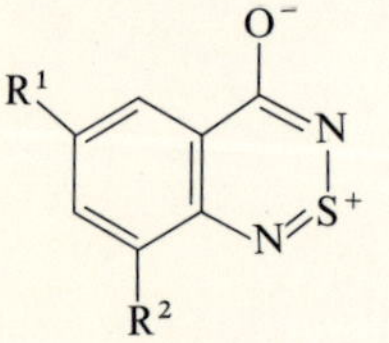

(278) 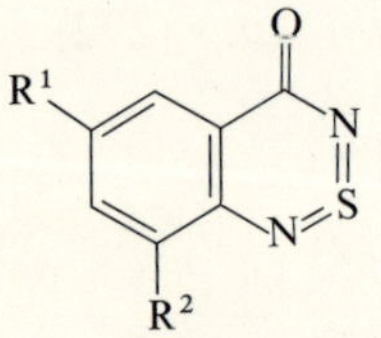 (279) 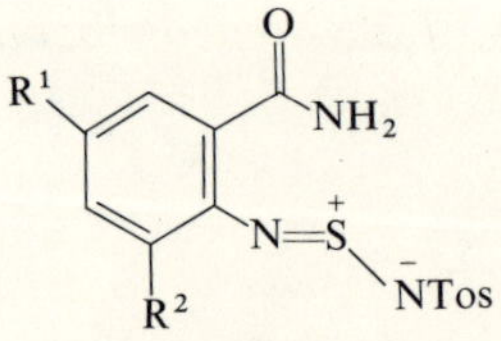 (280)

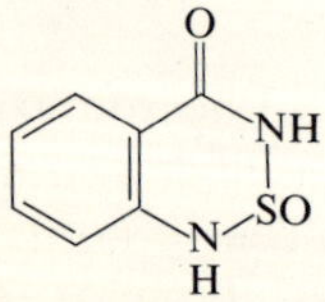

(281) 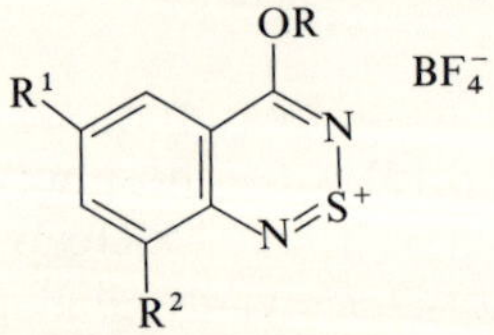 (282)

Reaction of 2-aminobenzamides with N,N'-bis(p-toluenesulfonyl)sulfurdiimide (Tos—N=S$^+$—N$^-$—Tos) gives betaines **278** via the acyclic intermediates **280**. Compound **278** (R^1 = R^2 = H) is a yellow crystalline solid (ν_{CO} 1640 cm^{-1}) which is hydrolyzed to compound **281**. Meerwein alkylation gives benzothiadiazinium tetrafluoroborates **282**.[201,202]

[199] A. Kakehi, S. Ito, K. Uchiyama, and Y. Konno, *Chem. Lett.*, 413 (1976).
[200] A. Kakehi, S. Ito, K. Uchiyama, Y. Konno, and K. Kondo, *J. Org. Chem.* **42**, 443 (1977).
[201] H. Grill and G. Kresze, *Justus Liebigs Ann. Chem.* **749**, 171 (1971).
[202] W. Kosbahn and H. Schäfer, *Angew. Chem., Int. Ed. Engl.* **16**, 780 (1977).

18. Quinolinium-3-olates (283)

3-Hydroxy-1-methylquinolinium iodide is deprotonated by tertiary base giving *N*-methylquinolinium-3-olate (**283**; R = Me) which can be trapped by dienes giving adducts **284**. These cycloadditions have been discussed in terms of frontier molecular orbital theory (see Section V,C,2).[203]

19. Cinnolinium-3-olates (285)

An orange compound (mp 280–283°C) obtained by methylation of 3-hydroxycinnoline with dimethyl sulfate and alkali has been formulated as the betaine **285** (R = Me).[204]

20. Quinolizinium-3-olates (286)

---

[203] K.-L. Mok and M. J. Nye, *J. C. S., Chem. Commun.*, 608 (1974).
[204] E. J. Alford and K. Schofield, *J. Chem. Soc.*, 1811 (1953).

Compound **287**, which is obtained from *N*-carbethoxymethyl-2-methylpyridinium bromide and diethyl mesoxalate, undergoes base-catalyzed cyclization to betaine **288** which is readily converted to the quinolizinium-3-olate **286** ($R^1 = CO_2Et$, $R^2 = CO_2H$). Decarboxylation of this compound (**286**; $R^1 = CO_2Et$, $R^2 = CO_2H$) by cupric oxide and quinoline at 180–190°C gives the unsubstituted betaine (**286**; $R^1 = R^2 = H$).[205] Quinolizinium-3-olate (**286**; $R^1 = R^2 = H$) has also been obtained by deprotonation of 3-hydroxyquinolizinium bromide[101] but a discrepancy between the melting points of the two samples, $C_9H_7NO \cdot 2H_2O$ (mp[101] 143–146°C) and $C_9H_7NO \cdot H_2O$ (mp[205] 75–80°C), is unsatisfactory.

21. *Quinolinium-6-olates* (**289**)

(**289**)

The methyl and benzyl derivatives (**289**; $R^1 = Me, PhCH_2$; $R^2 = H$) were described by Claus and Howitz in 1891. Their preparation involves *N*-alkylation of 6-hydroxyquinoline and base hydrolysis to the quaternary hydroxide which is then desiccated, giving the hygroscopic betaines **289**.[206,207] The 2-phenyl derivative (**289**; $R^1 = Me$, $R^2 = Ph$) has been similarly prepared.[208] With methyl iodide, the "phenol betaine" (**289**; $R^1 = Me$, $R^2 = H$) gives *N*-methyl-6-methoxyquinolinium iodide.[209]

C. Systems with Thirteen Conjugated Atoms

1. *2-(Pyridinium)phenolates* (**290**)

(**290**)   (**291**)   (**292**)

---

[205] V. Carelli, F. Liberatore, and G. Casini, *Ann. Chim.* (*Rome*) **57**, 269 (1967) [*CA* **67**, 54023 (1967)].
[206] A. Claus and H. Howitz, *J. Prakt. Chem.* **43**, 505 (1891).
[207] H. Decker and H. Engler, *Ber. Dtsch. Chem. Ges.* **36**, 1169 (1903).
[208] W. Schneider and A. Pothmann, *Ber. Dtsch. Chem. Ges. B* **74**, 471 (1941).
[209] A. Claus and H. Howitz, *J. Prakt. Chem.* **56**, 438 (1897).

The triaryl compounds **290** (R = Ar) are prepared by condensation of 2-aminophenol with triarylpyrylium salts followed by treatment with alkali. The triphenyl betaine (**290**; R = Ph) is obtained as a purple solid, mp 165°C (decomp), which shows large thermo/solvatochromic effects.[210,211] Oxidation of the betaine **290** (R = Ph) with hydrogen peroxide gives the triphenylpyridinium-3-olate **291** (R = Ph) (see Section III,A,2) and the pyrrole **292** (R = Ph).[45] The mechanism of this unusual reaction has not yet been established.

2. 6-(Aryl)-1,2,4-triazinium-5-olates (293)

(293) (294) (295) (296)

Examples of these betaines (**293**) are the bridged heterocycles **295**, which have been made by condensation of 1,2,4-triazin-5-one derivatives **294** with aldehydes or ketones.[212–215] The molecular structure of compound **295** (R^1 = H, R^2 = R^3 = Me, R^4 = SMe) has been determined by X-ray crystallography.[216] Acetylation gives derivatives **295** (R^1 = Ac) which with methyl iodide give the deep red salts **296** in high yield.[217]

3. 4-(Pyridinium)phenolates (297)

These compounds (**297**) have interesting thermochromic and solvatochromic effects. Ethanolic solutions of 4-(2′,4′,6′-triphenylpyridinium)-2,6-di-*tert*-butylphenolate (**297**; R^1 = R^2 = Ph, R^3 = *t*-Bu) are red (λ_{max} 513 nm) at −75°C, violet (λ_{max} 550 nm) at 25°C, and blue (λ_{max} 568 nm) at 78°C.[211] Dramatic changes are also observed when the solvent polarity is

---

[210] W. Schneider, W. Döbling, and R. Cordua, *Ber. Dtsch. Chem. Ges. B* **70**, 1645 (1937).
[211] K. Dimroth, C. Reichardt, and A. Schweig, *Justus Liebigs Ann. Chem.* **669**, 95 (1963).
[212] G. Doleschall and K. Lempert, *Tetrahedron* **29**, 639 (1973).
[213] G. Doleschall and K. Lempert, *Tetrahedron* **32**, 1735 (1976).
[214] G. Doleschall, *Acta Chim. Acad. Sci. Hung.* **90**, 419 (1976).
[215] G. Doleschall, *Tetrahedron* **32**, 2549 (1976).
[216] A. J. M. Duisenberg, A. Kálmán, G. Doleschall, and K. Lempert, *Cryst. Struct. Commun.* **4**, 295 (1975).
[217] G. Doleschall and K. Lempert, *Acta Chim. Acad. Sci. Hung.* **77**, 345 (1973).

(297) X = O
(298) X = S

varied. In water the pentaphenyl derivative (297; $R^1 = R^2 = R^3 = Ph$) has an absorption band at 453 nm which shifts to 810 nm in diphenyl ether.[218] This has been claimed to be the widest range of solvatochromism known and this property can be used to assess the polarity of solvents by eye and for monitoring the water content of solvents.[219,220]

The betaines 297 are easily prepared from a pyrylium salt and the appropriate 4-aminophenol followed by deprotonation. Typically, 2,4,6-triphenylpyridinium iodide and 4-aminophenol give the blue-black betaine (297; $R^1 = R^2 = Ph$, $R^3 = H$) which also occurs as a red hexahydrate. With methyl iodide this compound gives N-(p-methoxyphenyl)-2,4,6-triphenylpyridinium iodide.[210,221]

4. 4-(Pyridinium)thiophenolates (298)

Compounds of this type (298) have been made from 4-aminothiophenol and 2,4,6-triarylpyrylium salts.[211]

5. Benz[de]isoquinolines (Naphtho[1,8-cd]pyridines) (303)

The IUPAC-recommended name for this ring system tends to obscure its structural relationship to the others in succeeding sections (III,C,6–10), so an alternative, although strictly incorrect, form is included in the heading.

Benzisoquinolines (303) have not been isolated, but they have been generated *in situ* and trapped as 1,3-dipolar adducts. Treatment of the

---

[218] K. Dimroth, C. Reichardt, T. Siepmann, and F. Bohlmann, *Justus Liebigs Ann. Chem.* **661**, 1 (1963).
[219] K. Dimroth and C. Reichardt, *Z. Anal. Chem.* **215**, 344 (1966).
[220] C. Reichardt and K. Dimroth, *Fortschr. Chem. Forsch.* **11**, 1 (1968).
[221] W. Dilthey and H. Dierichs, *J. Prakt. Chem.* **144**, 1 (1936).

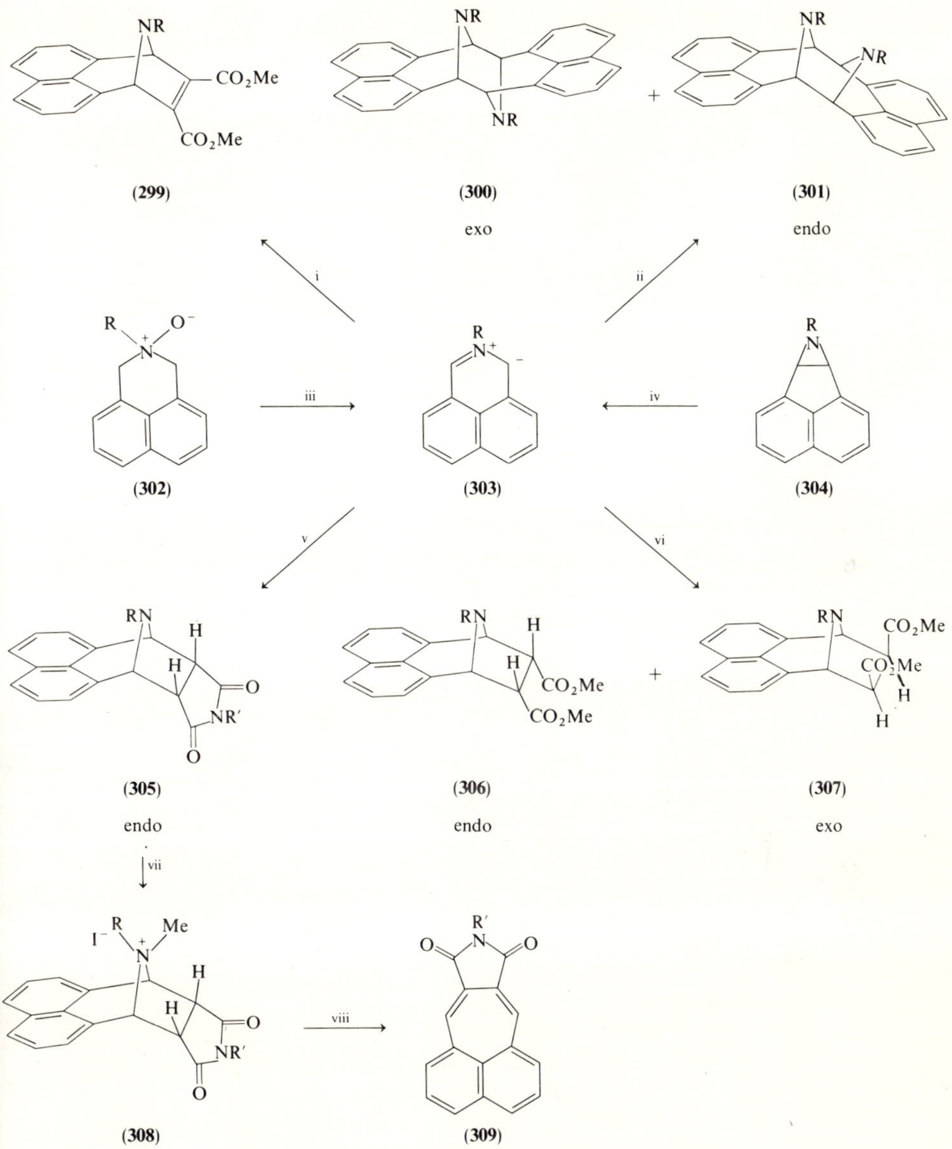

SCHEME 12. Reagents: i, MeO$_2$CC≡CCO$_2$Me; ii, ΔT; iii, Ac$_2$O/Et$_3$N; iv, hv or ΔT; v, N-methyl or -phenylmaleimide; vi, dimethyl maleate; vii, MeI; viii, KOH.

oxide (**302**; R = Me) with acetic anhydride and triethylamine in the presence of dimethyl acetylenedicarboxylate at $-10°C$ gives a high yield of adduct **299** (R = Me). Evidence for transient formation of the betaine **303** (R = Me) is afforded by the development of a green coloration. The betaine (**303**; R = Me) similarly gives the endo adducts (**305**; R = Me, R′ = Me,Ph) with N-methyl- and N-phenylmaleimide. The N-phenyl betaine (**303**; R = Ph) generated in the same way forms similar adducts, but reacts more slowly. With N-methylmaleimide a low yield of the exo adduct is also obtained. Quaternization of the compounds **305** (R = Me) followed by treatment of the methiodides (**308**; R = Me) with base gives cyclohepta[*de*]naphthalenes (**309**).[222]

An alternative method of generating the aryl betaines (**303**; R = Ar) involves thermolysis or photolysis of 7-arylacenaphtho[1,2-*b*]aziridines (**304**; R = Ar), which are made by reaction of acenaphthylene with arylazides. Heating the aziridines (**304**, R = Ph, *p*-MeC$_6$H$_4$, *p*-MeOC$_6$H$_4$, *p*-BrC$_6$H$_4$, *p*-NO$_2$C$_6$H$_4$, 3,5-Cl$_2$C$_6$H$_3$) with dimethyl acetylenedicarboxylate at 120°C gives the corresponding adducts (**299**) in high yield. Thermolysis or photolysis of the aziridines (**304**) in the presence of dimethyl maleate gives a mixture (~ 2:1) of endo and exo adducts (**306** and **307**), and similar adducts are formed with maleic anhydride. 1,2-Adducts are also formed with 4π- and 6π-addends such as 1,3-butadienes and cycloheptatriene. In the absence of 1,3-dipolarophiles, the betaine **303** (R = Ph) generated at 140°C, gives a mixture of the exo dimer (**300**) (47%) and the endo dimer (**301**) (41%).[223]

6. *Naphtho[1,8-cd]thiopyrans* (**310**)

(310)

(311)

(312)

(313) exo

(314) endo

(315)

---

[222] M. Ikeda, Y. Miki, S. Kaita, Y. Nishikawa, and Y. Tamura, *J. C. S., Perkin 1*, 44 (1977).
[223] A. C. Oehlschlager, A. S. Yim, and M. H. Akhtar, *Can. J. Chem.* **56**, 273 (1978).

(316)   (317)   (318)

These unusual sulfur heterocycles (310; R^1 = H,Ph) have been generated in situ by Pummerer dehydration of the sulfoxides 312 (R^1 = H, Ph). Isolation of the betaines has not been achieved, but they are trapped in high yield by N-phenylmaleimide giving the exo adducts (313; R^1 = H, Ph, R^2 = Ph).[224–226] This is in sharp contrast to the corresponding nitrogen betaines (303) (Section III,C,5) which give predominantly endo adducts. N-Methylmaleimide, however, does give a low yield of the endo adduct (314; R^1 = H, R^2 = Me) (13%) together with the exo adduct (313; R^1 = H, R^2 = Me) (52%).[222] Trapping with 1,4-naphthoquinone leads to adduct 315. Oxidation to the sulfone followed by base-catalyzed hydrolysis gives benzo[l]pleiadene-8,13-quinone (316) in moderate yield. Generation of compound 310 (R^1 = Ph) in the presence of oxygen gives mainly 1,8-dibenzoylnaphthalene which may well arise by initial formation of a peroxide adduct and subsequent extrusion of sulfur.[226]

When the diphenyl betaine 310 (R^1 = Ph) is generated in acetic anhydride at 120°C in the absence of dipolarophile, a quantitative yield of 1,2-diphenylacenaphthylene (318; R^1 = Ph) is obtained. This desulfurization occurs by initial valence tautomerism to the episulfide (317; R^1 = Ph) which can be isolated in 40% yield if the reaction is performed at 100°C.[226] The episulfide 317 (R^1 = Ph) and the hydrocarbon 318 (R^1 = Ph) are also produced by treating the sulfoxide 312 (R^1 = Ph) with phenyllithium. In the presence of oxygen, 1,8-dibenzoylnaphthalene is also a major product, and these observations suggest that the mesomeric betaine 310 (R^1 = Ph) is a common intermediate in these reactions.[227]

Although stable systems of the type 310 are so far unknown, the closely related species 319 (R^1 = Ph, R^2 = Br, Ph) have been isolated as dark-blue crystals.[228,229] The preparation and chemistry of these species closely

[224] M. P. Cava, N. M. Pollack, and D. A. Repella, J. Am. Chem. Soc. 89, 3640 (1967).
[225] R. H. Schlessinger and I. S. Ponticello, J. Am. Chem. Soc. 89, 3641 (1967).
[226] R. H. Schlessinger and A. G. Schultz, J. Am. Chem. Soc. 90, 1676 (1968).
[227] R. H. Schlessinger, G. S. Ponticello, A. G. Schultz, I. S. Ponticello, and J. M. Hoffman, Tetrahedron Lett., 3963 (1968).
[228] I. S. Ponticello and R. H. Schlessinger, J. Am. Chem. Soc. 90, 4190 (1968).
[229] J. M. Hoffman, Jr., and R. H. Schlessinger, J. Am. Chem. Soc. 91, 3953 (1969).

resemble those of the parent system, but these molecules (**319**) differ in that they can be represented by uncharged structures (**320**).[228-230]

(319)   (320)

Structure **311** has been widely used to represent the naphtho[1,8-*cd*]thiopyrans. The involvement of d-orbitals in the bonding of the molecules is briefly discussed in Section IV,A,4.

7. *Naphtho[1,8-de]triazines* (**321**)

2-Methyl-2*H*-naphtho[1,8-*de*]triazine (**321**; R = Me) is a blue compound which was first prepared by Sachs in 1909,[231] although the structure was not recognized until a reinvestigation by Perkins in 1964.[232] It is obtained, together with the red 1-methyl isomer (**322**; R = Me), by methylation of naphtho[1,8-*de*]triazine (**322**; R = H) using dimethyl sulfate and base.[232-234] A similar methylation gives 6,7-dihydro-2-methylacenaphtho[5,6-*de*]triazine (**324**; R = Me).[234-236] Other alkylating agents (e.g., Et_2SO_4, dodecyl tosylate, $PhCH_2Cl$) have also been used to give mixtures of the 1- and 2-alkylnaphthotriazines (**321** and **322**; R = alkyl) and *N*-aryl derivatives (**321**; R = aryl) have been obtained by reaction with activated chlorobenzenes.[234] Amination of naphtho[1,8-*de*]triazine (**322**; R = H) with cold ethereal chloramine gives mainly 1-aminonaphtho[1,8-*de*]triazine (**322**; R = NH_2) (43%), but a low yield of the isomeric betaine (**321**; R = NH_2) (2%) is also formed. A small amount of 2-amino-4-chloronaphtho[1,8-*de*]triazine (1%) is formed, presumably by electrophilic chlorination of the parent betaine.[237]

---

[230] R. H. Schlessinger and I. S. Ponticello, *Tetrahedron Lett.*, 4057 (1967).
[231] F. Sachs, *Justus Liebigs Ann. Chem.* **365**, 53 (1909).
[232] M. J. Perkins, *J. Chem. Soc.*, 3005 (1964).
[233] H. Beecken, *Angew. Chem., Int. Ed. Engl.* **6**, 360 (1967).
[234] P. Tavs, H. Sieper, and H. Beecken, *Justus Liebigs Ann. Chem.* **704**, 150 (1967).
[235] A. R. J. Arthur, P. Flowerday, and M. J. Perkins, *Chem. Commun.*, 410 (1967).
[236] P. Flowerday, M. J. Perkins, and A. R. J. Arthur, *J. Chem. Soc. C*, 290 (1970).
[237] C. W. Rees and R. C. Storr, *J. Chem. Soc. C*, 756 (1969).

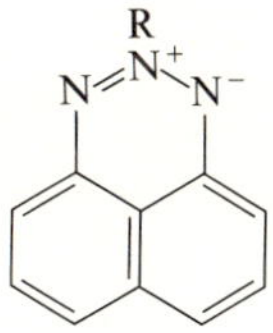

(321) 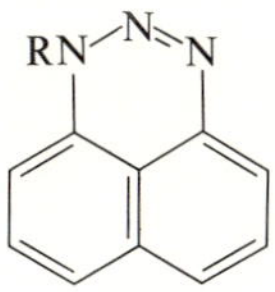 (322) 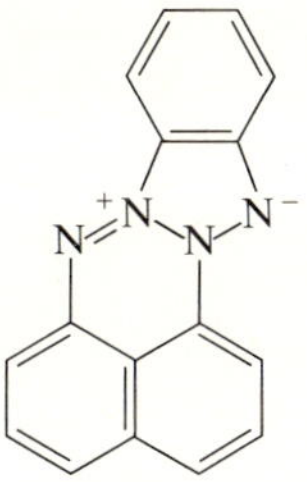 (323)

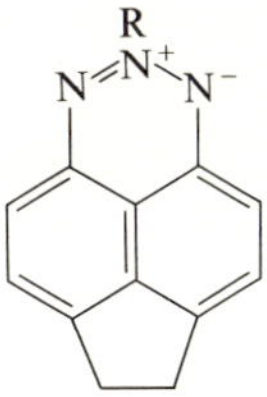

(324) 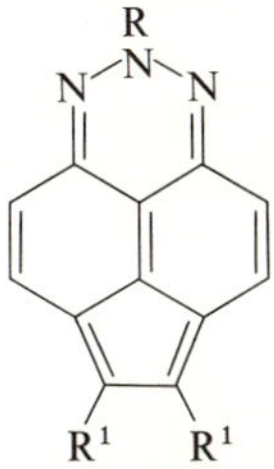 (325) 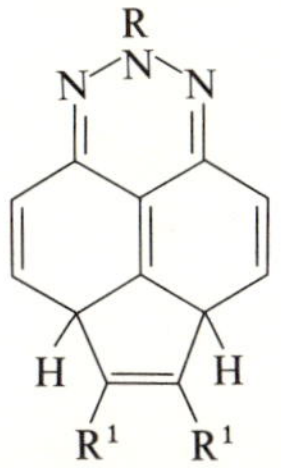 (326)

The chemistry of this class of compound (321) includes a number of novel transformations. The following reactions of the 2-methyl derivative (321; R = Me) have been reported: (i) Catalytic reduction gives 1,8-diaminonaphthalene and methylamine[232]; (ii) bromination (Br_2/AcOH) gives a tetrabromo derivative[232]; (iii) base-catalyzed condensation with benzaldehyde gives the conjugated derivatives 321 (R = CH=CHPh)[232]; and (iv) α-cyanoisopropyl or benzoyloxide radicals attack the naphthalene fragment, giving mainly substitution products.[238,239] Particularly interesting are the reactions of compound 321 (R = Me) with 1,3-dipolarophiles. Dimethyl acetylenedicarboxylate in hot o-dichlorobenzene gives moderate yields of the acenaphtho[5,6-de]triazine (325; R = Me, R¹ = CO_2Me).[240,241] [A derivative of this novel heterocyclic system (325; R = Me, R¹ = H) has alternatively been made by oxidation of the bridged betaine (324; R = Me)][235,236] The initial 1,11-cycloadduct (326) presumably undergoes dehydrogenation giving the observed product (325); the presence of sulfur as a dehydrogenating agent is advantageous. Similar additions have been observed with other alkynes. Reaction with alkenes has been observed but the nature of the products is unknown.[241]

---

[238] H. Beecken, P. Tavs, and H. Sieper, *Justus Liebigs Ann. Chem.* **704**, 166 (1967).
[239] H. Beecken and P. Tavs, *Justus Liebigs Ann. Chem.* **704**, 172 (1967).
[240] C. W. Rees, R. W. Stephenson, and R. C. Storr, *J. C. S., Chem. Commun.*, 1281 (1972).
[241] S. F. Gait, M. J. Rance, C. W. Rees, R. W. Stephenson, and R. C. Storr, *J. C. S., Perkin 1*, 556 (1975).

The 2-amino betaine **321** (R = NH$_2$) readily rearranges to 1-amino-8-azidonaphthalene (66%) in aqueous base. The same product is also obtained by treatment of compound **321** (R = NH$_2$), or the 1-amino isomer **322** (R = NH$_2$), with acid.[237]

The 2-nitrophenyl derivative (**321**; R = o-NO$_2$C$_6$H$_4$) on treatment with triethylphosphite gives the novel mesomeric betaine **323**.[242,243]

8. *Quinolino[8,1-ef][1,2,4]triazines* (**328**)

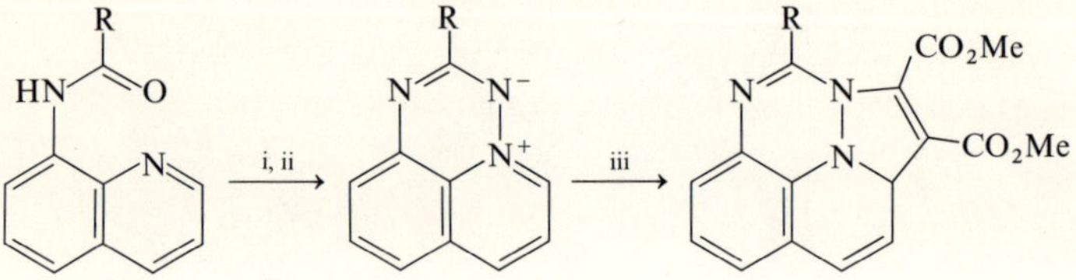

SCHEME 13. Reagents: i, ArSO$_2$ONH$_2$; ii, aq KOH; iii, MeO$_2$C—C≡C—CO$_2$Me.

The mesomeric betaines **328** (R = H, Me) have been prepared by reaction of 8-acylaminoquinolines (**327**) with *O*-mesitylenesulfonylhydroxylamine followed by aqueous alkali. The mechanism presumably involves N-amination and subsequent cyclodehydration. The red, hygroscopic betaine (**328**; R = H) reacts with dimethyl acetylenedicarboxylate giving the cycloadduct **329** (Scheme 13).[244]

9. *Naphtho[1,8-cd][1,2,6]thiadiazines* (**330**)

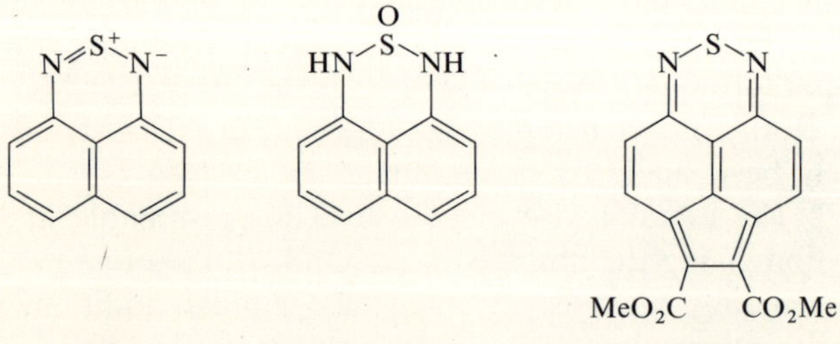

Lustrous blue-black crystals of naphtho[1,8-*cd*][1,2,6]thiadiazine (**330**) were first prepared by Dietz who obtained the sulfoxide (**331**) by reacting

---

[242] H. Sieper, *Tetrahedron Lett.*, 1987 (1967).
[243] H. Sieper and P. Tavs, *Justus Liebigs Ann. Chem.* **704**, 161 (1967).
[244] Y. Tamura, Y. Miki, H. Hayashi, Y. Sumida, and M. Ikeda, *Heterocycles* **6**, 281 (1977).

1,8-diaminonaphthalene with N-sulfinylaniline (PhNSO) and then converted this compound to the mesomeric betaine (**330**) by thermal dehydration.[245] Subsequently, other closely related conditions have been described for the preparation of compounds **330** and **331** from 1,8-diaminonaphthalene.[246] Use of N-sulfinylaniline or N-sulfinylbenzenesulfonamide (PhSO$_2$NSO) in the presence of triethylamine achieves the cyclization and dehydration in one step.[247,248] Treatment of compound **330** with phosphorus pentachloride or thionyl chloride gives a tetrachloro derivative.[246]

Compound **330** reacts with dimethyl acetylenedicarboxylate in hot o-dichlorobenzene giving the deep blue acenaphtho[5,6-cd]thiadiazine (**332**) in low yield (6%). Apparently, the primary adduct, which is not isolated, undergoes dehydrogenation giving the observed product **332**.[241]

10. *Naphtho[1,8-cd][1,2,6]selenadiazines* (**333**)

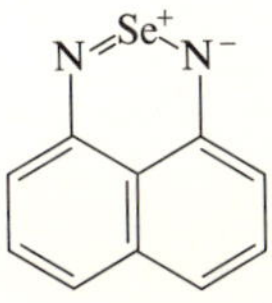

(333)

The preparation of derivative **333** was described in 1909 by Sachs who treated 1,8-diaminonaphthalene with selenous acid (H$_2$SeO$_3$). The chemistry of this molecule has not been investigated.[249]

D. Systems with Fifteen Conjugated Atoms

1. *Pyrido[1,2-b]cinnolinium-11-olates* (**338**)

Pyrolysis of the 2-(2-azidobenzoyl)pyridines **334** (R^1 and R^2 = H or Br) at 110°C gives the benzisoxazoles **335**, which rearrange to the mesomeric betaines **338** in good yield at higher temperatures (215°C). The yellow, crystalline betaines (v_{CO} 1640 cm^{-1}) are stable compounds whose structure is supported by an X-ray crystallographic study of the bromo derivative **338** (R^1 = H, R^2 = Br). The following reactions of the parent system (**338**; R^1 = R^2 = H) have been reported (Scheme 13); (i) Nitration gives the 2-nitro

---

[245] R. Dietz, *Chem. Commun.*, 57 (1965).
[246] H. Behringer and K. Leiritz, *Chem. Ber.* **98**, 3196 (1965).
[247] H. Beecken, *Chem. Ber.* **100**, 2164 (1967).
[248] H. Beecken, *Chem. Ber.* **100**, 2170 (1967).
[249] F. Sachs, *Justus Liebigs Ann. Chem.* **365**, 135 (1909).

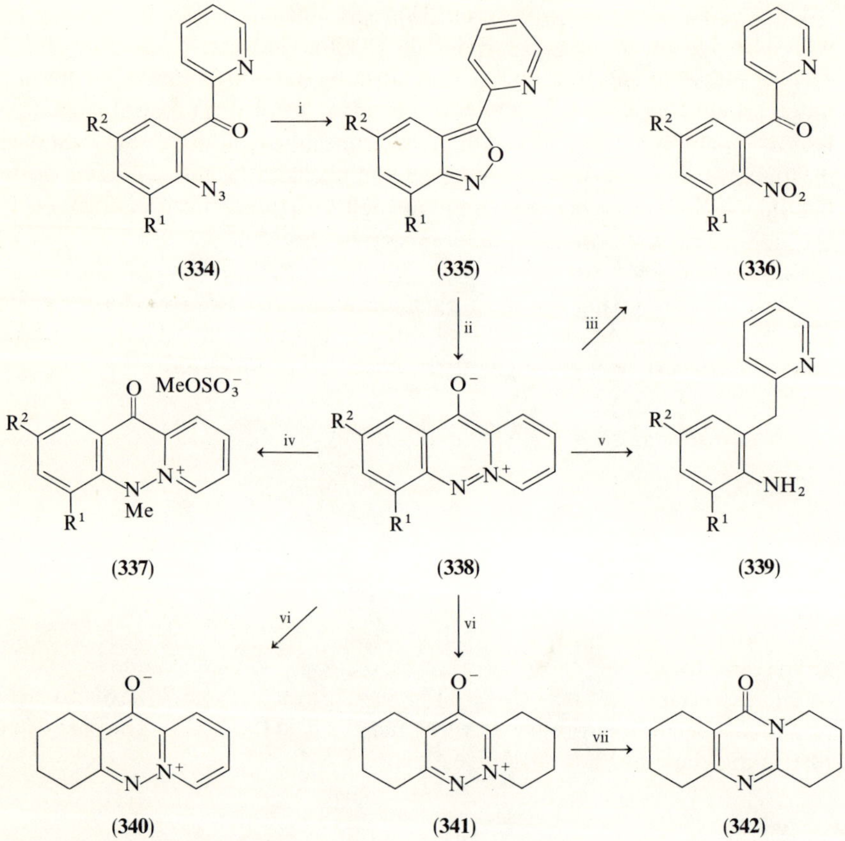

SCHEME 13. Reagents: i, 110°C; ii, 215°C; iii, m-ClC$_6$H$_4$CO$_3$H; iv, Me$_2$SO$_4$; v, Zn/AcOH; vi, H$_2$, Pd/C; vii, hν.

derivative (**338**; R^1 = H, R^2 = NO$_2$); (ii) iodination (ICl) gives the 2-iodo compound (**338**; R^1 = H, R^2 = I); (iii) methylation (Me$_2$SO$_4$) yields the N-methyl salt (**337**; R^1 = R^2 = H); (iv) treatment with phosphorus pentasulfide gives the corresponding thiolate (**348**) (Section III,D,4); (v) oxidation with m-chloroperbenzoic acid gives 2(2-nitrobenzoyl)pyridine (**336**; R^1 = R^2 = H); and (vi) reduction by zinc in acetic acid gives 2-o-aminobenzylpyridine (**339**; R^1 = R^2 = H). Catalytic hydrogenation affords the 1,9-naphthyridinium-4-olate (**340**) (Section III,B,12) and the pyridazinium-5-olate (**341**) (Section III,A,5), which upon photolysis gives the isomeric pyrimidinone (**342**) (Scheme 13). [250,251]

[250] R. Y. Ning, W. Y. Chen, and L. H. Sternbach, *J. Heterocycl. Chem.* **11**, 125 (1974).
[251] R. Y. Ning, J. F. Blount, W. Y. Chen, and P. B. Madan, *J. Org. Chem.* **40**, 2201 (1975).

Particularly interesting is the reaction of compound **338** ($R^1 = R^2 = H$) with dimethyl acetylenedicarboxylate to give the quinolin-4-one (**345**) (22%). The formation of this unexpected product can be rationalized by postulating initial generation of the 1,3-dipolar cycloadduct **343** followed by ring opening to the ketene derivative **344** and subsequent recyclization to the observed product **345**. This reaction is directly analogous to the observed thermal fragmentation of the cycloadducts of pyrylium-3-olates (**74**) (Section III,A,1).[251]

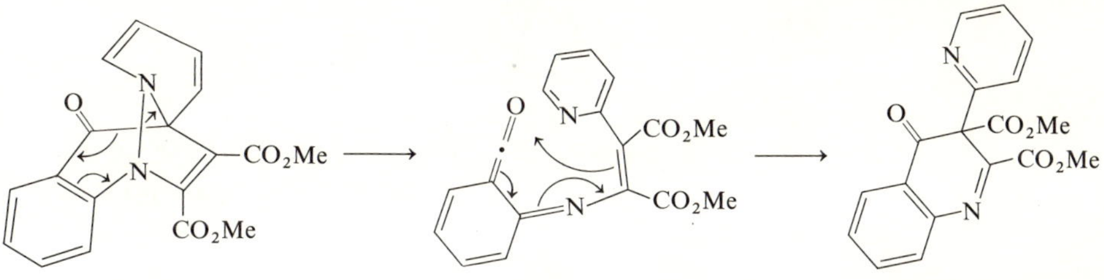

(343)   (344)   (345)

2. *Pyrido[1,2-b]cinnolinium-11-aminides* (**346**)

Reaction of the salts **349** (Section III,D,4) with a number of primary amines and hydrazines gives the 11-amino bromides (**350**). Treatment of two of these salts [**350**; $R^1 = R^2 = H$, $R = Me_2N(CH_2)_3$ and NH_2CSNH] with base gives the free betaines [**346**; $R^1 = R^2 = H$, $R = Me_2N(CH_2)_3$ and NH_2CSNH] as stable solids.[250]

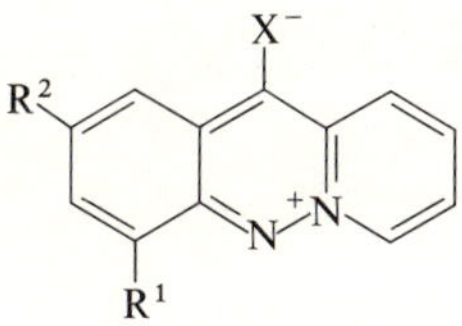

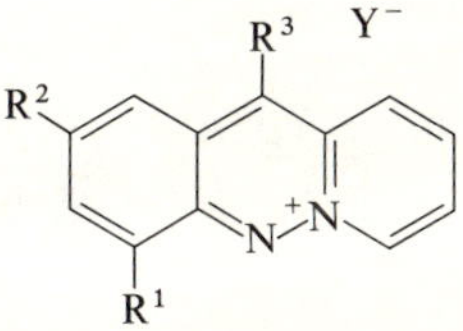

(346) X = NR
(347) X = CR$_2$
(348) X = S

(349) $R^3 = SCH_2CO_2Et$, Y = Br
(350) $R^3 = NHR$, Y = Br
(351) $R^3 = CH_3$, Y = Cl

3. *Pyrido[1,2-b]cinnolinium-11-methylates* (**347**)

In the presence of dimethyl malonate and aqueous sodium hydroxide, the salts **349** ($R^1 = H$, $R^2 = H$,Br) are converted to the purple betaines **347** ($R^1 = H$, $R^2 = H$,Br, $R = CO_2Me$). Acid hydrolysis of compound **347** ($R^1 = R^2 = H$, $R = CO_2Me$) generated the 11-methyl chloride **351** ($R^1 = R^2 = H$).[250]

4. Pyrido[1,2-b]cinnolinium-11-thiolates (348)

Compounds of this type are readily prepared from the pyrido[1,2-b]-cinnolinium-11-olates (338) (Section III,D,1) by reaction with phosphorus pentasulfide. Alkylation with ethyl bromoacetate gives the 11-ethoxycarboxylmethylthio salts (349) which are useful intermediates for preparing other derivatives (Sections III,D,2 and 3). Hydrolysis of 349 regenerates the betaines 348.[250]

5. Acridinium-4-olates (352)

N-Methylacridinium-4-olate (352; R = Me) has been obtained as a black solid which gives violet aqueous solutions. The method of preparation involves demethylation (AlCl$_3$) of a 4-methoxyacridinium salt followed by treatment of the resulting N-methyl-4-hydroxyacridinium salt with moist silver oxide.[252]

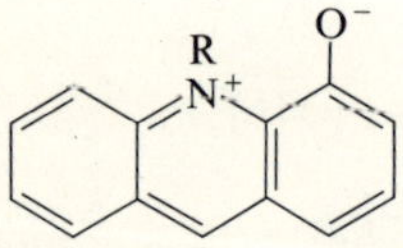

(352) 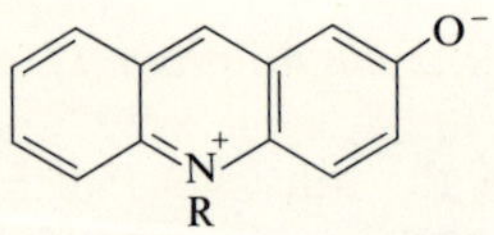   (353)

6. Acridinium-2-olates (353)

N-Methylacridinium-2-olate (353; R = Me) has been prepared in a manner analogous to that for the 4-isomer (352; R = Me) (Section III,D,5). This compound forms black crystals, mp 133°C, which give wine-red aqueous solutions.[252]

7. 5,8,10-Triazabenzo[a]quinolizinium-11-olates (354)

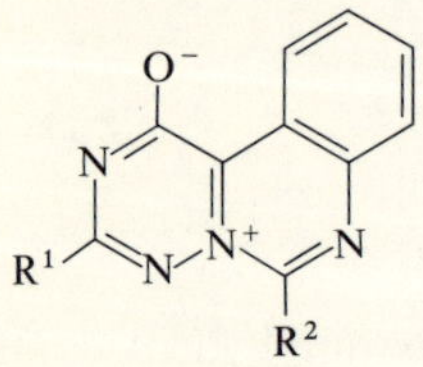

(354) 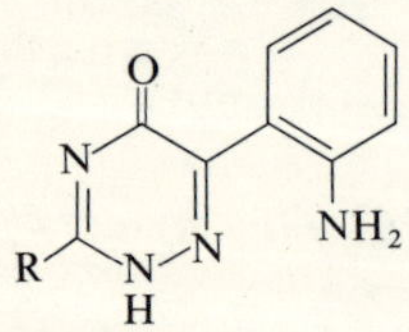   (355) 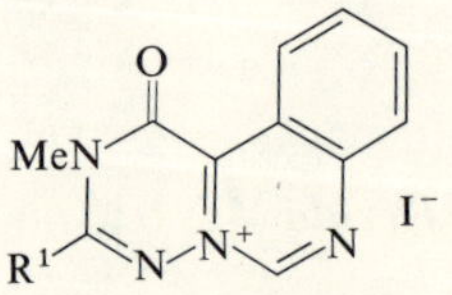   (356)

[252] S. Nitzsche, *Ber. Dtsch. Chem. Ges. B* **76**, 1187 (1943); **77**, 337 (1944).

Sec. III.E] HETEROCYCLIC BETAINES 65

Several compounds of this class (**354**) have been made by condensation of the 1,2,4-triazinone derivatives **355** (R = SMe) with orthoesters or acid chlorides. These betaines (**354**) form deeply colored methiodides (**356**) whose chemistry has been investigated.[212,213]

8. *Phenanthridinium-2-olates* (**357**)

(**357**)   (**358**)   (**359**)

Oxidation of the alkaloid lycorine (**359**) gives a red, hygroscopic solid which readily forms salts with mineral acids.[253] This product, which has been assigned the betaine structure **358**, has also been obtained from *Ungernia minor* and named ungeremine.[254,255] The *N*-methyl derivative (**357**; R^1 = Me, R^2R^3 = OCH_2O) has also been described.[253]

E. SYSTEMS WITH SEVENTEEN CONJUGATED ATOMS

1. *Neooxyberberines* (**364**)

The alkaloid berberine (**360**) is the source of two interesting mesomeric betaines (Scheme 14).[256,257] Reaction with acetone gives 8-acetonyldihydroberberine (**361**) which is oxidized to the bridged compound **362** (neooxyberberineacetone) by potassium permanganate.[258,259] Treatment of compound **362** with hot dilute mineral acid followed by base gives the betaine **364** which

---

[253] H. M. Fales, E. W. Warnhoff, and W. C. Wildman, *J. Am. Chem. Soc.* **77**, 5885 (1955).
[254] M. Normatov, K. A. Abduazimov, and S. Y. Yunusov, *Uzb. Khim. Zh.* **9**, 25 (1965) [*CA* **63**, 7061 (1965)].
[255] R. K.-Y. Zee-Cheng, S.-J. Yan, and C. C. Cheng, *J. Med. Chem.* **21**, 199 (1978).
[256] P. W. Jeffs, in "The Alkaloids" (R. H. F. Manske, ed.), Vol. 9, p. 41. Academic Press, New York, 1967.
[257] B. R. Pai, K. Nagarajan, H. Suguna, and S. Natarajan, *Heterocycles* **6**, 1377 (1977).
[258] F. L. Pyman, *J. Chem. Soc.* **99**, 1690 (1911).
[259] J. Iwasa and S. Naruto, *J. Pharm. Soc. Jpn.* **86**, 534 (1966) [*CA* **65**, 12247 (1966)].

has been named neooxyberberine.[258] Oxidation of berberine (**360**) by potassium ferricyanide gives an oxidation dimer (oxybisberberine) which with methanolic hydrogen chloride disproportionates to berberine chloride (**365**) and the 8-methoxybetaine **363**.[260] Catalytic reduction of this derivative (**363**) gives neooxyberberine (**364**) together with 13-hydroxytetrahydroberberine.[261]

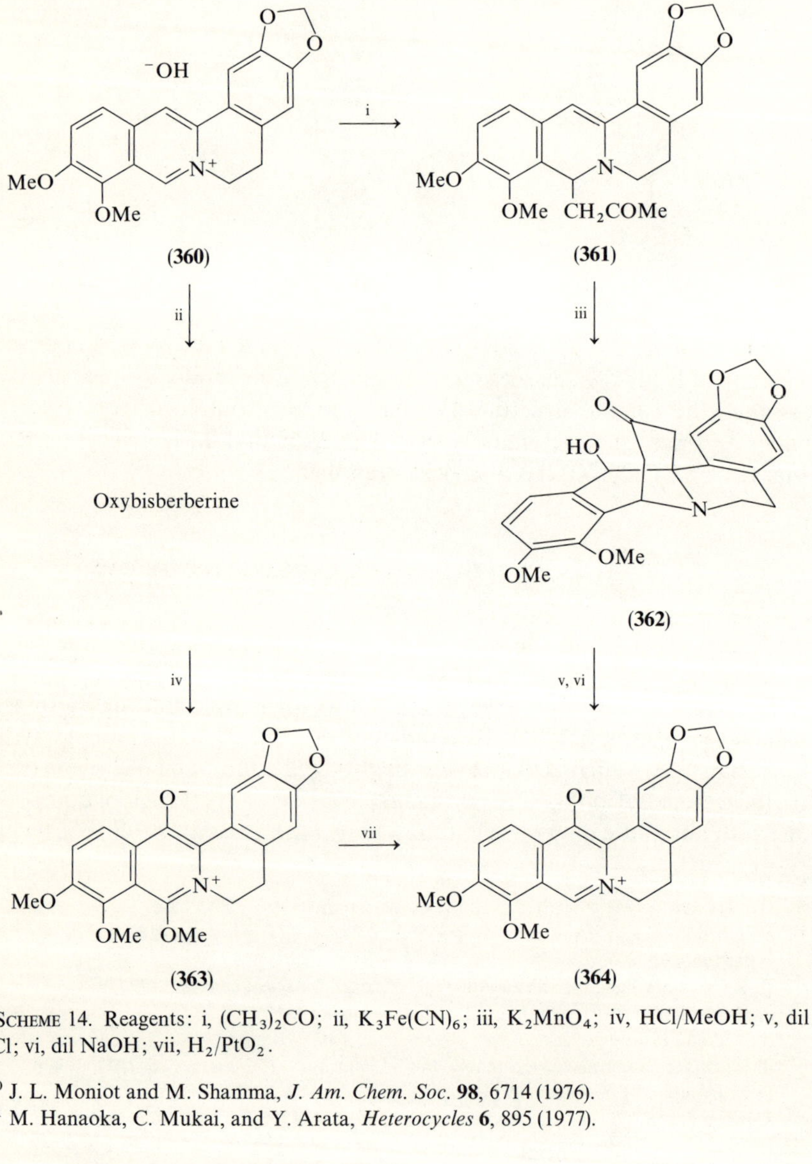

SCHEME 14. Reagents: i, $(CH_3)_2CO$; ii, $K_3Fe(CN)_6$; iii, K_2MnO_4; iv, HCl/MeOH; v, dil HCl; vi, dil NaOH; vii, H_2/PtO_2.

[260] J. L. Moniot and M. Shamma, *J. Am. Chem. Soc.* **98**, 6714 (1976).
[261] M. Hanaoka, C. Mukai, and Y. Arata, *Heterocycles* **6**, 895 (1977).

Neooxyberberine (**364**) has also been obtained by photooxidation of dihydroberberine (**366**; R = H).[262] In a closely related process, photooxidation of a methanolic solution of berberine chloride (**365**) containing sodium methoxide gives 8-methoxyneooxyberberine (**363**): Presumably addition of methoxide ion gives adduct **366** (R = OMe) which is oxidized by a process analogous to the transformation **366** (R = H) → **364**.[261]

(**365**)

(**366**)

(**367**)

(**368**)

Betaines **363** and **364** are yellow crystalline compounds. Methylation with methyl iodide gives methiodides (**367**)[258,263]: Acetylation of compound **363** with acetic anhydride and pyridine gives 13-acetoxyberberine **368**.[264] A number of reduction products have been reported. Sodium borohydride reduction of neooxyberberine (**364**) gives (±)ophiocarpine (**369**; R = H)[265] whereas reduction with zinc and acetic acid gives a mixture of tetrahydroberberine (**371**) and isomer **372**.[266,267] A ring-opening and recyclization sequence has been proposed to account for the formation of this interesting rearrangement product (**372**). Tetrahydroberberine (**371**) is formed in good yield by reducing neooxyberberine (**364**) with tin and hydrochloric acid.[258] Reduction of 8-methoxyneooxyberberine (**363**) with sodium borohydride

---

[262] Y. Kondo, H. Inoue, and J. Imai, *Heterocycles* **6**, 953 (1977).
[263] S. Pavelka and J. Kovar, *Collect. Czech. Chem. Commun.* **41**, 3654 (1976).
[264] J. L. Moniot, A. H. Abd el Rahman, and M. Shamma, *Tetrahedron Lett.*, 3787 (1977).
[265] T. Takemoto and Y. Kondo, *J. Pharm. Soc. Jpn.* **82**, 1413 (1962) [*CA* **59**, 689 (1963)].
[266] T. Takemoto, Y. Kondo, and K. Kondo, *J. Pharm. Soc. Jpn.* **83**, 162 (1963) [*CA* **59**, 3969 (1963)].
[267] C. Schöpf and M. Schweickert, *Chem. Ber.* **98**, 2566 (1965).

followed by acetylation (Ac$_2$O/pyridine) gives ($\pm$)-*O*-acetylophiocarpine (**369**; R = Ac) (87%) and ($\pm$)-*O*-acetyl-12-epi-ophiocarpine (**370**; R = Ac) (6%).[261]

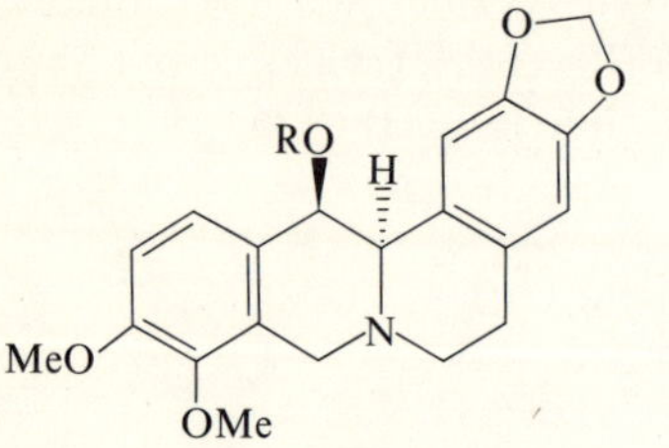

(369)

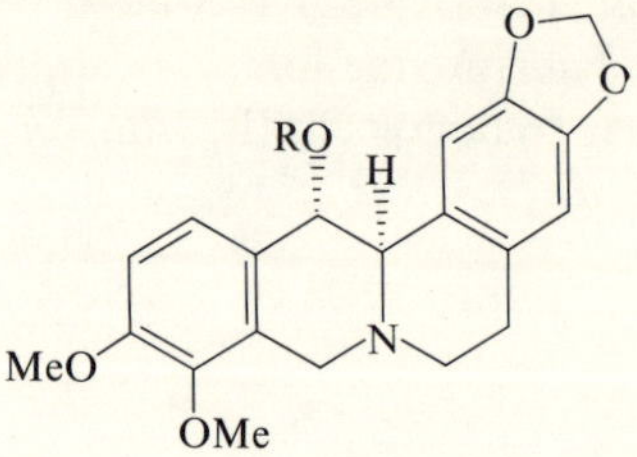

(370)

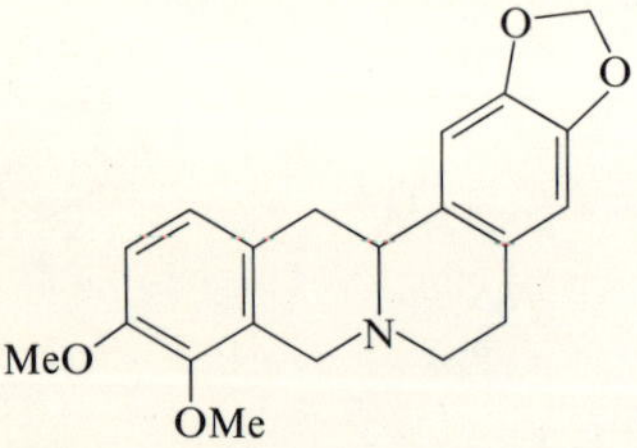

(371)

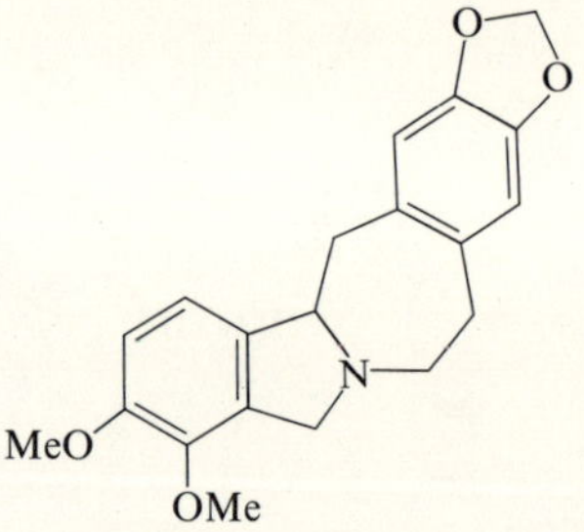

(372)

Interesting photooxidation products of betaines **363** and **364** have also been reported. Irradiation of methanol solutions (>0.1%) of compound **364** in the presence of oxygen gives a product (C$_{20}$H$_{17}$NO$_7$), mp 100.5–101.5°C, which has been formulated as adduct **373** (42%). In more dilute solutions the product is berberal (**374**; R = H) (56%).[262] Products **375**, **376**, and **374** (R = OMe) have been obtained by photooxygenation of the 8-methoxy-betaine **363**.[268]

Solutions of the betaine **363** in wet ether are slowly hydrolyzed to the methyl ester of dehydronorhydrastine (**377**).[260] In contrast, when hydrolysis

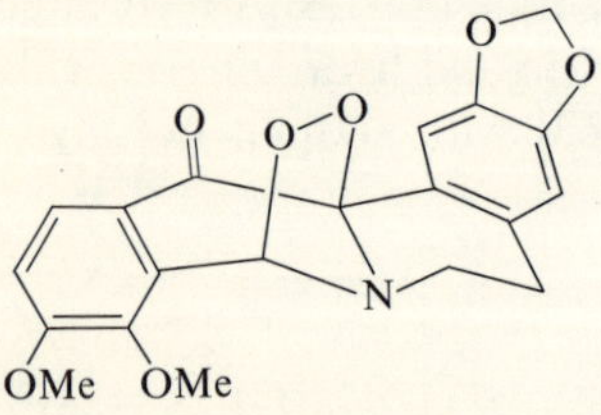

(373)

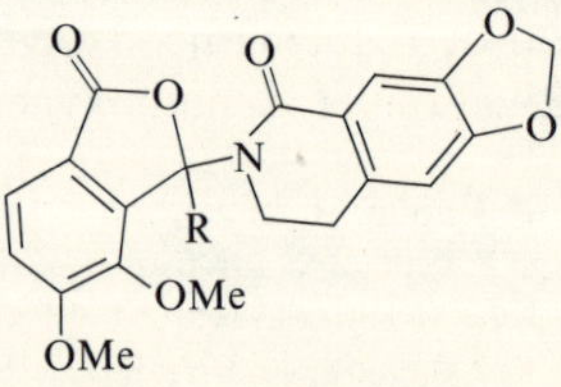

(374)

[268] M. Hanaoka and C. Mukai, *Heterocycles* **6**, 1981 (1977).

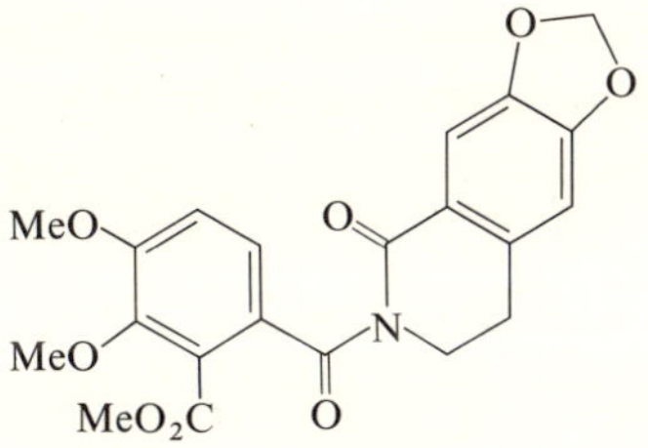

Sec. III.E] HETEROCYCLIC BETAINES 69

(375)

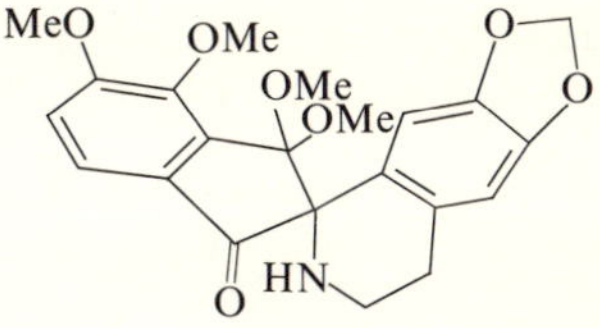

(376)

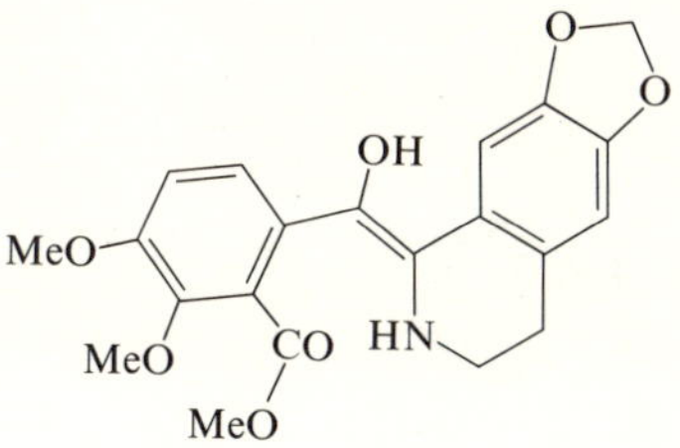

(377)

is carried out in wet tetrahydrofuran the product is methyl isoanhydroberberilate (375). A mechanism involving an aziridine intermediate has been proposed to account for this unusual rearrangement (363 → 375).[264]

2. *3-Arylcinnolinium-4-olates* (378)

Compound 378 has been made by oxidation of 6′-nitro-3,4-dihydropapaverine (379) to the *N*-oxide 380 using hot ethanolic iodine solution

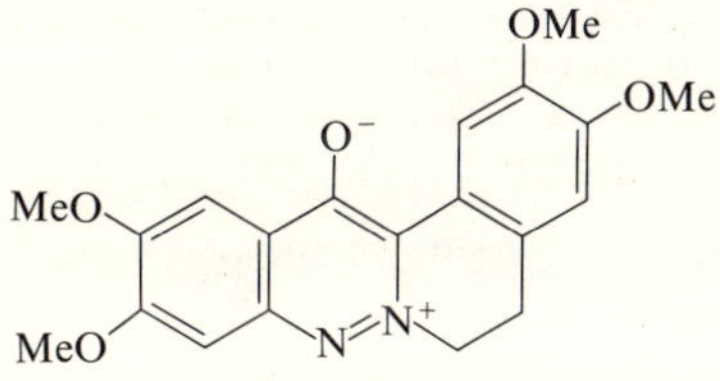

(378)

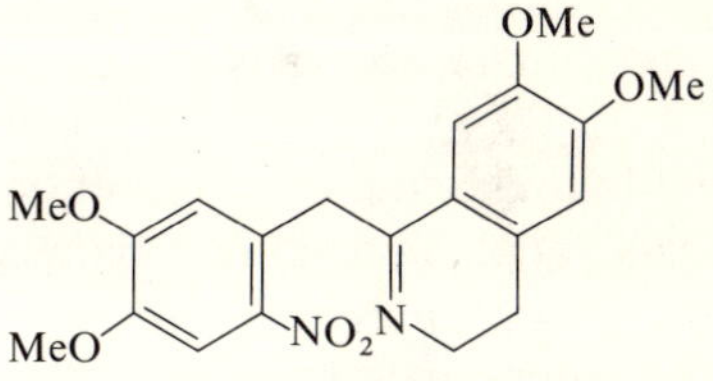

(379)

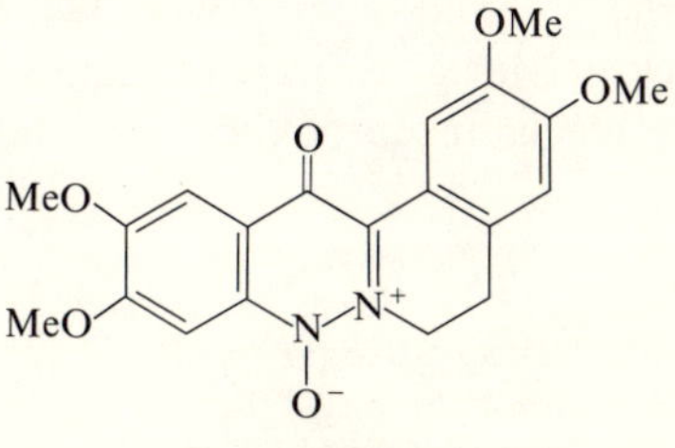

(380)

followed by reduction (Pd/H$_2$ or NaHSO$_3$). This orange-yellow betaine (378) is oxidized to the closely related betaine 388 (Section III,F,2) by dichlorodicyanoquinone.[269]

F. Systems with Nineteen Conjugated Atoms

1. *Dibenzo*[a,g]*quinolizinium-13-olates* (381)

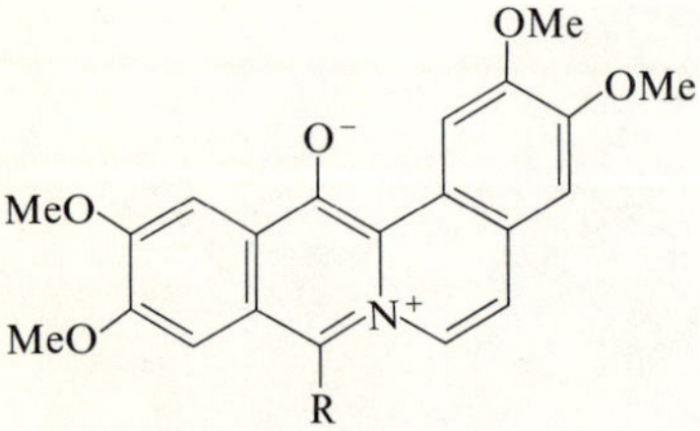

(381)

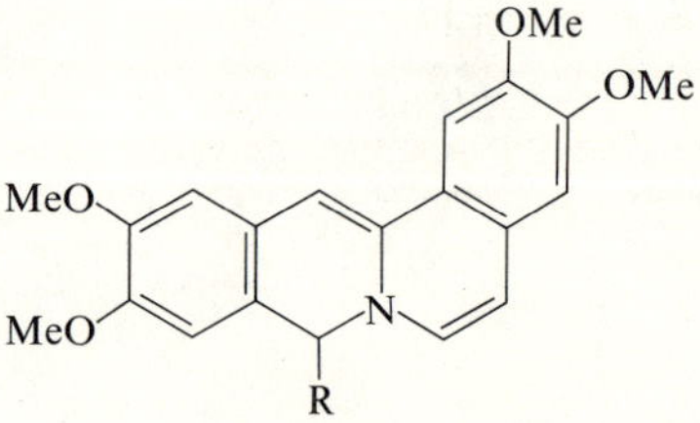

(382)

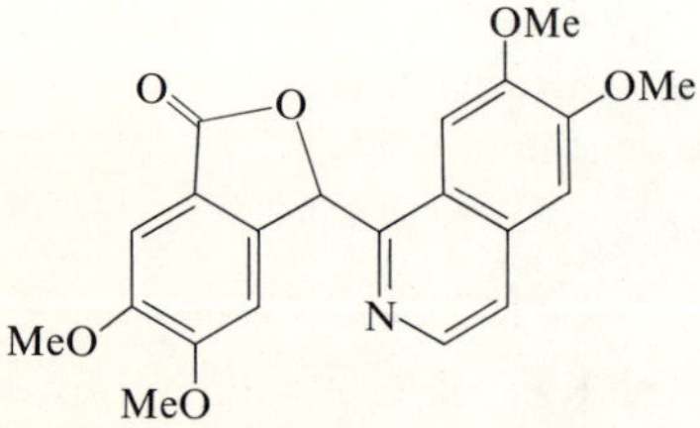

(383)

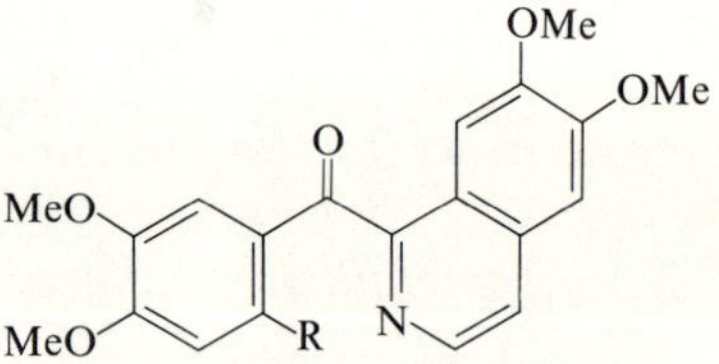

(384)

Preliminary accounts of the preparation and chemistry of betaines 381 (R = H, Me) have appeared. Compound 381 (R = H) is obtained in high yield by oxidation of dihydronorcoralyne (382; R = H) with *m*-chloroperbenzoic acid. In a similar manner, ethanolic solutions of dihydrocoralyne (382; R = Me) are oxidized to betaine 381 (R = Me) by air. The study of these compounds has so far been restricted to their oxidation products. Photooxidation of betaine 381 (R = H) followed by borohydride reduction gives a high yield of the phthalideisoquinoline 383. Photooxidation of the 8-methylbetaine 381 (R = Me) gives papaveraldine derivatives (384; R = COMe, CO$_2$Me). The mechanisms of these transformations are open to conjecture.[270-272]

[269] M. P. Cava, M. J. Mitchell, and D. T. Hill, *Chem. Commun.*, 1601 (1970).
[270] J. Imai and Y. Kondo, *Heterocycles* **5**, 153 (1976).
[271] J. Imai and Y. Kondo, *Heterocycles* **6**, 959 (1977).
[272] Y. Kondo, H. Inoue, and J. Imai, *Heterocycles* **7**, 45 (1977).

2. 8-Azadibenzo[a,g]quinolizinium-13-olates (388)

The yellow crystalline azaberbinone (388) has been prepared by several, closely related methods. Dehydrogenation of betaine 378 (Section III,E,2) with dichlorodicyanoquinone gives compound 388 directly. Alternatively, this compound (388) has been prepared in three ways from 6'-nitropapaverine (385) (Scheme 15): (i) Treatment of compound 385 with triethyl phosphite gives a low yield of betaine 388 directly; (ii) compound 385 in methanolic KOH at reflux temperature gives the anthranil 386 which rearranges to the betaine 388 in hot triethyl phosphite; and (iii) iodine oxidation of 6'-nitropapaverine (385) generates the N-oxide 387 which is reduced to the betaine 388 by sodium bisulfite. The chemistry of compound 388 remains unexplored.[269,273]

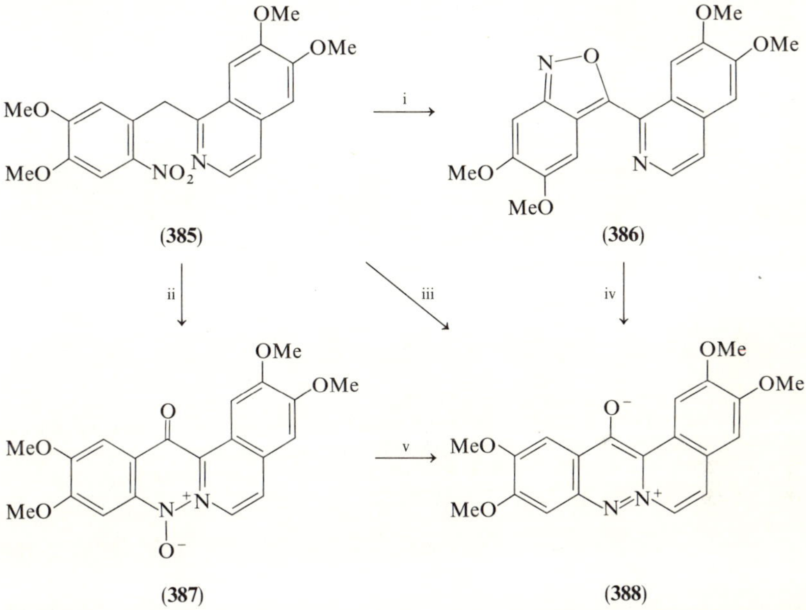

SCHEME 15. Reagents: i, KOH/MeOH; ii, I_2/EtOH; iii, (EtO)$_3$P; iv, heat; v, NaHSO$_3$/MeCO$_2$H.

3. Phenanthro[4,10-bc]azinium-6,11-diolates (389)

Betaines of this class (389) are of particular interest in that they have been isolated from natural sources. The alkaloid corunnine (389; $R^1 = H$, $R^2 = R^3 = OMe$) is obtained as violet needles, mp 255–257°C, from *Glaucium*

[273] T. Kametani, T. Yamanaka, K. Ogasawara, and K. Fukumoto, *J. Chem. Soc. C*, 380 (1970).

flavum Cr. var. *vestitum*[274]; nandazurine (**389**; $R^1 = H$, $R^2R^3 = OCH_2O$), which is a dark green alkaloid, has been ioslated from *Nandina domestica* Thumb. (Berberidaceae)[275,276]; and the green alkaloid PO-3 (**389**; $R^1 = $ OMe, $R^2 = R^3 = H$) has been extracted as its perchlorate from *Papaver orientale*.[277]

Corunnine (**389**; $R^1 = H$, $R^2 = R^3 = $ OMe) is also obtained in small amounts by oxidation of glaucine (**390**; $R^1 = R^3 = R^4 = $ OMe, $R^2 = H$).[278] Similarly, oxidation of isothebaine (**390**; $R^1 = $ OH, $R^2 = $ OMe, $R^3 = R^4 = H$) gives PO-3 (**389**; $R^1 = $ OMe, $R^2 = R^3 = H$)[279,280] and photooxidation of dehydronuciferine (**391**) gives the blue-green betaine **389** ($R^1 = R^2 = R^3 = H$).[281]

Reduction of corunnine (**389**; $R^1 = H$, $R^2 = R^3 = $ OMe) by zinc and acetic acid gives ($\pm$)-thalicmidine (**390**; $R^1 = $ OH, $R^2 = H$, $R^3 = R^4 = $ OMe). Similar reduction of nandazurine (**389**; $R^1 = H$, $R^2R^3 = OCH_2O$)

(**389**)

(**390**)

(**391**)

(**392**)

[274] I. Ribas, J. Sueiras, and L. Castedo, *Tetrahedron Lett.*, 3093 (1971).
[275] Z. Kitazato, *J. Pharm. Soc. Jpn.* **45**, 695 (1925).
[276] J. Kunitomo, M. Ju-ichi, Y. Yoshikawa, and H. Chikamatsu, *Experientia* **29**, 518 (1973).
[277] V. L. Preininger and F. Santavy, *Acta Univ. Palacki. Olomuc.*, 5 (1966) [*CA* **67**, 54290 (1967)].
[278] L. Castedo, R. Suau, and A. Mourino, *An. Quim.* **73**, 290 (1977) [*CA* **87**, 201836 (1977)].
[279] V. Preininger, J. Hrbek, Z. Samek, and F. Santavy, *Arch. Pharm.* (*Weinheim, Ger.*) **302**, 808 (1969) [*CA* **72**, 55710 (1970)].
[280] V. Preininger, J. Hrbek, Z. Samek, and F. Santavy, *Acta Univ. Palacki. Olomuc., Fac. Med.* **52**, 5 (1969) [*CA* **76**, 25454 (1972)].
[281] J. M. Saa, M. J. Mitchell, and M. P. Cava, *Tetrahedron Lett.*, 601 (1976).

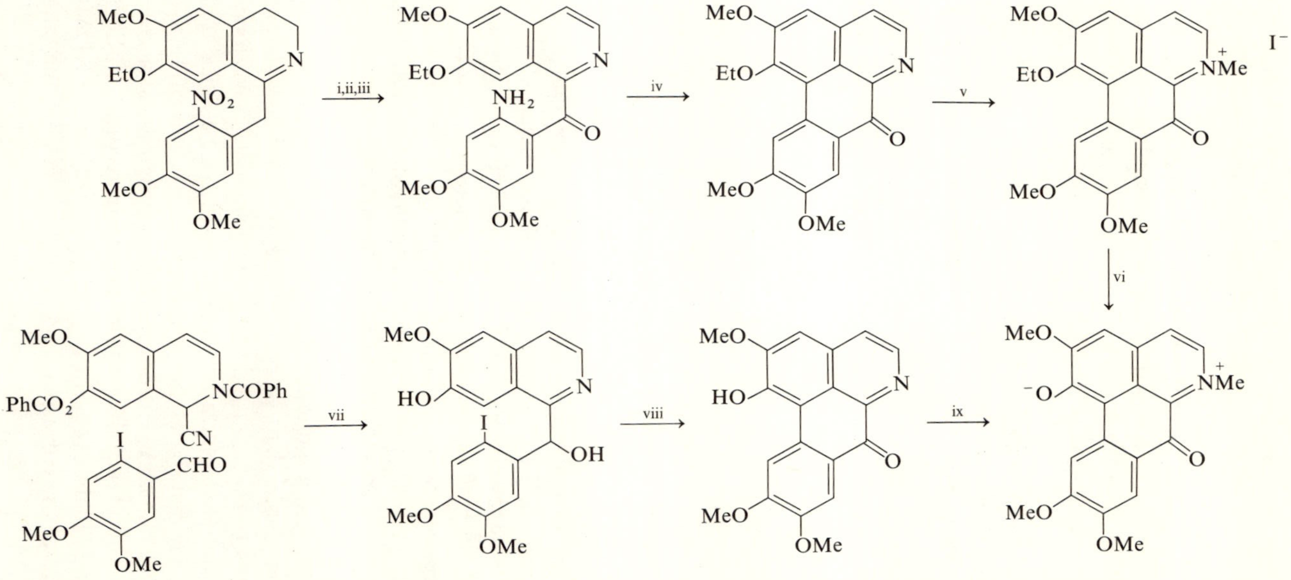

SCHEME 16. The synthesis of corunnine. Reagents: i, CrO$_3$/AcOH; ii, OH$^-$/EtOH; iii, H$_2$/Raney Ni; iv, HNO$_2$/Cu; v, MeI; vi, heat; vii, NaH/DMF; viii, O$_2$/$h\nu$; ix, MeI.

gives (±)-domesticine (**390**; $R^1 = OH$, $R^3R^4 = OCH_2O$) whereas reduction by sodium borohydride gives hexahydronandazurine (**392**).[276,282]

Two unambiguous methods of synthesizing corunnine (**389**; $R^1 = H$, $R^2 = R^3 = OMe$) have been reported and are shown in Scheme 16. The second route has also been used to synthesise nandazurine (**389**; $R^1 = H$, $R^2R^3 = OCH_2O$). The total synthesis of these naturally occurring mesomeric betaines seems a suitable high point upon which to complete this section.[282,283]

IV. The Electronic Structure of Mesomeric Betaines

A. A Perturbation Molecular Orbital (PMO) Model

1. *Ionization Potentials*

If a carbon atom in an odd AH anion is replaced by a heteroatom, then assuming no change in π-bond strengths (i.e., $\Delta\beta = 0$), the change in energy (ΔE_m) of a π-molecular orbital (ψ_m) is given to a first-order approximation by

$$\Delta E_m \simeq C_{m,r}^2 \cdot \Delta\alpha_r \tag{1}$$

where $C_{m,r}$ is the orbital coefficient at the position of substitution (r) and $\Delta\alpha_r$ is the corresponding change in the Coulomb integral.[10,11]

Mesomeric betaines are generated from odd AH anions by a conceptual replacement of inactive carbon atoms (CH) by hetero-cations (N^+R, O^+, S^+) (Section II). Because the HOMO of the AH anion is a NBMO which vanishes on inactive atoms (i.e., $C_{m,r} = 0$), this substitution will not perturb the HOMO ($\Delta E_m = 0$; Eq. 1) and mesomeric betaines are expected to be associated with a HOMO that is high in energy. The low ionization potentials of pyridinium-3-olates (**393**) and isoquinolinium-4-olates (**395**) (Table III) lend support to this conclusion. Note that values for the pyridinium-3-olates (**393**) are significantly less (~0.75 eV) than those for isomeric 2-pyridones (**394**). It is also significant that ionization potentials of the betaines are relatively insensitive to the nature of N-substituents; para-substitution of phenyl groups has little influence. Introduction of heteroatoms at active positions of the betaines will reduce the HOMO energy in accord with Eq. (1). In the betaines **393** and **395** the exocyclic oxygen atoms perturb the HOMO. On the basis of the size of the corresponding orbital coefficients, the magnitude of

---

[282] S. M. Kupchan and P. F. O'Brien, *J. C. S., Chem. Commun.*, 915 (1973).
[283] I. Ribas, J. Saa, and L. Castedo, *Tetrahedron Lett.*, 3617 (1973).

TABLE III
IONIZATION POTENTIALS (IP)

(393)   (394)   (395)

Compound	IP^a (eV)	Reference
393; R^1 = Me, R^2 = H	7.9	284
393; R^1 = Ph, R^2 = H	7.55, 7.48	63, 285
393; R^1 = Ph, R^2 = Me	7.27	63
393; R^1 = p-MeOC$_6$H$_4$, R^2 = H	7.4	285
393; R^1 = p-ClC$_6$H$_4$, R^2 = H	7.6	285
394; R^1 = Ph, R^2 = H	8.26	63
394; R^1 = Ph, R^2 = Me	8.01	63
395; R = PhCH$_2$	7.10	95
395; R = Ph	7.10	286
395; R = p-MeOC$_6$H$_4$	6.93	286
395; R = p-NO$_2$C$_6$H$_4$	7.29	286

^{a} Measured by electron impact.

this perturbation is expected to be greater for pyridinium-3-olates (**393**) than for isoquinolinium-4-olates (**395**) and this may account for the difference in their ionization potentials (Table III).[63,95,284–286]

2. Absorption Spectra

The effect of heteroatoms on the LUMO energy of mesomeric betaines can be treated similarly, and this leads to an estimate of the HOMO–LUMO splitting. If E is the splitting in an AH anion, the change (ΔE) upon substitution of a heteroatom at position r is given by Eq. (2).

$$\Delta E \simeq (C^2_{m+1,r} - C^2_{m,r}) \cdot \Delta \alpha_r \qquad (2)$$

where $C_{m,r}$ and $C_{m+1,r}$ are the orbital coefficients of the anion HOMO and LUMO at position r.[10,11]

---

[284] T. Groenneberg and K. Undheim, *Org. Mass Spectrum.* **6**, 225 (1972).
[285] K. Undheim and P. E. Hansen, *Org. Mass Spectrom.* **7**, 635 (1973).
[286] P. E. Hansen and K. Undheim, *Acta Chem. Scand. Ser. B* **29**, 221 (1975).

Using Eq. (2) we can estimate the HOMO–LUMO splitting of the nitrogen heterocycles **396**–**398** (Fig. 1). Assuming that perturbations due to exocyclic oxygen are constant, the separation of the frontier orbitals (E_{HET}) is given by

$$E_{\text{HET}} \simeq E_o + (C^2_{m+1,r} - C^2_{m,r}) \cdot \Delta\alpha_{\text{N}} \quad (3)$$

where E_o is the orbital splitting in the phenoxide anion, r is the position of the nitrogen atom, and $\Delta\alpha_{\text{N}}$ is the change in the Coulomb integral due to the nitrogen atom.

Calculated values of $(C^2_{m+1,r} - C^2_{m,r})$ for a number of nitrogen heterocycles are given in Table IV. The splitting for betaine **396** is predicted to be substantially smaller than for the phenoxide ion (i.e., ΔE is large and negative since $\Delta\alpha_{\text{N}}$ is negative). In N-methylpyrid-2-one (**397**) the splitting increases relative to **396** and an even larger splitting is estimated for N-methylpyrid-4-one (**398**). Some measure of the reliability of these estimates can be obtained by considering the frequency of the first $\pi \rightarrow \pi^*$ transition of these species (in comparable solvents) since this corresponds to a transition between HOMO and LUMO. The UV and visible spectra of relevant heterocycles are given in Table IV.[33,101,106,287–304]

The frequency of the first $\pi \rightarrow \pi^*$ absorption band (v) is related to frontier orbital separation by the expression

$$hv = E_o + (C^2_{m+1,r} - C^2_{m,r}) \cdot \Delta\alpha_{\text{N}} \quad (4)$$

or

$$v = (E_o/h) + (C^2_{m+1,r} - C^2_{m,r}) \cdot (\Delta\alpha_{\text{N}}/h) \quad (5)$$

---

[287] S. F. Mason, *J. Chem. Soc.* 1253 (1959).
[288] L. C. Anderson and N. V. Seeger, *J. Am. Chem. Soc.* **71**, 343 (1949).
[289] H. Specker and H. Gawrosch, *Ber. Dtsch. Chem. Ges, B* **75**, 1338 (1942).
[290] K. G. Cunningham, G. T. Newbold, F. S. Spring, and J. Stark, *J. Chem. Soc.*, 2091 (1949).
[291] M. L. W. Chang and B. C. Johnson, *J. Biol. Chem.* **234**, 1817 (1959).
[292] S. F. Mason, *J. Chem. Soc.*, 5010 (1957).
[293] D. J. Brown and S. F. Mason, *J. Chem. Soc.*, 3443 (1956).
[294] C. A. Ramsden, unpublished work.
[295] B. S. Thyagarajan, K. Rajagopalan, and P. V. Gopalakrishnan, *J. Chem. Soc. B*, 300 (1968).
[296] C. F. H. Allen, H. R. Beilfuss, D. M. Burness, G. A. Reynolds, J. F. Tinker, and J. A. Van Allan, *J. Org. Chem.* **24**, 779 (1959).
[297] V. Boekelheide and J. P. Lodge, *J. Am. Chem. Soc.* **73**, 3681 (1951).
[298] J. M. Hearn, R. A. Morton, and J. C. E. Simpson, *J. Chem. Soc.*, 3318 (1951).
[299] G. W. Ewing and E. A. Steck, *J. Am. Chem. Soc.* **68**, 2181 (1946).
[300] D. W. Jones, *J. Chem. Soc. C*, 1729 (1969).
[301] D. A. Evans, G. F. Smith, and M. A. Wahid, *J. Chem. Soc. B*, 590 (1967).
[302] A. Fozard and G. Jones, *J. Chem. Soc.*, 2760 (1964).
[303] H. Ley and H. Specker, *Ber. Dtsch. Chem. Ges. B* **72**, 192 (1939).
[304] R. D. Brown and F. N. Lahey, *Aust. J. Sci. Res., Ser. A* **3**, 615 (1950) [*CA* **45**, 9369 (1951)].

TABLE IV: Ultraviolet and Visible Spectra

Compound	$(C_{m+1,r}^2 - C_{m,r}^2)$	Solvent	$\lambda_{max}(\log \varepsilon)$ (nm)	References
396 (Fig. 1)	0.25	H_2O (pH 7)	320(3.75), 250(3.84), 215(4.40)	287
		H_2O (pH 10)	320(3.70)	33
		EtOH	328(—)	287
		Dioxane	356(3.70)	287
397 (Fig. 1)	0.11	H_2O (pH 7)	298(3.92)	288
		MeOH	300(3.64), 227(3.64)	289
		EtOH	305(3.69), 230(3.78)	290
			300(—)	287
398 (Fig. 1)	−0.14	H_2O (pH 7)	262(—)	291
		MeOH	263(4.26)	289
		EtOH	262(—)	287
399 (Fig. 2; see also Table V)	0.24	H_2O (pH 10)	442(3.20), 346(3.10), 334(3.10), 273(4.60)	292
		EtOH	484(2.70), 355(2.85), 280(4.00), 250(3.50)	106
		$CHCl_3$	554(2.50), 375(3.00), 290(3.80), 240(3.50)	106
400 (Fig. 2)	0.17	H_2O (pH 8.5)	462(3.58), 332(3.06), 318(3.09), 273(4.52), 237(4.06)	293
401 (Fig. 2)	0.10	H_2O (pH 10)	408(3.60), 349(3.80), 269(4.30), 233(4.20)	292
402 (Fig. 2)	0.07	H_2O (pH 7)	370(3.78)	294
		H_2O (pH 8.5)	364(3.99), 320(3.56), 248(3.96)	292
		EtOH	384(3.87)	294
403 (Fig. 2)	0.00	EtOH	400(3.94), 374sh(3.91), 266(3.91)	101
		$CHCl_3$	431(3.85), 382(3.85), 278(3.85), 250(3.85)	101
404 (Fig. 2)	0.00	H_2O (pH 10)	400(3.76), 334(3.69), 257(4.30)	292
405 (Fig. 2)	−0.02	H_2O (pH 7)	380(4.05), 245(4.05)	295
		MeOH	380(4.20), 277(3.20), 230(4.10), 220(4.20)	296
		EtOH	378(4.15), 275(3.15), 255(4.10), 247(4.15)	297
406 (Fig. 2)	−0.14	EtOH	338(4.12), 331(4.06), 325(4.11), 261(2.97), 237(4.27)	298
407 (Fig. 2)	−0.20	EtOH	325(3.70), 275(3.95), 285(3.95)	299
418 (Fig. 3)	0.19	H_2O (pH 10)	408(3.54), 325(3.60), 316(3.60), 270(4.43)	292
419 (Fig. 3)	0.18	H_2O (pH 10)	384(3.85), 320(3.23), 308(3.23), 255(4.16)	292
420 (Fig. 3)	0.11	EtOH	410(3.65)	300
		$CHCl_3$	424(3.27), 308(2.88), 295(3.20)	301
421 (Fig. 3)	0.11	H_2O (pH 10)	408(3.47), 298(3.91), 261(4.70)	292
422 (Fig. 3)	0.095	H_2O (pH 8)	406(4.00), 311(3.19), 261(4.51)	293
423 (Fig. 3)	0.005	$CHCl_3$	362(4.18), 278(4.36), 269(4.36), 250(4.38)	101
424 (Fig. 3)	−0.03	H_2O (pH 10)	358(4.09), 267(4.36), 230(4.54)	292, 293
425 (Fig. 3)	−0.04	H_2O (pH 7)	325sh(3.75), 299(4.00), 226(4.50)	302
426 (Fig. 3)	−0.15	H_2O (pH 7)	325(3.81), 272(3.84), 245(4.01), 228(4.53)	292
		MeOH	329(3.76)	303
		EtOH	344sh(3.61), 330(3.77), 279(3.81), 271(3.82), 301(4.61)	304

A plot of absorption frequency against the factor $(C_{m+1,r}^2 - C_{m,r}^2)$ should be linear if the approximation is realistic. The plot of $v/(C_{m+1,r}^2 - C_{m,r}^2)$ for the heterocycles **396–398** in ethanol solution (Fig. 1) does indeed show remarkably good linearity. A similar analysis of heterocycles isoconjugate with the α-naphthylmethyl anion is shown in Fig. 2 and again the correlation is good. A bathochromic shift relative to the systems **396–398** is due to narrowing of the frontier orbitals by extension of the conjugation. Note how this simple PMO treatment accounts for the large bathochromic shift of N-methylquinolinium-8-olate (**399**).

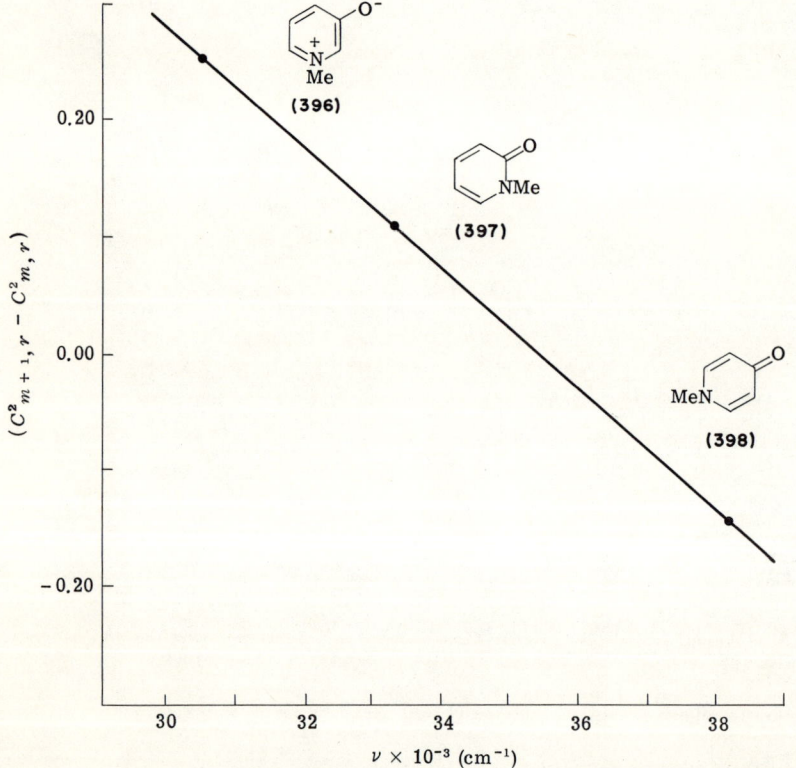

FIG. 1. Calculated change in HOMO–LUMO splitting of heterocycles isoconjugate with the benzyl anion vs frequency of their first $\pi \to \pi^*$ absorption in ethanol solution.

The correlations shown on Figs. 1 and 2 are particularly remarkable when other factors which influence the spectra are considered. A primary complication is the effect of solvent polarity. Ideally, UV spectra should be recorded in nonpolar hydrocarbon solvents to minimize the effect of

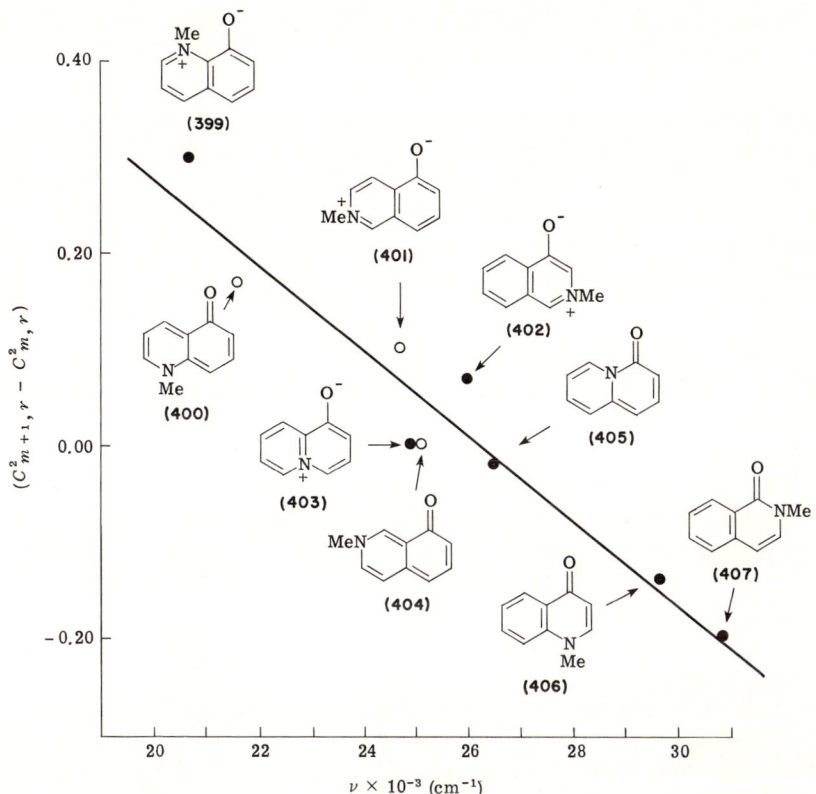

Fig. 2. Calculated change in HOMO–LUMO splitting of heterocycles isoconjugate with the α-naphthylmethyl anion vs frequency of their first $\pi \to \pi^*$ absorption band in (i) ethanol (●) or (ii) water buffered at pH 10 (○).

solvation, which is greatest for polar solvents and is particularly pronounced if the solute molecule has a large dipole moment. A hypsochromic shift is observed if the ground state is more polar than the excited state, and if the electron distribution of the two states differs, the excited state may be produced so rapidly that the solvent cage has no time to rearrange and relatively poor solvation results. Some comment on the charge distribution in mesomeric betaines is therefore relevant.

The formal π-electron density at each atom in an odd AH radical (e.g., **408**) is unity, and to a first approximation it will be the same in isoelectronic radical cations (e.g., **409**). Approximately (because inductive effects are neglected), a unit positive charge is localized on the heteroatom. Introduction of an additional electron into these cations (e.g., **409**) gives mesomeric betaines (e.g., **410**). Because the electron enters a NBMO, it is restricted to

(408) (409) (410) (411)

(412) ⟷ (413) ⟷ (414) ⟷ (415)

active positions and will result in charge distributions exemplified by structure **411**. Separation of charge is expected and the description of these compounds as betaines is justified. Inductive effects will of course modify this idealized model, especially when additional heteroatoms are present, but on an approximate basis the distribution shown can be expected. Indeed, this is predicted by resonance theory (i.e., **412–415**) and is also consistent with the observed ^{1}H- and ^{13}C-chemical shifts of several betaines.[305–307] It must be remembered, however, that a similar treatment of nonbetaine isomers also suggests substantial polarity (cf. dipole moments of pyridones). Charge separation as such is not, therefore, a property which necessarily distinguishes mesomeric betaines from their isomers.

Although the HOMOs of mesomeric betaines vanish at the position of the positively charged heteroatom, this is not generally the case with the LUMO, which usually has a substantial orbital coefficient at this position. It follows that promotion of an electron from HOMO to LUMO will result in migration of electron density toward the positive pole, and the first excited state will be less polar than the ground state. The first absorption band should, therefore, be displaced to shorter wavelengths by polar solvents. The effect of solvent on the spectra of two betaines has been studied and the results are shown in Table V. For both compounds the expected solvent effect is observed and because these solutions are colored the effect is clearly visible. The bands are displaced to shorter wavelengths as the polarity of the solvent increases (negative solvatochromism). Compound **417** (Section III,C,3) gives an orange solution in water (λ_{max} 453 nm) whereas in diphenyl ether the solution is blue-green (λ_{max} 810 nm). Dimroth et al.[218–220] have used the transition energy of compound **417** expressed in kilocalories per

---

[305] U. Vögeli and W. von Philipsborn, *Org. Magn. Reson.* **5**, 551 (1973).
[306] Y. Takeuchi and N. Dennis, *Org. Magn. Reson.* **8**, 21 (1976).
[307] L. Stefaniak, *Tetrahedron*, 1065 (1976).

TABLE V
SOLVATOCHROMISM OF THE FIRST $\pi \to \pi^*$ TRANSITIONS OF TWO MESOMERIC BETAINES[a]

(416)   (417)

Solvent	Dielectric constant (ε)	Compound 416 λ_{max} (nm)	Compound 417 λ_{max} (nm)	E_T
Diphenyl ether	3.65	—	810	35.3
p-Dioxane	2.21	—	795	36.0
Benzene	2.28	568	—	—
Chloroform	4.81	554	730	39.7
Ethyl acetate	6.02	539	750	38.1
Benzonitrile	25.2	—	680	42.0
Nitrobenzene	34.8	530	680	42.0
Acetone	20.7	535	677	42.2
tert-Butanol	10.9	502	652	43.9
Acetonitrile	38.8	512	622	46.0
Nitromethane	20.8	388	618	46.3
Ethanol	24.3	484	550	51.9
Methanol	32.6	—	515	55.5
Water	78.5	443	453	63.1

[a] See Saxena et al.[106] and Dimroth et al.[218-220]

mole as a polarity parameter (E_T) (Table V). Similarly, compound **416** (Section III,B,5) varies from purple in benzene (λ_{max} 568 nm) to red in water (λ_{max} 443 nm).[106]

Considering the large influence of solvent on the position of absorption bands, the correlations shown in Figs. 1 and 2, which contain no solvent term, are surprisingly good. Can it be that the degree of solvation varies little among the isomeric species and that the effect can be included in the constants of the equation? It would be interesting to examine these correlations for a range of solvents. A plot of $(C_{m+1,r}^2 - C_{m,r}^2)/v$ for heterocycles isoconjugate with the β-naphthylmethyl anion is shown in Fig. 3. In this case a greater variation of solvent is unavoidable owing to shortage of data, and although correlation is clearly discernible, the scatter is much greater.

Various other factors may influence the spectra. Protonation in neutral, aqueous solution may cause anomalous results for heterocycles with large

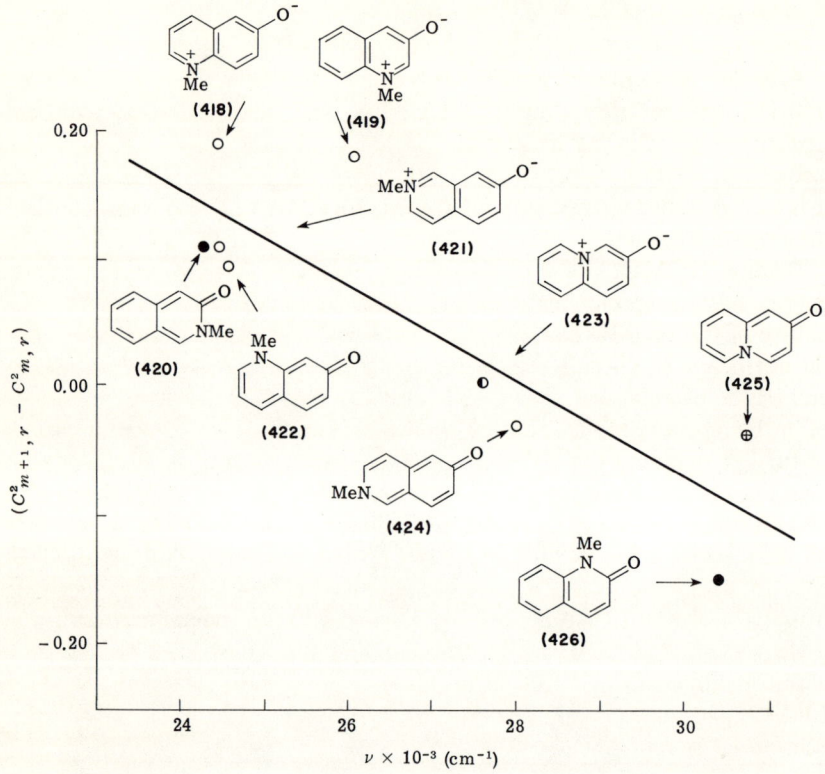

Fig. 3. Calculated change in HOMO–LUMO splitting of heterocycles isoconjugate with the β-naphthylmethyl anion vs frequency of their first $\pi \to \pi^*$ absorption band in (i) ethanol (●), (ii) water (⊕), (iii) water at pH > 7 (○), or (iv) chloroform (◐).

pK_a values. Solutions buffered at pH 8–10 are preferred. Monocentric perturbations can cause changes in orbital ordering relative to AH anions. Mason has demonstrated that this may occur in 4-pyridones,[308] and the large extinction coefficient for 4-methylpyridone (**398**, Table IV) possibly indicates the close proximity of the first two absorption bands.

A detailed analysis of the spectra of the compounds shown in Figs. 1–3 and in related species is beyond the scope of this review. The simple model described here, however, does provide some insight into the electronic structure of these molecules and enables some general conclusions to be made about the nature of their frontier orbitals.

[308] S. F. Mason, in "Physical Methods in Heterocyclic Chemistry" (A. R. Katritzky, ed.), Vol. 2, p. 1. Academic Press, New York, 1963.

3. Frontier Orbitals

Frontier molecular orbital (FMO) theory has been successful in rationalizing the reactivity, electroselectivity, and regioselectivity of many heterocycles, including mesomeric betaines. To apply FMO theory, some knowledge of the frontier orbital coefficients and energies is necessary, and it is useful to draw some general conclusions about the frontier orbitals in mesomeric betaines.

Betaine HOMOs are high in energy (Section IV,A,1), but the LUMO energy has no special features and varies with individual examples. The nature of the HOMO causes the HOMO–LUMO gap to be narrower than the norm, and this is indicated by grouping of betaines at the long wavelength end of the correlations shown in Figs. 1–3. This is the reason for betaines often being more highly colored than their isomers. An ingenious method of further reducing the HOMO–LUMO separation in pyridinium-3-olates (and thereby increasing reactivity) has been demonstrated by Katritzky. The HOMO is relatively insensitive to the nature of the N-substituent (Section VI,A,1), because the two fragments are cross-conjugated, but this is not the case for the betaine LUMO. A conjugated N-substituent (e.g., aryl) will usually lower the LUMO energy by extending the conjugation, and further lowering is achieved if the substituent contains heteroatoms or electron-withdrawing groups. The effect of this type of modification of the HOMO–LUMO gap on the dimerization of pyridinium-3-olates is described in Section III,A,2.

Since the betaine HOMO is treated as a perturbed NBMO, the shape, as well as the energy, of this orbital might be expected to be close to that of a NBMO. Hückel calculations on three betaines **427–429** (Fig. 4) lend support to this view. The following features of the calculated HOMOs are notable: (i) 90% of the charge density is restricted to active positions; (ii) the only

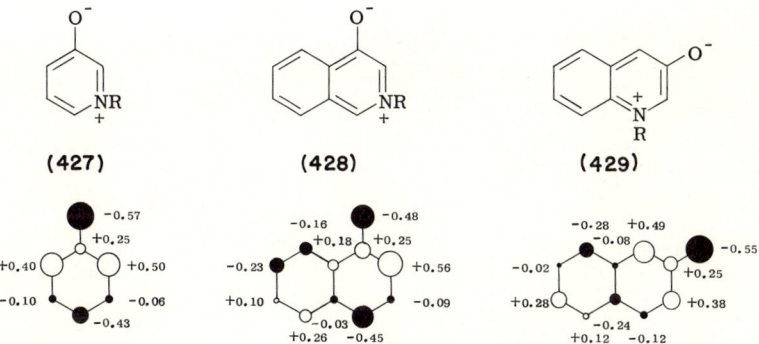

FIG. 4. HOMO coefficients of three betaines calculated by the Hückel method.

inactive positions associated with significant charge density ($\sim 6\%$) are those adjacent to oxygen atoms; and (iii) the phase relationships between orbital coefficients at active positions are the same as in the corresponding NBMO. Similar results are found using more sophisticated MO methods (Section VI,B). For qualitative applications of FMO theory, therefore, betaine HOMOs can be represented to a first approximation by the appropriate NBMO. Calculations also indicate that the shape of betaine LUMOs are closely related to those of the corresponding AH, but since the perturbations of these orbitals are greater, generalizations should be treated cautiously.

In summary, the following generalizations can be made:

1. The HOMO of a mesomeric betaine is similar in shape and energy to a NBMO. The energy of this pseudo-NBMO is lowered by replacing active carbon atoms by heteroatoms or by introducing electron-withdrawing substituents at active positions.

2. The HOMO–LUMO separation of mesomeric betaines is relatively small. Heteroatoms and conjugated substituents at inactive positions will usually reduce this separation by lowering the energy of the LUMO without perturbing the HOMO.

4. *Sulfur d-Orbitals*

Structures depicting tetracovalent sulfur (e.g., **431**) are often used to represent sulfur-containing mesomeric betaines (e.g., **430**). Because the betaine HOMO is similar to a NBMO, it has symmetry suitable for overlap with a sulfur d_{yz}-orbital (e.g., **432**) and the relative importance of the canonical forms **430** and **431** depends upon the extent of this d_{yz}–HOMO mixing. The situation is essentially the same as for heteropentalene mesomeric betaines, and a discussion of the main factors influencing this orbital mixing can be found in an earlier review.[2] At present, the importance of structures involving the participation of sulfur d-orbitals is not clear. A recent study of the photoelectron spectra of naphtho[1,8-*cd*][1,2,6]thiadiazines (**330**) (Section III,C,9) and naphtho[1,8-*de*]triazines (**321**) (Section III,C,7) suggests that 3d-orbitals are not important.[309]

(430)   (431)   (432)

---

[309] R. Bartetzko and R. Gleiter, *Angew. Chem. Int. Ed. Engl.* **17**, 468 (1978).

5. Thermodynamic Stability

Consider the replacement of a carbon atom in an odd AH anion by a heteroatom. Dewar has shown that if the heteroatom contributes one π-electron (e.g., **434 → 435**), the change in total π-bond energy ΔE_π is given to a first-order approximation by Eq. (6).[10,11]

$$\Delta E_\pi \simeq - C_{o,r}^2 \cdot \Delta \alpha_r \qquad (6)$$

where $C_{o,r}$ is the orbital coefficient of the NBMO at position r.

If the heteroatom contributes two π-electrons (e.g., **434 → 433**) the difference in total π-energy between the AH anion and the neutral heteroconjugated system is given by Eq. (7).

$$\Delta E_\pi \simeq (1 - C_{o,r}^2) \cdot \Delta \alpha_r \qquad (7)$$

The effects are opposed. Introduction of a 2π-electron heteroatom reduces the bond energy whereas a 1π-electron heteroatom increases the bond energy. By taking the total π-bond energy of an odd AH anion as reference point, Eq. (7) can be used to estimate the relative π-bond energy (and therefore the relative thermodynamic stability) of mesomeric betaines and their "nonpolar" isomers. Since the heteroatom in mesomeric betaines occupies an inactive position, the decrease in π-bond energy relative to the isoconjugate AH anion (Eq. 7) will always be the maximum value ($\Delta \alpha_r$). For isomeric species, a smaller change is estimated since the factor $(1 - C_{o,r}^2)$ (Eq. 7) is usually less than unity. We therefore tentatively predict that mesomeric betaines are thermodynamically less stable than their isoconjugate isomers (in the gas phase). Substitution of 1π-electron heteroatoms at active positions will increase the stability (Eq. 6). The Rowe rearrangement (**213 → 215**) (see Section III,B,7) may well reflect the predicted difference in energy between isomers. It is certainly clear from Section III that mesomeric betaines are associated with reactive (i.e., high energy) π-systems and that those with only one heteroatom in the ring system are often intrinsically unstable at room temperature.

In the absence of precise thermochemical data it is difficult to draw firm conclusions about the relative stability of mesomeric betaines and their isomers (e.g., **213** and **215**), but some estimate can be obtained by comparing their pK_a values.

(433)   (434)   (435)

6. pK_a Values

The pK_a values of a number of betaines and related heterocycles are shown in Table VI.[310–312]

TABLE VI
pK_a VALUES OF SOME BETAINES AND RELATED HETEROCYCLES[a]

Compound	pK_a	Reference
396 (Fig. 1)	4.96	310
397 (Fig. 1)	0.32	310
398 (Fig. 1)	3.33	310
399 (Fig. 2)	6.81	311
400 (Fig. 2)	6.12	311
401 (Fig. 2)	6.90	311
402 (Fig. 2)	4.93	311
404 (Fig. 2)	5.81	311
406 (Fig. 2)	2.36[b]	312
407 (Fig. 2)	−1.8	310
418 (Fig. 3)	7.15	311
419 (Fig. 3)	5.42	311
421 (Fig. 3)	7.09	311
422 (Fig. 3)	5.56	311
424 (Fig. 3)	6.02	311
426 (Fig. 3)	−0.71	310

[a] In water, at 20°C, unless otherwise indicated.
[b] At 30°C.

Suppose that the free energy difference between these heterocycles (Table VI) and their conjugate acids can be equated to a constant term, which represents the dissociation energy of the OH bond (ΔE_{OH}), plus the difference in the total π-bond energies of the two species (ΔE_π). For convenience, we will restrict attention to heterocycles isoelectronic with the α-naphthylmethyl anion. The equilibrium **436** ⇌ **437** is a representative example.

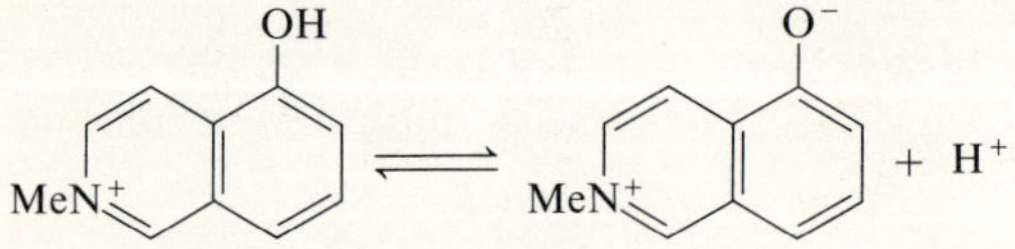

(**436**)   (**437**) + H⁺

---

[310] A. Albert and J. N. Phillips, *J. Chem. Soc.*, 1294 (1956).
[311] S. F. Mason, *J. Chem. Soc.*, 674 (1958).
[312] G. F. Tucker and J. L. Irvin, *J. Am. Chem. Soc.* **73**, 1923 (1951).

For the general case, if E_{π_1} is the total π-bond energy of the cation and E_{π_2} is the total π-bond energy of the neutral heterocycle, then using Eq. (7) (Section IV,A,5), the difference in total π-bond energy (ΔE_π) is given by

$$\Delta E_\pi = E_{\pi_2} - E_{\pi_1}$$
$$= E_{\pi_0} + (1 - C_{o,r}^2) \cdot \Delta\alpha - E_{\pi_1} \qquad (8)$$

where E_{π_0} is the π-bond energy of the 1-naphtholate ion.

We now assume that π-bond energies of the conjugate acids (e.g., **436**) are independent of the position of the N-atom (i.e., E_{π_1} is constant). This is reasonable if we consider the cations as monocentric perturbations of 1-naphthol. Equation (8) now reduces to the form

$$\Delta E_\pi = A + B \cdot C_{o,r}^2 \qquad (9)$$

where A and B are constants.

If we adhere to our assumption that the free energy difference can be equated to the difference in π-bond energies (ΔE_π) plus a constant term (ΔE_{OH}), then

$$\Delta E_\pi + \Delta E_{\text{OH}} = -2.3\, RT \cdot pK_a \qquad (10)$$

Combining equations (9) and (10)

$$pK_a = M - N \cdot C_{o,r}^2 \qquad (11)$$

where M and N are constants.

If our approximations are realistic, then variation of the pK_a values in Table VI should be governed by Eq. (11). This is not in fact the case. So far we have found it convenient to neglect an important factor. If the heterocyclic nitrogen atom is located close to the position of the acidic OH group, the positive charge localized on nitrogen weakens the OH bond—presumably by a field and/or inductive effect. This proximity effect can be expected to be a function of the separation of the two groups and to decrease fairly rapidly with distance. When the heteroatom is located at a remote region of the heterocycle, we can assume that these proximity effects are unimportant and that the pK_a is governed by Eq. (11). If we consider those systems in Table VI in which the OH group and nitrogen atom are in different rings, then we find that the betaines **399, 401, 418,** and **421** [pK_a (average) = 7.0] are consistently less acidic than their isomers **400, 404, 422,** and **424** [pK_a (average) = 5.9]. This fluctuation of about one pK_a unit as the heteroatom moves round the ring is consistent with Eq. (11), since $C_{o,r}^2$ is zero for betaines. This result, therefore, provides indirect experimental support for the conclusion (Section IV,A,5) that the π-bond energies of betaines are less than those of their isomers.

TABLE VII

Molecule	pK_a(obs)^a	pK_a(calc)^b	Difference
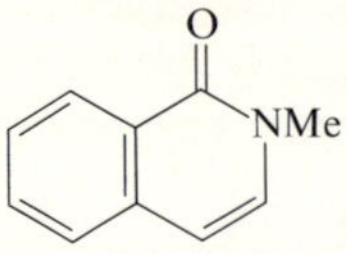 (isoquinolin-1(2H)-one, N-Me)	−1.8	3.3	5.1
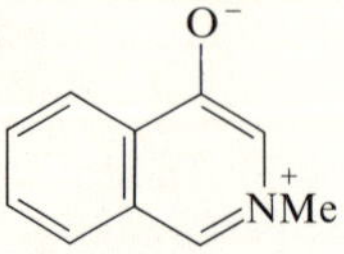 (isoquinolin-4-olate, N-Me)	4.9	6.9	2.0
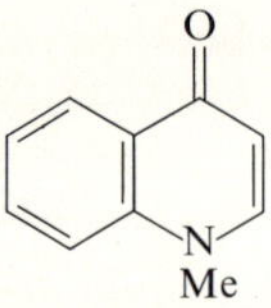 (quinolin-4(1H)-one, N-Me)	2.4	3.3	0.9

^{a} Taken from Table VI.
^{b} Calculated using Eq. (11).

Some measure of the magnitude of the proximity effect can be obtained by calculating the pK_a values using Eq. (11) and comparing these with observed values. The difference is a measure of the proximity effect. The constants M and N in Eq. (11) can be estimated using pK_a values for systems where the proximity effect can be regarded as negligible. Using this approach the values shown above (Table VII) are obtained. The actual calculated values in themselves are of no great significance. It does appear significant, however, that the difference between calculated and observed pK_a values is larger when the two functional groups are close together and falls off rapidly as they separate.

B. Molecular Orbital Calculations

Considering the proliferation of MO calculations in recent years, it is remarkable how few calculations have been reported for these heterocycles. Frontier orbitals of several betaines have been calculated using the Hückel method and electroselectivity correctly predicted on the basis of the orbital symmetry.[139,203] Similar results have been obtained for pyridinium-3-olates (**427**) using the Pariser–Parr–Pople (PPP) method.[32,313,314] The CNDO

[313] L. Paoloni, M. Cignitti, and M. L. Tosato, *Org. Magn. Reson.* **6**, 469 (1974).
[314] G. Berthier, B. Lévy, and L. Paoloni, *Theor. Chim. Acta* **16**, 316 (1970).

method has been used to investigate the properties of pyridinium-3-olates (**427**)[55] and N-methylquinolinium-8-olate (**416**).[315]

Recently, a detailed CNDO/2 study of pyridinium-3-olates (**427**) has been used to rationalize their kinetic rates and peri-, site-, regio-, and stereoselectivity of cycloaddition.[55a] These calculations clearly indicate that the energy of the betaine HOMO is insensitive to the N-substituent, whereas the LUMO energy varies considerably with the N-substituent and its orientation. Using the calculated eigenvalues and eigenvectors of the frontier orbitals, Eq. (12) (Section V) successfully predicts the preferred regioisomers for dimerization and cycloaddition of alkenes and alkynes. Stereoselectivity of these reactions is explained by a monopole repulsion treatment. However, the relative rates of cycloaddition are rationalized by a simplified version of Eq. (12) in which the numerators are equated to a constant (Eq. 13). Some justification for this simplification has been presented. It is interesting that these studies[55a] indicate that addition of acrylonitrile across the O-2 or O-4 positions of the pyridinium-3-olates (**427**) is kinetically preferred to the observed 2,6-addition. The O-2 and O-4 adducts are probably not observed because they do not lead to thermodynamically stable adducts. A notable exception is the addition of chloroketenes which do give products formed via O-2 and O-4 adducts (see **119**, Section III,A,2).

V. Pericyclic Reactions of Mesomeric Betaines

The structural requirements of the mesomeric betaines described in Section III endow these molecules with reactive π-electron systems whose orbital symmetries are suitable for participation in a variety of pericyclic reactions. In particular, many betaines undergo 1,3-dipolar cycloaddition reactions giving stable adducts. Since these reactions are moderately exothermic, the transition state can be expected to occur early in the reaction and the magnitude of the frontier orbital interactions, as 1,3-dipole and 1,3-dipolarophile approach, can be expected to influence the energy of the transition state—and therefore the reaction rate and the structure of the product. This is the essence of frontier molecular orbital (FMO) theory, several accounts of which have been published.[316,317] The application of the FMO method to the pericyclic reactions of mesomeric betaines has met with considerable success. The following section describes how the reactivity, electroselectivity, and regioselectivity of these molecules have been rationalized.

---

[315] I. Zuika, Z. Bruveris, and A. Jurgis, *Khim. Geterotsikl. Soedin.*, 1524 (1976) [*CA* **86**, 105698 (1977)].
[316] I. Fleming, "Frontier Orbitals and Organic Chemical Reactions." Wiley, New York, 1976.
[317] K. Fukui, *Acc. Chem. Res.* **4**, 57 (1971).

Equation (12) is the mathematical basis of the FMO method and describes the energy change due to frontier orbital interactions when two molecules M and N interact (M + N → MN). ΔE_{FMO} is a measure of transition-state stabilization (or destabilization). The first term describes the interaction between the HOMO of molecule M and the LUMO of N whereas the second term describes the alternative interaction.[316]

$$\Delta E_{\text{FMO}} = \frac{2[(C_{m,r} \cdot C_{n+1,s} + C_{m,r'} \cdot C_{n+1,s'})\beta]^2}{E_m - E_{n+1}}$$
$$+ \frac{2[(C_{m+1,r} \cdot C_{n,s} + C_{m+1,r'} \cdot C_{n,s'})\beta]^2}{E_n - E_{m+1}} \quad (12)$$

where m and n are the HOMOs of molecules M and N; $m + 1$ and $n + 1$ are the LUMOs of M and N; E_m, E_{m+1}, E_n, and E_{n+1} are the corresponding orbital energies; β is the resonance integral; $C_{m,r}$, $C_{n+1,s}$, etc. are the HOMO and LUMO coefficients at the reaction centers, which are atoms r and r' on M and s and s' on N.

A simplified form of Eq. (12), which has been used to predict rates of cycloaddition, is Eq. (13) (K is a constant here).[55a]

$$\Delta E_{\text{FMO}} = K^2 \left[\frac{1}{E_m - E_{n+1}} + \frac{1}{E_n - E_{m+1}} \right] \quad (13)$$

A. Valence Tautomerism

Betaines have been generated by photochemical or thermal ring opening of isomeric aziridines or oxiranes. Photolysis of the epoxide derivatives **438** gives pyrylium-3-olates **439** (Section III,A,1). Similarly, betaines **443**, **445**, and **447** have been generated by photochemical or thermal ring opening of their valence tautomers **442**, **443**, and **446** (Sections III,B,1–3) and transformation **448** → **449** (Section III,C,5) has been achieved thermally. These processes are often reversed photochemically: Irradiation of N-phenylpyridinium-3-olate (**441**; R = Ph) gives isomer **440** (R = Ph) (Section III,A,2).

Geometrical constraints require that these electrocyclic reactions proceed by disrotatory mechanisms. In a disrotatory thermal cleavage the σ-bond

(**438**) X = O   (**439**) X = O
(**440**) X = NR   (**441**) X = NR

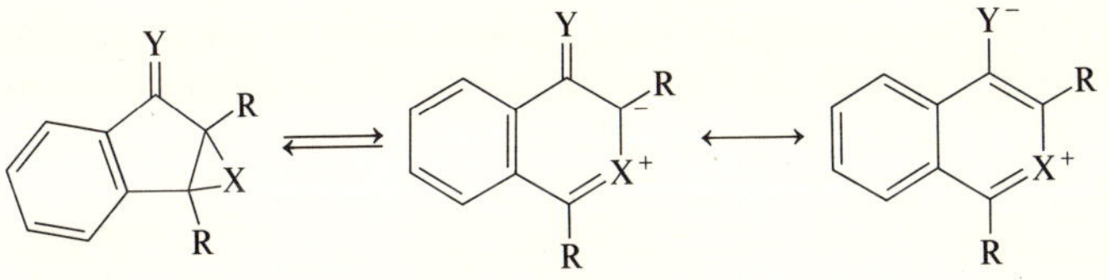

(442) X = Y = O
(444) X = NR, Y = O
(446) X = Y = NR

(443) X = Y = O
(445) X = NR, Y = O
(447) X = Y = NR

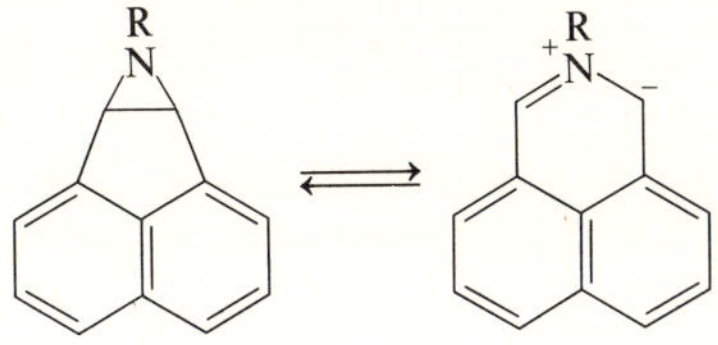

(448)   (449)

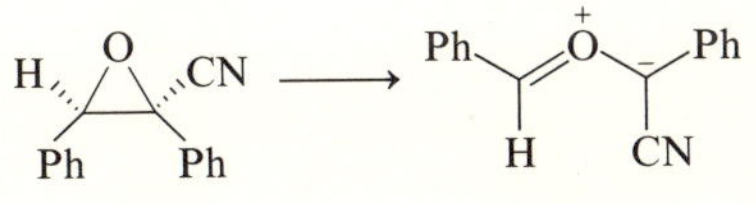

(450)   (451)

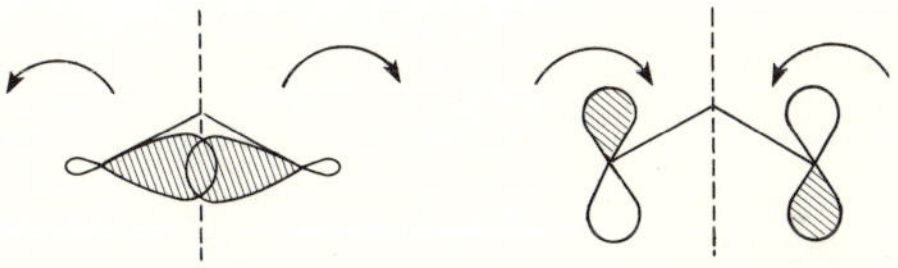

σ-bond
(a)

betaine HOMO
(b)

FIG. 5. Thermal disrotatory valence tautomerism; the orbitals do not correlate.

(Fig. 5a) does not correlate with the betaine HOMO (Fig. 5b), and it can be shown, therefore, that the thermal reactions are symmetry forbidden whereas photochemical reactions are symmetry allowed. Mechanistically, these reactions are analogous to the disrotatory ring opening of α-cyano-*cis*-stilbene oxide (450 → 451) which Huisgen and Markowski have demonstrated takes place thermally with a free energy of activation of 35.6 kcal mol^{-1} at 130°C.[318–320] Clearly, although forbidden, these reactions are not prevented. Ullman and Milks have found the activation energy for ring

[318] V. Markowski and R. Huisgen, *J. C. S., Chem. Commun.*, 439 (1977).
[319] R. Huisgen and V. Markowski, *J. C. S., Chem. Commun.*, 440 (1977).
[320] R. Huisgen, *Angew. Chem. Int. Ed. Engl.* **16**, 572 (1977).

opening of 2,3-diphenylindenone-2,3-oxide (**442**; R = Ph) to be 29.4 kcal mol^{-1}.[89] The enthalpy of activation for the thermal valence tautomerism (**448** → **449**; R = Ph) is 25.1 kcal mol^{-1}.[223]

B. Dimerization

Two modes of thermal dimerization of these heterocycles now seem to be firmly established. The first type of dimer is formed by pyridinium-3-olates (Section III,A,2). Adducts having the exo-syn structures (**452**) are kinetically favored, but over a longer period the thermodynamically preferred exo-anti isomers (**453**) often accumulate.[9] Using the FMO method, the relationship between structure, reactivity, and regiospecificity has been investigated by the East Anglia group.[55–56] Figure 6 shows the betaine frontier orbitals as calculated by the CNDO/2 method. The second order HOMO–LUMO interactions for thermal dimerization (Fig. 6a) are indicated, and clearly the orbital symmetry is suitable for bonding interactions. For thermal dimerization, both denominators in Eq. (12) are equal. Small HOMO–LUMO gaps increase the transition-state stabilization [ΔE_{FMO}; Eq. (12)] and

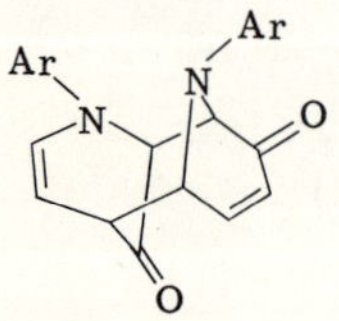

(**452**) exo-syn 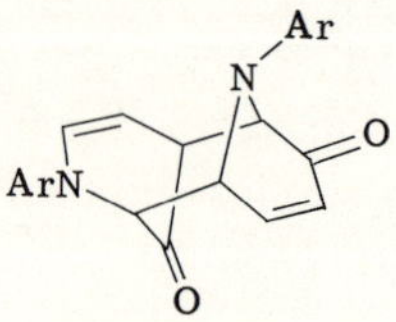   (**453**) exo-anti 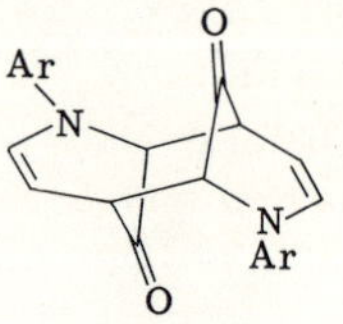   (**454**)

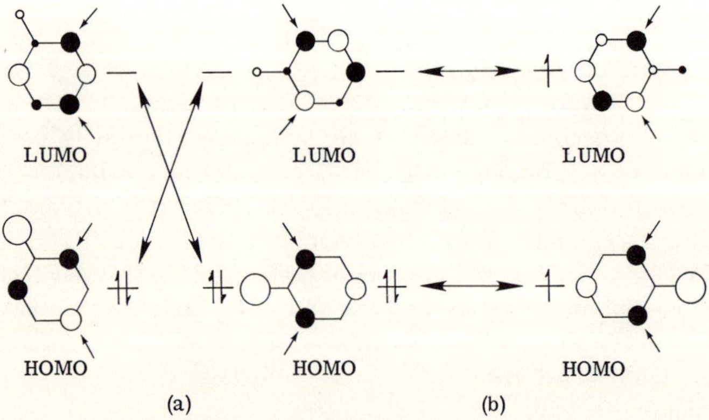

Fig. 6. Frontier orbital interactions in (a) thermal and (b) photochemical dimerization of pyridinium-3-olates. Sites of addition are indicated by arrows.

accordingly increase the rate of reaction. When the frontier orbital separation of pyridinium-3-olates is reduced by the method discussed in Section IV,A,3, dimerization is facilitated and is convincingly demonstrated by the substituent effects described in Section III,A,2. Computation of the frontier orbital coefficients of 1-methylpyridinium-3-olate enabled Katritzky and co-workers to calculate the FMO interaction energy (ΔE_{FMO}) for both syn and anti approach and correctly to predict that the syn regioisomer (**452**) is kinetically favored.[55,55a] The absence of any endo dimers can be attributed to unfavorable secondary interactions in the transition state, but it is interesting to note that a similar dimerization of thiopyrylium-3-olates gives predominantly endo dimers (see Section III,A,4).

Irradiation of pyridinium-3-olates gives a dimer of different structure (**454**).[61] In a photochemical process the important frontier orbital interactions are those between HOMO–HOMO and LUMO–LUMO.[316] Figure 6b demonstrates that for photodimerization, a different orientation favors transition-state stabilization and leads to the novel adduct **454**.

The second type of thermal dimerization is that shown by benzopyrylium-4-olates (Section III,B,1), benzothiopyrylium-4-olates (Section III,B,6), and benz[*de*]isoquinolines (Section III,C,5) which give adducts **455–460**. Theory predicts these reactions to be forbidden and it is possible that these dimers are formed by a two-step mechanism. Other 1,3-dipoles are known to undergo the same reaction: The 1,3-dipole **461** (Ar = p-NO$_2$C$_6$H$_4$) generated *in situ* from *N*-benzylisoquinolinium bromide gives the piperazine derivative **462** (Ar = p-NO$_2$C$_6$H$_4$).[321] A similar dimerization of mesoionic 1,3-dithiol-4-ones (**463 → 464**) has recently been described.[322]

(**455**) X = O
(**457**) X = S

(**456**) X = O
(**458**) X = S

(**459**)

(**460**)

[321] G. Müller, K. H. Duchardt, and F. Kröhnke, *Chem. Ber.* **110**, 3224 (1977).
[322] H. Gotthardt, C. M. Weisshuhn, O. M. Huss, and D. J. Brauer, *Tetrahedron Lett.*, 671 (1978).

(461) (462) (463) (464)

C. Cycloadditions

1. 2π-Electron Addends

Mesomeric betaines (high-energy HOMO) are particularly reactive toward electron-deficient 1,3-dipolarophiles (low-energy LUMO). Ample evidence of this reactivity is afforded by the large number of adducts that have been formed between betaines and dimethyl acetylenedicarboxylate (Sections III,A,1 and 2, III,B,1–3,7, and 8, III,C,5,7,8, and 9 and III,D,1). The reactions of benz[*de*]isoquinolines (Fig. 7a) are typical.[222,223] A large bonding interaction between the betaine HOMO and the acetylene LUMO (Fig. 7a) leads to transition-state stabilization. Since all betaine HOMOs have the symmetry of a NBMO, this favorable interaction is an important general feature of their cycloadditions. Equally important, this mode of addition usually gives thermodynamically stable adducts. Lowering the energy of betaine HOMOs can be expected to reduce reactivity by reducing the betaine HOMO/alkyne LUMO interaction, and this possibly accounts for the lack of reactivity of pyridazinium-3-olates (Section III,A,5) toward 1,3-dipolarophiles.[78]

Cycloaddition is not necessarily restricted to the active positions flanking the heteroatom. *N*-Methylnaphtho[1,8-*de*]triazine and dimethyl acetylenedicarboxylate give an alternative adduct (Fig. 7b)[241]—an addition which is also facilitated by the frontier orbital interactions.

Similar reactivity is observed between betaines and electron-deficient alkenes (E·CH=CH$_2$) and a high degree of regiospecificity is often observed. Pyridinium-3-olates (Section III,A,2) react with acrylonitrile or methyl-

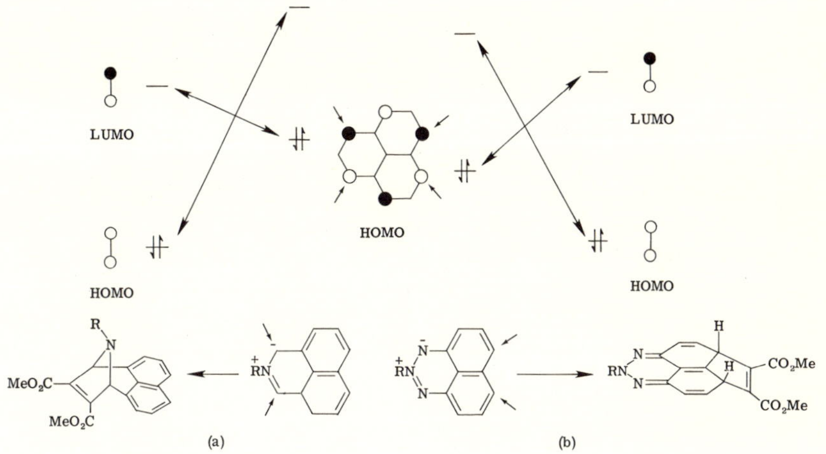

Sec. V.C] HETEROCYCLIC BETAINES 95

FIG. 7. Exemplifying two distinct modes of symmetry allowed cycloaddition of mesomeric betaines: (a) conventional 1,3-dipolar cycloaddition; (b) addition across peri positions.

acrylate giving the exo and endo adducts (**465**) in good yield, but significant amounts of the regioisomers (**466**) are not encountered. This selectivity has been rationalized by Katritzky and co-workers who have calculated the frontier orbital energy levels and coefficients of *N*-methylpyridinium-3-olate, acrylonitrile, and methylacrylate using the CNDO/2 method and have shown, using Eq. (12), that for both dipolarophiles the total interaction energy (ΔE_{FMO}) is greatest for the observed mode of addition.[73] As expected, the term representing the betaine HOMO/alkene LUMO interaction is dominant. As a general rule,[316] kinetically preferred regioisomers are those in which the largest coefficient of the HOMO interacts with the largest coefficient of the LUMO, and this is indeed found to be the case for reactions of pyridinium-3-olates (Fig. 8).

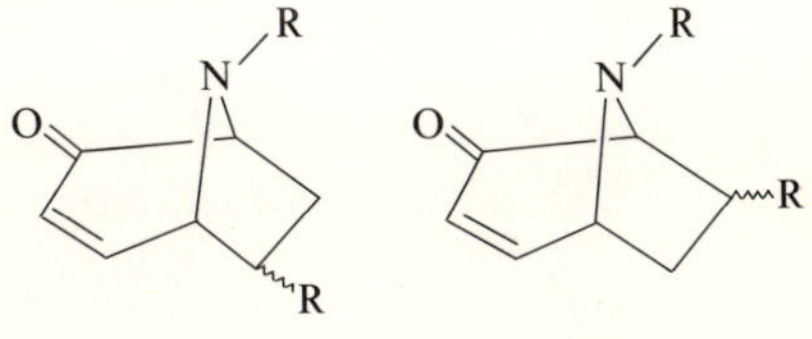

(**465**)   (**466**)

Cycloadditions of betaines are not restricted to electron-deficient alkenes. Pyridinium-3-olates also react with conjugated olefins (e.g., styrenes) and with electron-rich olefins (e.g., ethyl vinyl ether). In the latter case, the betaine LUMO/alkene HOMO interaction becomes dominant and reaction is only observed with pyridinium-3-olates having a low-energy LUMO

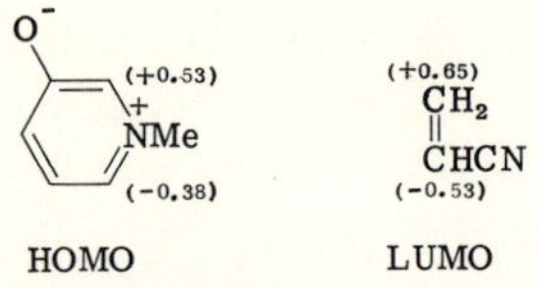

FIG. 8. Calculated orbital coefficients and preferred geometry of approach for 1,3-dipolar cycloaddition.[73]

(e.g., *N*-nitropyridyl derivatives). With conjugated olefins both HOMO–LUMO interactions are equally important and reaction occurs only with *N*-arylpyridinium-3-olates. Regiospecificity giving the adducts **465** is also a feature of these cycloadditions, and these results have also been rationalized by the FMO method. The additions of styrenes give predominantly endo adducts—presumably due to favorable secondary orbital overlap.[9,73]

2. *4π-Electron Addends*

The study of reactions between mesomeric betaines and 1,3-dienes has so far been restricted to pyridinium-3-olates (Section III,A,2) and quinolinium-3-olates (Section III,B,18). The mode of addition of dienes differs from that of 2π-electron addends in two respects: (i) Dienes add across a different

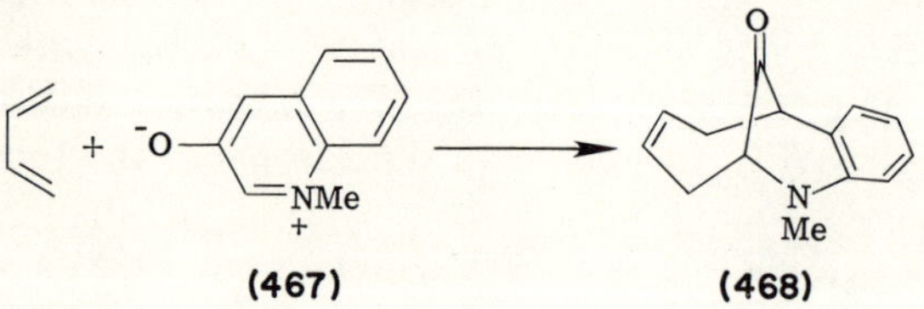

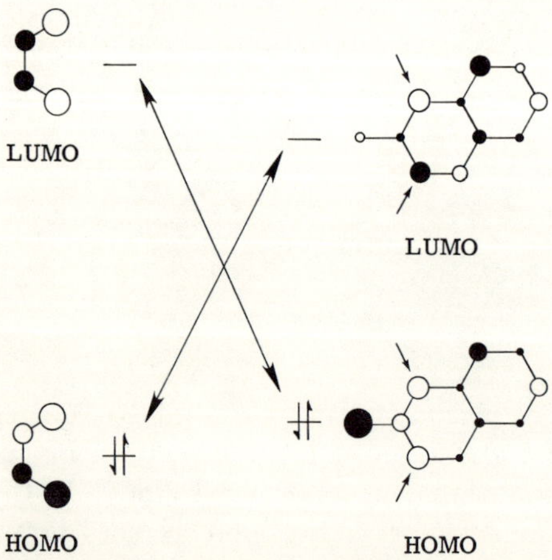

FIG. 9. Frontier orbital mixing for the thermal addition of 1,3-butadiene to *N*-methylquinolinium-3-olate (**467** → **468**).[203]

Sec. VI.A] HETEROCYCLIC BETAINES 97

fragment of the betaine ring giving novel adducts (e.g., **468**) and (ii) the betaine acts as an electron acceptor and electron-rich dienes are most reactive. The betaine LUMO/diene HOMO interaction dominates and the most reactive heterocycles are those with a low-lying LUMO.

Dennis, Ibrahim, and Katritzky demonstrated the addition of 1,3-dienes to mesomeric betaines.[54] They prepared adducts from N-pyridyl and N-pyrimidinylpyridinium-3-olates and various dienes (including cyclopentadiene) and have analyzed the frontier orbital interactions in some detail.[73] Almost simultaneously, Mok and Nye discovered that N-methylquinolinium-3-olate (**467**), generated *in situ*, gives adducts (e.g., **468**) with several 1,3-dienes, and using the HMO method they showed that the frontier orbital interactions during addition (Fig. 9) are favorable for transition state stabilization.[203]

VI. Heterocyclic Betaines Isoelectronic with Even Alternant Hydrocarbon Dianions

A. Kekulé and Non-Kekulé Alternant Hydrocarbons

So far, this review has described betaine molecules that are isoconjugate with odd AH anions. We now consider betaines isoconjugate with even AH anions.

Two types of even AH have been recognized. Those which can be represented as classical polyenes (in which all the atoms are linked in pairs by double bonds) have been described as Kekulé hydrocarbons. Benzene is a Kekulé hydrocarbon and so are the *o*- and *p*-quinodimethanes (**469** and **470**).[10,11]

Not all even AHs can be represented by Kekulé structures. The *m*-quinodimethane **471** can only be represented by diradical structures, and molecules

(469)   (470)   (471)

(472)   (473)

of this type are described as non-Kekulé hydrocarbons. Other examples are 2,2′-biallyl (**472**) and triangulene (**473**).[10,11] Non-Kekulé hydrocarbons can be regarded as being formed by union of pairs of odd AH through inactive positions (Scheme 17). Because union occurs through inactive positions, the NBMOs of the odd AH fragements are unperturbed, and it follows that non-Kekulé molecules are associated with a pair of NBMOs. Conversely, combinations of the NBMOs of non-Kekulé hydrocarbons can be chosen which are localized on cross-conjugated odd AH fragments, and properties that depend upon total energy or total electron distribution (collective properties) can conveniently be discussed in terms of independent odd AH fragments. Introduction of two additional electrons into the NBMOs gives a dianion. We now consider neutral heterocyclic systems isoconjugate with non-Kekulé AH dianions.

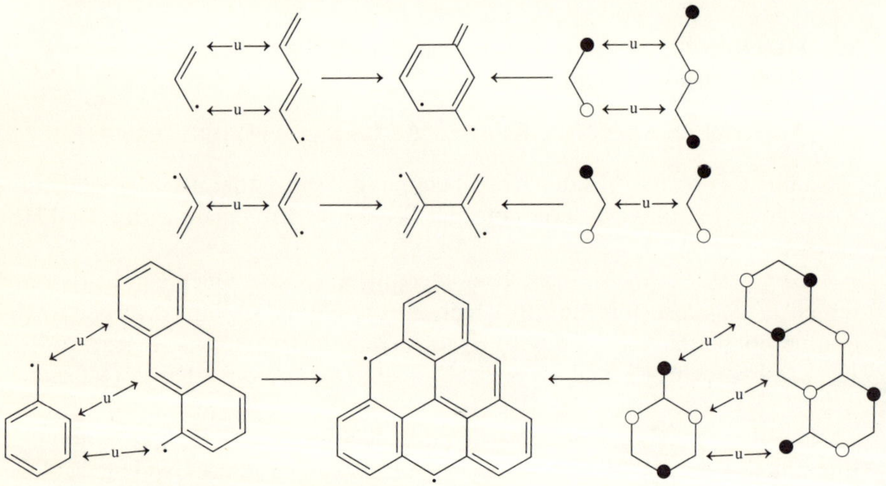

SCHEME 17

B. Heterocyclic Betaines Isoelectronic with Non-Kekulé Dianions

Heterosystems isoconjugate with non-Kekulé dianions are characterized by two lone pairs of electrons which are donated by heteroatoms. Consider heterocycles derived from the non-Kekulé dianion **474**, which can be treated as independent allyl anion ($\bar{a}$—b=c), and pentadienyl anion (d=e—f=g—h$^-$) fragments—providing that we are discussing collective properties. [This division is not unique: division into a methyl anion (f$^-$) and a heptatrienyl anion fragment (d=e—a=b—c=g—h$^-$) is also possible and leads to similar conclusions.] If one lone pair originates on each of the cross-conjugated fragments, then the rule described in Section II,B can be applied

Sec. VI.B] HETEROCYCLIC BETAINES 99

independently to each fragment. The condition that a molecule can be represented by a nonpolar structure is that the lone pair on each fragment must be donated to the π-electron system by a starred heteroatom. Uracil (**475**) is a molecule of this type. Alternatively, if one lone pair originates at an unstarred position of the odd AH fragment, then the molecule can only be represented by dipolar structures. The pyrazinium-2,6-diolate (**476**) conforms to this type and may be recognized as representative of a large class of mesomeric betaines that are isoelectronic with non-Kekulé AH dianions and that can be expected to have properties similar to the betaines described in Section III.

(474)   (475)   (476)

If both lone pairs originate at unstarred positions the molecule can only be represented by tetrapolar structures. Such a molecule is dinitrogen tetroxide (**477**) in which the lone pairs originate on the nitrogen atoms, each of which is located at the inactive position of an allyl fragment (**478**). The structure and bonding of dinitrogen tetroxide is interesting and most textbooks agree that no satisfactory explanation of its structure is available.[323–326] In particular, its planar structure and the exceptionally long N–N bond (1.75 Å) have been difficult to rationalize. Since the two nitro functions are linked through inactive positions, the molecule can be expected to behave as independent (cross-conjugated) nitro groups linked by a N–N σ-bond with no π-character. Because the nitrogen atoms are each associated with substantial positive charge, the N–N σ-bond is probably abnormally long due to electrostatic repulsion. Comparison with the N–N bond length in hydrazine dihydrochloride (H_3N^+–NH_3^+·2Cl$^-$) (1.42 Å) is not strictly valid since in the latter molecule a large proportion of the positive charge probably resides on the hydrogen atoms. The planar structure of O_2N–NO_2 is probably favored by weak second-order interactions between the first π-molecular orbital of one nitro group and the π*-orbital of its partner. This

---

[323] F. A. Cotton and G. Wilkinson, "Advanced Inorganic Chemistry. A Comprehensive Text," 3rd ed., p. 358. Wiley (Interscience), New York, 1972.

[324] K. Jones, in "Comprehensive Inorganic Chemistry" (A. F. Trotman-Dickenson, ed.), Vol. 2, p. 342. Pergamon, Oxford, 1973.

[325] J. D. Lee, "A New Concise Inorganic Chemistry," 3rd ed., p. 216. Van Nostrand-Reinhold, New York, 1977.

[326] C. T. Rawcliffe and D. H. Rawson, "Principles of Inorganic and Theoretical Chemistry," 2nd ed., p. 252. Heinemann, London, 1974.

weak π-bond may well be sufficient to compensate for the unfavorable O...O repulsions which are maximized in the planar structure.

(477)   (478)

Let us now consider a different type of molecule isoconjugate with non-Kekulé dianions. If the heteroatoms donating the lone pairs are both associated with the same odd AH fragment (and there is no alternative way of subdividing the molecule so that they are on different cross-conjugated fragments) then a different type of dipolar structure is produced. An example of these structural types is the 1,3-diazinium-4,6-diolates **479**, which are isoconjugate with the dianion **474**. In these species the positive and negative charges are restricted to different cross-conjugated fragments of the molecule, and these molecules can be recognized as examples of the cross-conjugated dipoles which were described in Section II,C. The betaines **479** can be regarded as inner salts, and their chemical properties (Section VII,B) are quite different from those of the mesomeric betaines (e.g., **476**) (Section VII,A).

(479)

VII. The Chemistry of Heterocyclic Betaines Derived from Even Alternant Hydrocarbon Dianions

A. Mesomeric Betaines

Only one system of this type appears to be known. The stable yellow pyrazinium-2,6-diolates **482** ($R^1 = R^3 = $ SAr, R^2 and $R^4 = $ Ar or Me) have been obtained by reaction of 2,6-dioxopiperazines (**480**) with arylsulfonyl chlorides in pyridine.[327–329] 3,4-Unsubstituted derivatives (**482**; $R^1 = R^3 = $ H) are not isolated, but the 1,4-diphenyl compound (**482**; $R^1 = R^3 = $ H,

[327] J. Honzl and M. Sorm, *Tetrahedron Lett.*, 3339 (1969).
[328] J. Honzl, M. Sorm, and V. Hanus, *Tetrahedron* **26**, 2305 (1970).
[329] M. Sorm and J. Honzl, *Tetrahedron* **28**, 603 (1972).

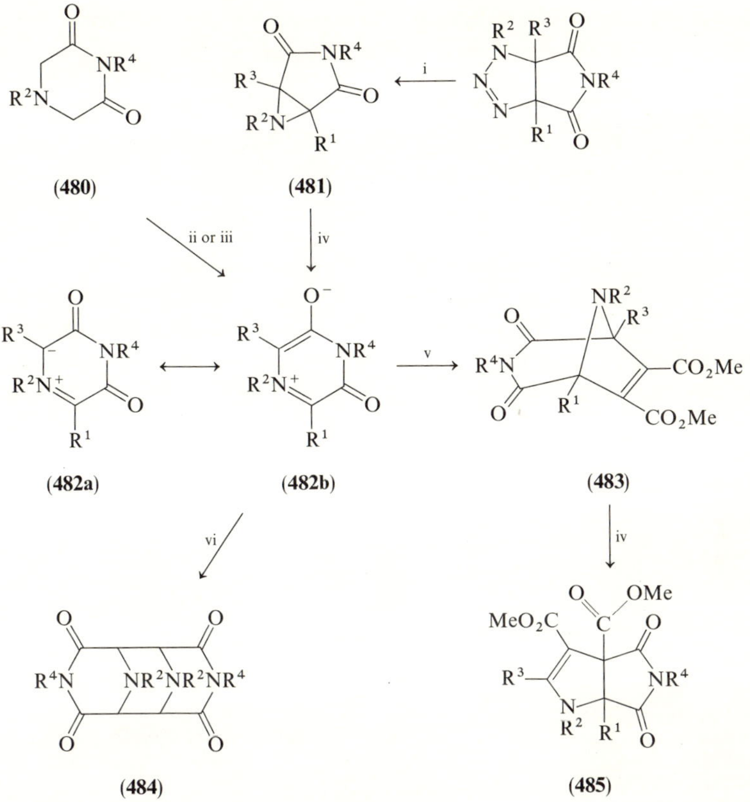

Sec. VII.A] HETEROCYCLIC BETAINES 101

SCHEME 18. Reagents: i, heat; ii, $PhSO_2Cl$/pyridine; iii, $C_6H_5NO_2$ at 120°C; iv, $h\nu$; v, $MeO_2C-C\equiv C-CO_2Me$; vi, spontaneous at 120°C.

$R^2 = R^4 = Ph$) has been generated *in situ* simply by heating the 2,6-dioxopiperazine **480** ($R^2 = R^4 = Ph$) in nitrobenzene. Under these conditions the betaine **482** ($R^1 = R^3 = H$, $R^2 = R^4 = Ph$) gives the dimer **484** ($R^2 = R^4 = Ph$), or alternatively if the reaction is carried out in the presence of *N*-phenylmaleimide a 1,3-dipolar cycloadduct is obtained.[330] Similarly, the 3,5-disubstituted derivative (**482**; $R^1 = R^3 = SPh$, $R^2 = R^4 = Ph$) gives cycloadducts with maleic anhydride and formaldehyde.[327,328]

An alternative method of generating the betaines **482** in solution involves photolysis of the bicyclic aziridines **481**.[331–333] Irradiation in the presence

[330] T. Tanaka, H. Yamazaki, and M. Ohta, *Bull. Chem. Soc. Jpn.* **50**, 1821 (1977).
[331] M. Ohta and H. Kato, *Nippon Kagaku Zasshi* **78**, 1400 (1957).
[332] S. Oida and E. Ohki, *Chem. Pharm. Bull.* **16**, 764 (1968).
[333] R. Huisgen and H. Mäder, *Angew. Chem., Int. Ed. Engl.* **8**, 604 (1969).

of dimethyl acetylenedicarboxylate gives adducts **483**. Under these conditions the adduct **483** ($R^1 = R^3 = H$, $R^2 = PhCH_2$, $R^4 = p\text{-MeOC}_6H_4$) undergoes photorearrangement to the bicyclic product **485** ($R^1 = R^3 = H$, $R^2 = PhCH_2$, $R^4 = p\text{-MeOC}_6H_4$) (Scheme 18).[332,333]

In their chemical reactions these betaines **482** are clearly behaving like the mesomeric betaines described in Section III.

B. Cross-Conjugated Betaines

Examples of this class of betaine molecule are well-known. The general types that have been prepared, together with references, are shown in structures **486–492**. Because this review is primarily concerned with meso-

(**486**)[334–352] (**487**)[348,353,354] (**488**)[355]

---

[334] G. M. Kheifets, N. V. Khromov-Borisov, and A. I. Kol'tsov, *Zh. Org. Khim.* **2**, 1516 (1966) [*CA* **66**, 46392 (1967)].

[335] M. Prystas and F. Sorm, *Collect. Czech. Chem. Commun.* **32**, 1298 (1967).

[336] M. Prystas, *Collect. Czech. Chem. Commun.* **32**, 4241 (1967).

[337] R. A. Coburn, *J. Heterocycl. Chem.* **8**, 881 (1971).

[338] T. Kappe and W. Lube, *Angew. Chem., Int. Ed. Engl.* **10**, 925 (1971).

[339] A. Kotarska, S. Staniszewska, and L. Kociszewski, *Soc. Sci. Lodz., Acta Chim.* **16**, 89 (1971) [*CA* **77**, 34448 (1972)].

[340] K. T. Potts and M. Sorm, *J. Org. Chem.* **36**, 8 (1971).

[341] Y. Maki, M. Sako, and M. Suzuki, *J. C. S., Chem. Commun.*, 999 (1972).

[342] K. T. Potts and M. Sorm, *J. Org. Chem.* **37**, 1422 (1972).

[343] R. A. Coburn, R. A. Carapellotti, and R. A. Glennon, *J. Heterocycl. Chem.* **10**, 479 (1973).

[344] R. A. Coburn and R. A. Glennon, *J. Heterocycl. Chem.* **10**, 487 (1973).

[345] R. A. Coburn and R. A. Glennon, *J. Pharm. Sci.* **62**, 1785 (1973).

[346] K. T. Potts and R. K. C. Hsia, *J. Org. Chem.* **38**, 3485 (1973).

[347] R. A. Coburn, R. A. Glennon, and Z. F. Chmielewicz, *J. Med. Chem.* **17**, 1025 (1974).

[348] R. A. Coburn and R. A. Carapellotti, *J. Pharm. Sci.* **65**, 1505 (1976).

[349] P. Dvortsak, G. Resofszki, M. Huhn, L. Zalantai, and A. I. Kiss, *Tetrahedron* **32**, 2117 (1976).

[350] T. Kappe and W. Golser, *Chem. Ber.* **109**, 3668 (1976).

[351] G. Schindler, D. Furtunopulos, and T. Kappe, *Z. Naturforsch., Teil B* **31B**, 500 (1976).

[352] E. Ziegler, W. Steiger, and C. Strangas, *Z. Naturforsch., Teil B* **32**, 1204 (1977).

[353] R. A. Coburn and B. Bhooshan, *J. Org. Chem.* **38**, 3868 (1973).

[354] W. Stadlbauer and T. Kappe, *Chem. Ber.* **109**, 3661 (1976).

[355] K. T. Potts, F. Huang, and R. K. Khattak, *J. Org. Chem.* **42**, 1644 (1977).

(489)[356]   (490)[350,357]   (491)[355]

(492)[354,358]

meric betaines, a comprehensive account of the chemistry of the cross-conjugated species **486–492** is not included. However, their cycloaddition reactions are briefly discussed in order to demonstrate that they are quite different in character from mesomeric betaines. Whereas mesomeric betaines participate in 1,3-dipolar cycloaddition reactions, the cross-conjugated betaines undergo 1,4-dipolar cycloadditions.

The 1,3-thiazinium-4,6-diolates **493**, which are made by condensation of thiobenzamides (PhCSNHR^2) with malonyl chlorides [CHR(COCl)$_2$] or carbon suboxide, react with aryl isocyanates giving the pyrimidinium-4,6-diolates **495** by initial formation of the bicyclic adducts (**494**) followed by elimination of carbonyl sulfide (Scheme 19).[350,357] This 1,4-dipolar cycloaddition (**493 → 494**) probably involves a dipolar intermediate which cyclizes to adduct **494**. These reactions are quite different to the cycloadditions of mesomeric betaines (Section V,C). The betaines **495** also undergo 1,4-dipolar cycloaddition reactions: Heating with dimethyl acetylenedicarboxylate gives the pyridones **497** (Scheme 19).[338,342] Using compound **495** (R^1 = Me, R^2 = R^4 = Ph, R^3 = H) the intermediate adduct **496** (R^1 = Me, R^2 = R^4 = Ph, R^3 = H) (mp 188–189°C) can be isolated in 94% yield and transformed to the pyridone **497** (R^1 = Me, R^2 = Ph, R^3 = H) by heating above its melting point.[342] Reaction of compound **495** (R^1 = R^2 = R^4 = Ph, R^3 = CH$_2$Ph) with maleic anhydride similarly gives a stable bicyclic adduct (61%; mp 231–233°C).[338]

These elegant transformations (Scheme 19) clearly demonstrate that cross-conjugated betaines are in no way the poor relations of the betaine family.

---

[356] G. Nöhammer and T. Kappe, *Monatsh. Chem.* **107**, 859 (1976).
[357] T. Kappe and W. Golser, *Synthesis*, 312 (1972).
[358] H. Hagemann and K. Ley, *Angew. Chem., Int. Ed. Engl.* **11**, 1012 (1972).

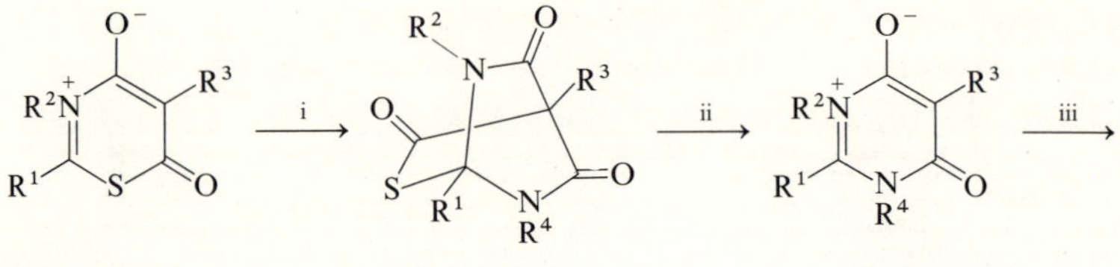

(493) (494) (495)

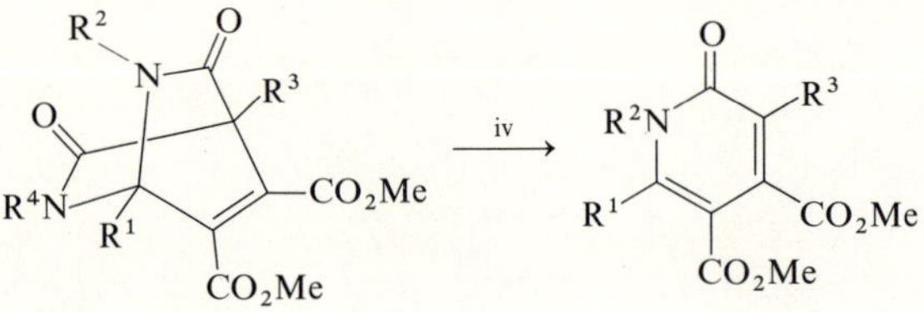

(496) (497)

SCHEME 19. Reagents: i, R^4NCO; ii, heat; iii, $MeO_2C—C{\equiv}C—CO_2Me$; iv, heat.

VIII. Conclusion

This series of three reviews has surveyed the chemistry of the major types of heterocyclic mesomeric betaine.[1,2] Examples of other heterocyclic betaines that do not fall into any of the categories that have been discussed are known. These include, for example, compounds **498**[359] and **499**,[360] but so far systems of this type have received little attention.

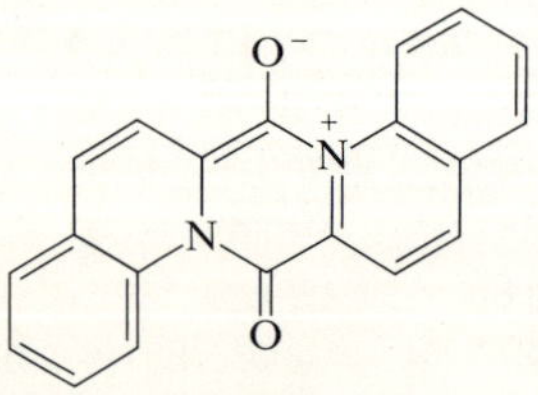

(498)

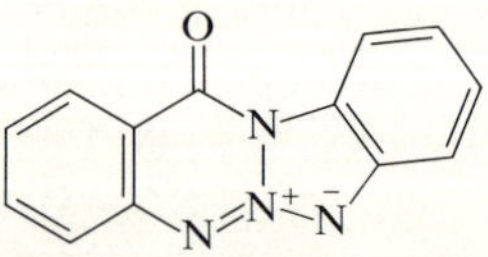

(499)

In addition to these systems, application of the PMO method leads to the prediction of the possible existence of many new and interesting classes of betaines. This predictive approach can be demonstrated by considering perturbations of the 1,3-dipoles **500**, which are isoconjugate with the

[359] B. R. Brown and D. L. Hammick, *J. Chem. Soc.*, 628 (1950).
[360] A. W. Murray and K. Vaughan, *Chem. Commun.*, 1282 (1967).

heptatrienyl anion **501**: (i) Intramolecular union between the 2 and 6 positions leaves the NBMO (**502**) unperturbed and leads to reactive betaine molecules **503**; (ii) intermolecular union with a benzene ring similarly gives a mesomeric betaine (**504**). Examples of betaines of types **503**[361,362] and **504**[363-366] have already been encountered as reaction intermediates that undergo transformations characteristic of mesomeric betaines. Many other systems can now be predicted. We hope that future studies, whether synthetic or spectroscopic, theoretical or therapeutic, will be rewarding.

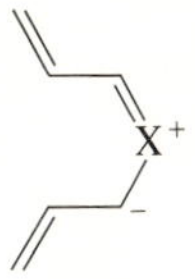

(500) 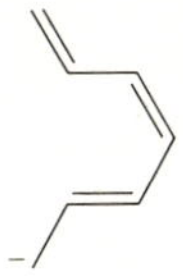   (501)    (502)

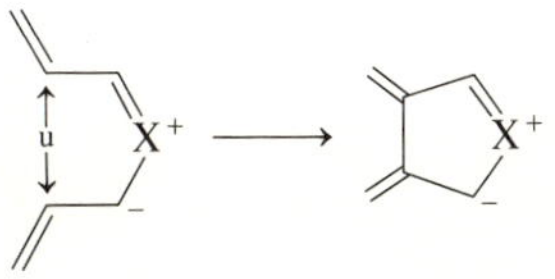

(503)

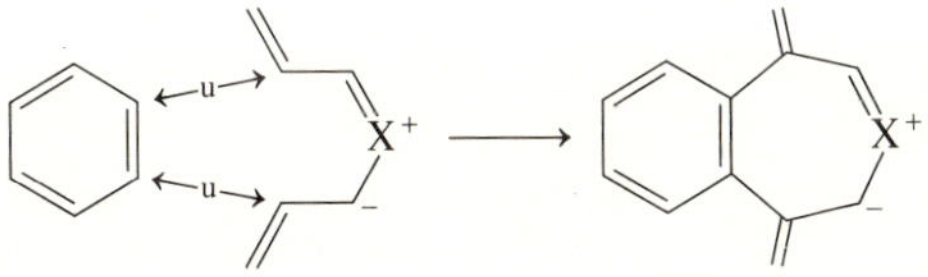

(504)

IX. Appendix Added in Proof

This appendix extends the literature coverage to October, 1979. Section headings are used for easy reference to the main text.

[361] P. Vogel and M. Hardy, *Helv. Chim. Acta* **57**, 196 (1974).
[362] Y. S. P. Cheng, E. Dominguez, P. J. Garratt, and S. B. Neoh, *Tetrahedron Lett.*, 691 (1978).
[363] H. Kato, K. Yamaguchi, and H. Tezuka, *Chem. Lett.*, 1089 (1974).
[364] H. Kato, H. Tezuka, K. Yamaguchi, K. Nowada, and Y. Nakamura, *J. C. S., Perkin 1*, 1029 (1978).
[365] K. Maruyama, S. Arakawa, and T. Oysuki, *Tetrahedron Lett.*, 2433 (1975).
[366] S. Arakawa, *J. Org. Chem.* **42**, 3800 (1977).

Section III,A

2. *Pyridinium-3-olates* (**86**) (page 16)

Further studies on the 1,3-dipolar cycloadditions of these molecules (**86**) have been reported. Addition of allyl alcohol gives endo adducts (**505**) which are not isolated but spontaneously cyclize to tricyclic products (**506**). Similar tricyclic products were also obtained using *N*-allylbenzenesulfonamide, triethylammonium acrylate, and vinylpyridines as dipolarophiles.[367] It has previously been shown that the pyridinium-3-olates with chloroketenes (RCCl=C=O) give 2-oxofuro [2,3-*c*]pyridine **507** (see p. 22). Further studies demonstrate that when bromoketenes (RCBr=C=O) are used as dipolarophiles, a mixture of 2-oxofuro[2,3-*c*]pyridines (**507**) and isomeric 2-oxofuro[3,2-*b*]pyridines (**508**) is obtained.[368]

(**505**) (**506**) (**507**)

(**508**) (**509**)

Treatment of the 1-(4,6-dimethylpyrimidin-2-yl)pyridinium-3-olate dimer **510** with 1-dimethylaminobuta-1,3-diene and subsequent oxidation leads to the benzo dimer **511** which is converted to the regioisomer **512** by heating in an inert solvent. Thermal dissociation of the adduct **511** into the two betaines **513** and **514** was demonstrated by trapping these species with several 1,3-dipolarophiles (Scheme 20).[369]

---

[367] A. R. Katritzky, N. Dennis, G. J. Sabongi, and L. Turker, *J. C. S., Perkin 1*, 1525 (1979).
[368] A. R. Katritzky, A. T. Cutler, N. Dennis, S. Rahimi-Rastgoo, G. J. Sabongi, I. J. Fletcher, and G. W. Fischer, *Z. Chem.* **19**, 20 (1979).
[369] A. R. Katritzky, N. Dennis, and H. A. Dowlatshahi, *J. C. S., Chem. Commun.*, 316 (1978).

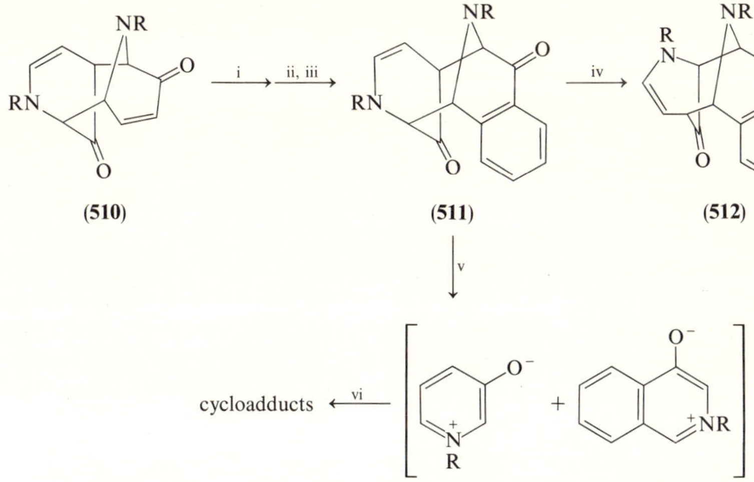

Sec. IX]   HETEROCYCLIC BETAINES   107

(510)   (511)   (512)

cycloadducts ← (513) + (514)

SCHEME 20. Reagents: i, Me$_2$NCH=CHCH=CH$_2$; ii, KOH/EtOH; iii, CrO$_3$/pyridine; iv, 80°C in ClCH$_2$CH$_2$Cl; v, heat; vi, dipolarophiles.

Polymethylene-bridged bis(pyridinium-3-olates) 509 and their bis-adducts have been prepared.[370] The ^{13}C NMR spectra of a number of pyridinium-3-olate cycloadducts have been discussed.[371]

The rates of cycloaddition of methyl acrylate with six 1-substituted pyridinium-3-olates have been studied. A Hammett plot of the second-order rate constants against σ values of the substituents gave satisfactory correlation.[372,373]

The photochemical and thermal transformations of 5-methyldihydrothiazolo[3,2-a]pyridinium-8-olate (515) have been investigated. Irradiation using a Hanovia mercury lamp and Pyrex filter gives the isomeric pyridone 516 (4%). When a Rayonet reactor was used, a product which appears to be the valence tautomer 517 (6%) was isolated. Further irradiation of 517 gave

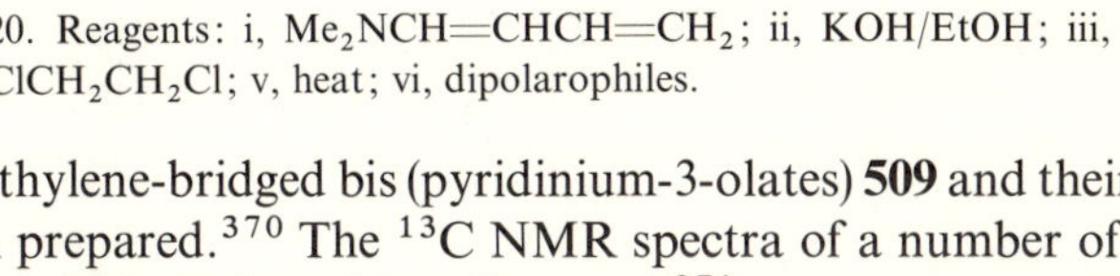

(515)   (516)   (517)

[370] N. Dennis, H. A. Dowlatshahi, and A. R. Katritzky, *J. Chem. Res.* (*M*), 1182 (1977).
[371] A. R. Katritzky, N. Dennis, and G. J. Sabongi, *Org. Magn. Resron.* **12**, 357 (1979).
[372] G. Guiheneuf, C. Laurence, and A. R. Katritzky, *J. C. S. Perkin 2*, 1829 (1976).
[373] A. R. Katritzky, B. El-Osta, G. Musumarra, and C. Ögretir, *J. Chem. Res.* (*M*), 4074 (1978).

the pyridone (**516**).[374] Vacuum pyrolysis of compound **515** in a quartz tube at 400°C yields the following five products: **518** (23%), **519** (18%), **520** (4%), **521** (29%), and **522** (26%).[375]

3. *Pyridinium-3-aminides* (**130**) (page 24)

Two betaines of type **130** (R^2 = CONHR^3) have been prepared by quaternization of the corresponding pyridine derivative followed by base treatment.[376]

5. *Pyridazinium-3-olates* (**137**) (page 25)

A full account of the photochemical behavior of 1,6-dimethylpyridazinium-3-olate **137** ($R^1 = R^2$ = Me, R^3 = R = H) has now been published.[377]

6. *Pyridazinium-5-olates* (**142**) (page 26)

---

[374] T. Laerum and K. Undheim, *J. C. S., Perkin 1*, 1150 (1979).
[375] T. Laerum, T. Ottersen, and K. Undheim, *Acta Chem. Scand., Ser. B* **33**, 299 (1979).
[376] W. H. Guendel, *Z. Naturforsch., Teil B* **33**, 84 (1978).
[377] Y. Maki, M. Suzuki, T. Furuta, M. Kawamura, and M. Kuzuya, *J. C. S., Perkin 1*, 1199 (1979).

Dieckmann condensation of the diester **523** gives compound **524** which is readily oxidized to the pyridazinium-5-olate **525** (R = CO_2Et). This betaine (**525**; R = CO_2Et) has been converted into a number of derivatives (**525**; R = H, CO_2H, Br, NHPh, SCH_2Ph).[378] Photolysis of the betaines **525** (R = H, CO_2Et) results in rearrangement to isomeric 4(3H)-pyrimidinones.[379] Full details of another study of this type of photochemical rearrangement of pyridazinium-5-olates (**142**) have now been published.[377]

Reaction of the 2-acetoxy-3(2H)-furanones (**526**) with monosubstituted hydrazines gives good yields of the pyridazinium-5-olates (**527**) together with varying amounts of isomeric products. Alkyl derivatives (**527**; R = alkyl) have also been prepared by base-catalyzed alkylation (MeI, Me_2SO_4, $PhCH_2Cl$) of 3-methyl-6-phenyl-5-ethoxycarbonyl-4(1H)-pyridazinone. Reduction of the diphenyl compound **527** (R = Ar = Ph) by zinc and hydrochloric acid gives 3-ethoxycarbonyl-5-hydroxy-5-methyl-1,2-diphenyl-2-pyrrolin-4-one (**528**; R = Ar = Ph) (Scheme 21).[380]

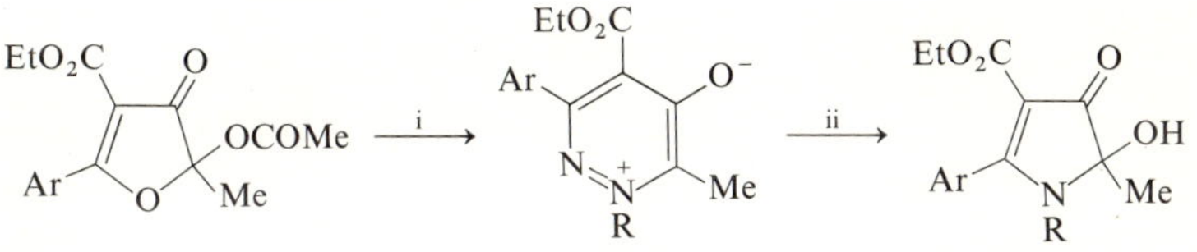

SCHEME 21. Reagents: i, $RNHNH_2$; ii, Zn/HCl.

SECTION III,B

1. *2-Benzopyrylium-4-olates* (**150**) (page 27)

Further examples of these betaines (**150**) have been prepared by photolysis of indenone oxides (**151**).[381]

7. *Phthalazinium-1-olates* (*Pseudophthalazones*) (**196**) (page 35)

Full details of the photochemical rearrangement of these molecules (**196**) have now been published.[377]

---

[378] T. Yamazaki, M. Nagata, F. Nohara, and S. Urano, *Chem. Pharm. Bull.* **19**, 159 (1971).
[379] T. Yamazaki, M. Nagata, S. Hirokami, and S. Miyakoshi, *Heterocycles* **8**, 377 (1977).
[380] S. Gelin, *J. Org. Chem.* **44**, 3053 (1979).
[381] V. M. Zolin, N. D. Dmitrieva, A. V. Zubkov, and Y. E. Gerasimenko, *Khim. Geterotsikl. Soedin.*, 605 (1978).

13. 1,2,3-Benzotriazinium-4-olates (261) (page 47)

(261) (529) (530)

Irradiation of acetonitrile solutions of the *N*-aryl betaines **261** has been shown to result in high yields of the isomeric 3-arylbenzotriazinones (**529**). These species (**529**) are also formed in moderate yield by irradiation of the hydrazonyl bromides (**530**). The betaines (**261**) and their *N*-oxides (**262**) are believed to be intermediates in this process (**530** → **529**).[377,382]

An X-ray structural analysis of the 2,4-dibromophenyl derivative (**261**; R = 2,4-Br$_2$C$_6$H$_3$) has been reported.[383]

Section III,C

7. Naphtho[1,8-de]triazines (321) (page 58)

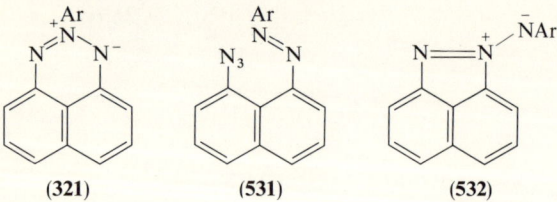

(321)   (531)   (532)

Photolysis of the 8-azido-1-arylazonaphthalene (**531**; Ar = 2-methoxy-1-naphthyl) gives a low yield (5%) of the betaine (**321**; Ar = 2-methoxy-1-naphthyl). Under the same conditions five other aryl derivatives (**531**) failed to give this type of product. In all cases the main products of this reaction were benz[*cd*]indazole-*N*-arylimines (**532**) which are also formed by thermolysis of the azides **531**. Irradiation of compound **321** (Ar = 2-methoxy-1-naphthyl) resulted in partial conversion into the isomer **532**.[384]

---

[382] Y. Maki and T. Furuta, *Synthesis*, 382 (1978).
[383] M. A. Hamid, *Libyan J. Sci.* **8B**, 75 (978) [*CA* **90**, 186907 (1979)].
[384] P. Spagnolo, A. Tundo, and P. Zanirato, *J. Org. Chem.* **43**, 2508 (1978).

Sec. IX]   HETEROCYCLIC BETAINES   111

9. Naphtho[1,8-cd][1,2,6]thiadiazines (330) (page 60)

An ^{15}N NMR study of compound **330** and some related species has been published.[385]

10. Naphtho[1,8-cd][1,2,6]selenadiazines (333) (page 61)

Recent attempts to repeat the preparation of compound **333** from 1,8-diaminonaphthalene have been unsuccessful.[386]

SECTION III,E

1. Neooxyberberines (364) (page 65)

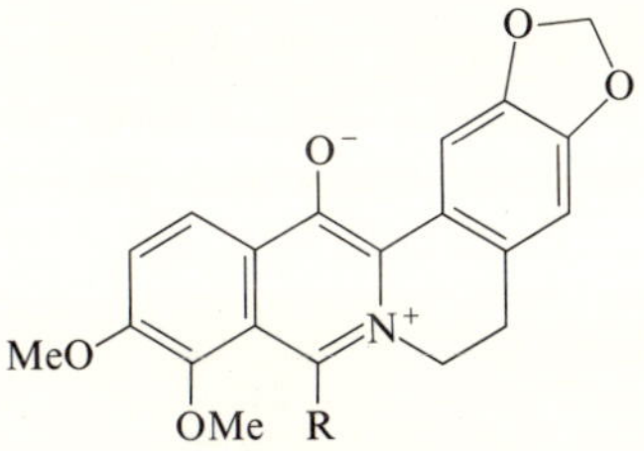

(**363**) R = OMe
(**364**) R = H

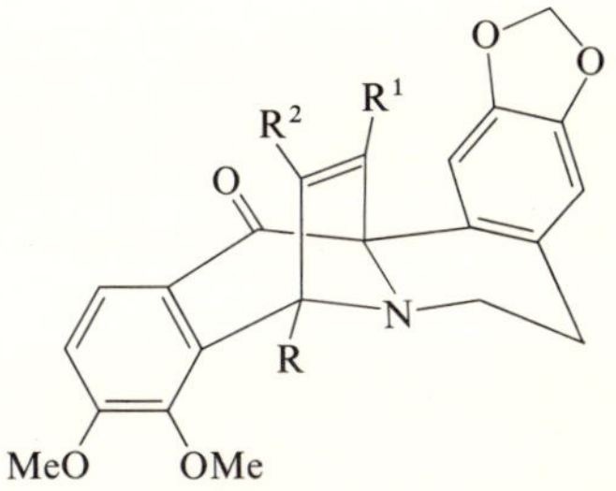

(**533**)

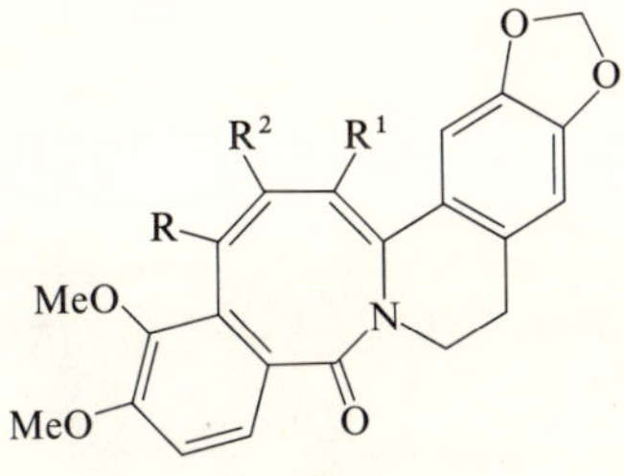

(**534**)

The betaines **363** and **364** react with acetylenes giving the cycloadducts **533** which rearrange to the benzazocines **534** in hot ethanol.[387]

---

[385] I. Yavari, R. E. Botto, and J. D. Roberts, *J. Org. Chem.* **43**, 2542 (1978).
[386] M. L. Kaplan, R. C. Haddon, F. C. Schilling, J. H. Marshall, and F. B. Bramwell, *J. Am. Chem. Soc.* **101**, 3306 (1979).
[387] M. Hanaoka, A. Wada, S. Yasuda, C. Mukai, and T. Imanishi, *Heterocycles* **12**, 511 (1979).

Section VII,B

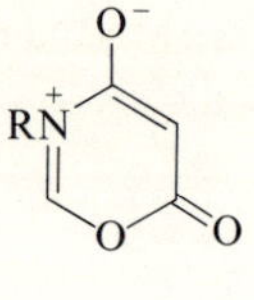

(535)[388-390] 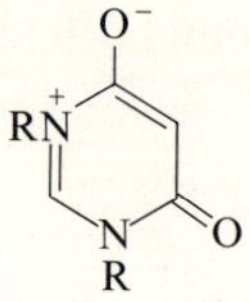   (536)[390-393] 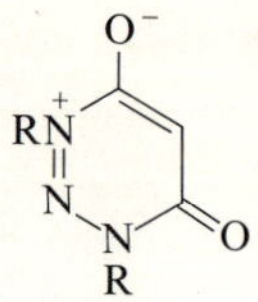   (537)[394]

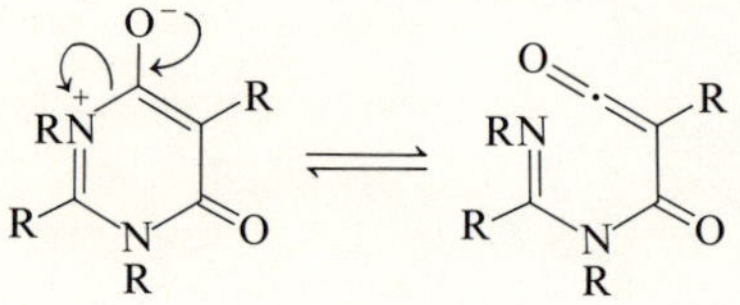

(538)   (539)

Further examples of cross-conjugated betaines (535–537) have been reported.[388-394] Particularly notable is the thermal ring opening of compounds of the general type 538 to give the ketene intermediates 539.[391]

Section VIII

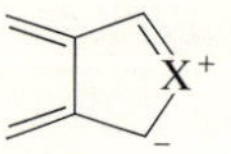

(540) 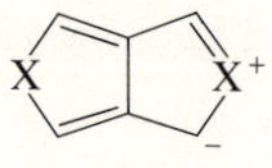   (541) 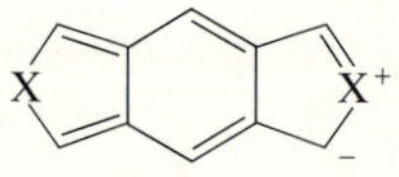   (542)

Further studies on molecules of the type 540 (X = O, S, NR) have been described.[395,396] Although the workers in this field favor a diradical representation, simple MO considerations of these molecules (540) and their close structural relationship to heteropentalenes (541)[2] and related species

---

[388] W. Friedrichsen, E. Kujath, G. Liebezeit, R. Schmidt, and I. Schwarz, *Justus Liebigs Ann. Chem.*, 1655 (1978).
[389] W. Friedrichsen, C. Krüger, E. Kujath, G. Liebezeit, and S. Mohr, *Tetrahedron Lett.*, 237 (1979).
[390] T. Kappe, W. Golser, M. Hariri, and W. Stadlbauer, *Chem. Ber.* **112**, 1585 (1979).
[391] T. Kappe and W. Lube, *Chem. Ber.* **112**, 3424 (1979).
[392] F. Mercer, L. Hernandez, and H. W. Moore, *Heterocycles* **12**, 45 (1979).
[393] R. A. Glennon, M. E. Rogers, R. G. Bass, and S. B. Ryan, *J. Pharm. Sci.* **67**, 1762 (1978).
[394] T. Kappe, W. Golser, and W. Stadlbauer, *Chem. Ber.* **111**, 2173 (1978).
[395] Y. S. P. Cheng, E. Dominguez, P. J. Garratt, and S. B. Neoh, *Tetrahedron Lett.*, 691 (1978).
[396] P. J. Garratt and S. B. Neoh, *J. Org. Chem.* **44**, 2667 (1979).

(**542**) suggest that 1,3-dipolar representations should also be considered. The known chemistry of these species (**540**) is very similar to that of the heteropentalenes (**541**) and entirely consistent with a betaine structure. Future studies will presumably lead to a clearer understanding of the bonding in these interesting molecules.

References Added in Proof

The following references (397–410) extend the coverage of the literature available to the author up to December 31, 1979. A few references which were inadvertently omitted from the main text are also included.

397. Section III,A,2. J. C. Craig and S. D. Hurt, *J. Org. Chem.* **44**, 1108 (1979); see also, E. C. Kornfeld, E. J. Fornefeld, G. B. Kline, M. J. Mann, D. E. Morrison, R. G. Jones, and R. B. Woodward, *J. Am. Chem. Soc.* **78**, 3087 (1956).
398. Section III,A,2. G. A. Ulsaker, H. Breivik, and K. Undheim, *J. C. S., Perkin 1*, 2420 (1979).
399. Section III,A,2. A. R. Katritzky, S. I. Bayyuk, N. Dennis, G. Musumarra, and E.-U. Wurthwein, *J. C. S. Perkin 1*, 2535 (1979).
400. Section III,A,2. A. R. Katritzky, J. Banerji, N. Dennis, J. Ellison, G. J. Sabongi, and E.-U. Wurthwein, *J. C. S. Perkin 1*, 2528 (1979).
401. Section III,B,3. A. Padwa and E. Vega, *J. Org. Chem.* **40**, 175 (1975).
402. Section III,B,3. B. Bhat and M. V. George, *J. Org. Chem.* **44**, 3288, (1979).
403. Section III,E,1. J. L. Moniot and M. Shamma, *J. Org. Chem.* **44**, 4337 (1979).
404. Section III,E,1. J. L. Moniot, D. M. Hindenlang, and M. Shamma, *J. Org. Chem.* **44**, 4343 (1979).
405. Section III,E,1. J. L. Moniot, D. M. Hindenlang, and M. Shamma, *J. Org. Chem.* **44**, 4347 (1979).
406. Section III,E,1. M. Hanaoka, S. Yasuda, K. Nagami, K. Okajima, and T. Imanishi, *Tetrahedron Lett.*, 3749 (1979).
407. Section III,F,3. L. Castedo, D. Dominguez, J. M. Saa, and R. Suau, *Tetrahedron Lett.*, 4589 (1979).
408. Section IV,A,2. V. Kokars and V. Kampars, *Tezisy Dokl.-Resp. Konf. Molodykh Uch.-Khim.*, 2, **1**, 39 (1977) [*CA* **89**, 197286 (1978)].
409. Section IV,A,2. S. P. Tandon, M. P. Bhutra, P. C. Mehta, J. P. Saxena, and K. Tandon, *Indian J. Pure Appl. Phys.* **6**, 694 (1968) [*CA* **70**, 62656 (1969)].
410. Section VIII. K. Maruyama and A. Osuka, *Chem. Lett.*, 77 (1979).

Thiocoumarins

O. METH-COHN AND B. TARNOWSKI

Department of Chemistry and Applied Chemistry,
University of Salford, Salford, England

I. Introduction 115
II. Synthetic Methods 116
   A. From Thiophenols 116
      1. By Reaction with a β-Dicarbonyl Compound 116
      2. By Reaction with α,β-Unsaturated Carboxylic Acid Derivatives 117
   B. From 2-Mercaptobenzaldehydes and Related Compounds 118
   C. From 2-Mercaptocinnamic Acid and Its Derivatives 120
   D. From Other Heterocycles 121
III. Applications of Thiocoumarins 122
IV. Spectroscopic and Physical Properties of Thiocoumarins 123
V. Reactions of Thiocoumarins 124
   A. Reactions at the Carbonyl Group 124
   B. Reactions at the 3,4-Double Bond 124
   C. Benzenoid Ring and Substituent Reactivity 125
VI. Dithiocoumarins 131
VII. 3,4-Dihydrothiocoumarins 132

I. Introduction

The chemistry of thiocoumarin (**1**) has for the most part been neglected, compared to that of coumarins and related systems. Apart from Elderfield's extremely brief coverage in 1950[1] and recent surveys in the Chemical Society's Specialist Periodical Reports,[2-5] no reviews of the chemistry of

---

[1] R. C. Elderfield, "Heterocyclic Compounds," Vol. 2, Chapter 14. Wiley, New York, 1950.
[2] R. J. S. Beer, "Organic Compounds of Sulphur, Selenium and Tellurium" (Sr. Reporter, D. H. Reid), Vol. 1, p. 351. Chemical Society, London, 1970.
[3] R. J. S. Beer, "Organic Compounds of Sulphur, Selenium and Tellurium," Vol. 2, p. 537. Chemical Society, London, 1973.
[4] R. J. S. Beer, "Organic Compounds of Sulphur, Selenium and Tellurium," Vol. 3, p. 534. Chemical Society, London, 1975.
[5] U. Eisner, "Organic Chemistry of Sulphur, Selenium and Tellurium" (Sr. Reporter D. R. Hogg), Vol. 4, p. 321. Chemical Society, London, 1977.

this system have appeared. It is the authors' hope that this review will not only close this gap but will stimulate renewed interest in this overlooked system, from both an academic and application viewpoint.

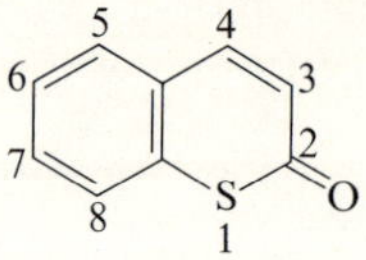

(1)

The structure and numbering are shown (1) and alternative names include 2H-1-benzothiopyran-2-one and benzothiopyrone.

II. Synthetic Methods

A. From Thiophenols

1. *By Reaction with a β-Dicarbonyl Compound*

This method, being the best source of the important 4-hydroxythiocoumarins (3), has numerous variations. Thus, esterification of thiophenol with malonic acid yields the diester (2) which is readily cyclized in good yield with Lewis acids[6–8] (Scheme 1). The reaction may also be accomplished in

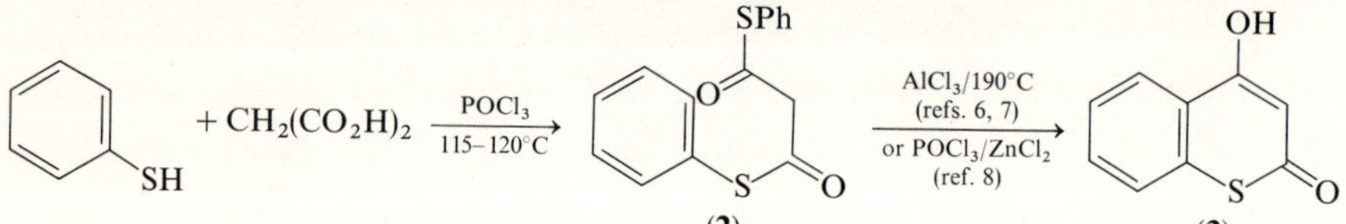

Scheme 1

one step from thiophenol by treatment with malonic acid, phosphoryl and zinc chlorides at 30–110°C.[9] Various substituted thiophenols,[10,11] and substituted malonic acids[12,13] have been similarly reacted giving the correspondingly substituted 4-hydroxythiocoumarins in 50–90% yields. Poly-

---

[6] E. Ziegler and H. Junek, *Monatsh. Chem.* **86**, 29 (1955).
[7] J. R. Geigy A.-G., British Patent 755,162 (1956) [*CA* **51**, 8146 (1957)].
[8] J. L. Bose, R. C. Shah, and V. R. Shah, *Chem. Ind. (London)*, 623 (1960).
[9] V. R. Shah, J. L. Bose, and R. C. Shah, Indian Patent 62,890 (1960) [*CA* **54**, 12161 (1960)].
[10] P. S. Jamkhandi and S. Rajagopal, *Monatsh. Chem.* **94**, 1271 (1963).
[11] P. S. Jamkhandi and S. Rajagopal, *Symp. Synth. Heterocycl. Compd. Physiol. Interest* [*Proc.*], *1964*, 83 (1966) [*CA* **69**, 10329 (1968)].
[12] P. S. Jamkhandi and S. Rajagopal, *Arch. Pharm. (Weinheim, Ger.)* **300**, 561 (1967).
[13] A. Ruwet, C. Draguet, and M. Renson, *Bull. Soc. Chim. Belg.* **79**, 639 (1970).

phosphoric acid is also an effective agent for the one-step reaction.[13] The high-temperature transesterification of malonic esters with thiophenol has been observed to give low yields of thiocoumarins.[14]

Ethyl acetoacetate or diketene undergo an acid-catalyzed conversion into 4-methylthiocoumarin (5) by way of the thioester (4), which need not be isolated.[15]

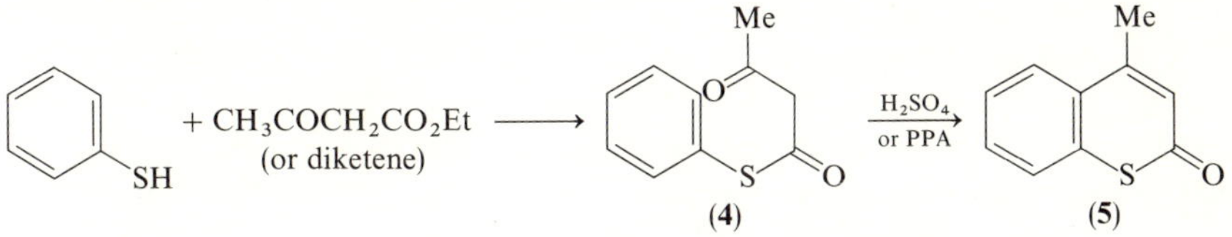

SCHEME 2

2. By Reaction with α,β-Unsaturated Carboxylic Acid Derivatives

Formation of cinnamoyl esters of thiophenols is readily accomplished. These esters cyclize, with elimination of benzene, by the action of aluminum chloride (Scheme 3).[16] An alternative approach, exemplified in Scheme 4, utilizes the ready thermal isomerization of thiochromones to thiocoumarins (see p. 121), the former being derived from phenylthioacrylic esters.[17,18]

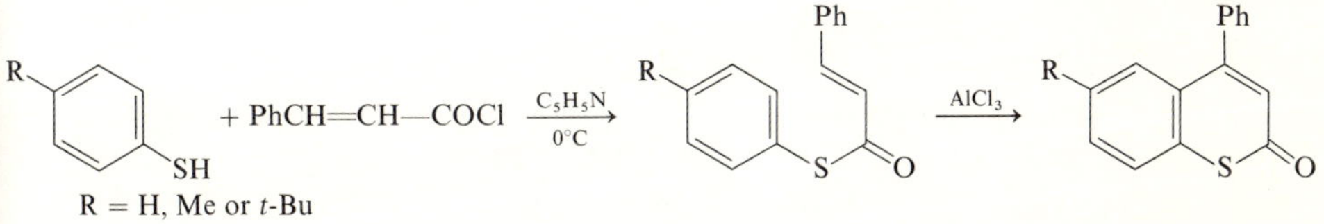

SCHEME 3

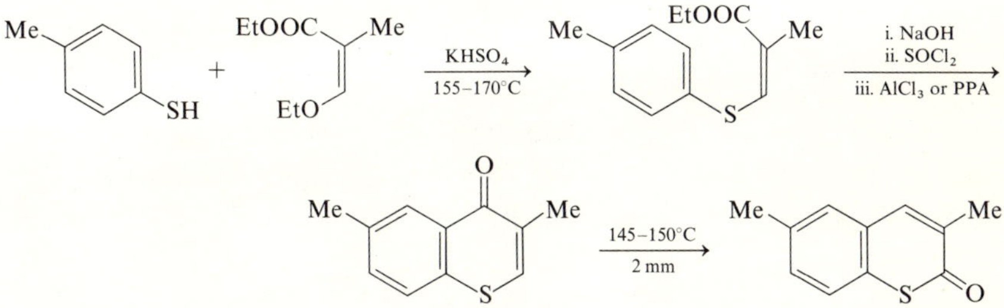

SCHEME 4

[14] C. Mentzer, A. Ville, and J. Faure-Bouvin, Bull. Soc. Chim. Fr., 92 (1963).
[15] K. Konishi, H. Umemoto, M. Yamamoto, and T. Kitao, Nippon Kagaku Kaishi 1, 118 (1973).
[16] T. Manimaran, T. K. Thiruvengadam, and V. T. Ramakrishnan, Synthesis, 739 (1975).
[17] L. Legrand and N. Lozac'h, Bull. Soc. Chim. Fr., 953 (1958).
[18] L. Legrand, N. Lozac'h, and N. Bignebat, Bull. Soc. Chim. Fr., 3247 (1964).

Thiocoumarin itself has been made in this way from phenylthioacrylic acid, although in only 10% yield.[19] Some thiocoumarin is also formed in the the aluminum chloride-catalyzed cyclization together with the thiochromone[18] (see p. 121).

B. From 2-Mercaptobenzaldehydes and Related Compounds

Although 2-mercaptobenzaldehyde (6) is an unstable intermediate, prone to resinification on storage, a surprising amount of research has been conducted using it as a source of thiocoumarins, with many workers preferring to use it directly or store it in solution at 0°C or below. It was first synthesized by Friedländer and Lenk in 1912 (Scheme 5).[20]

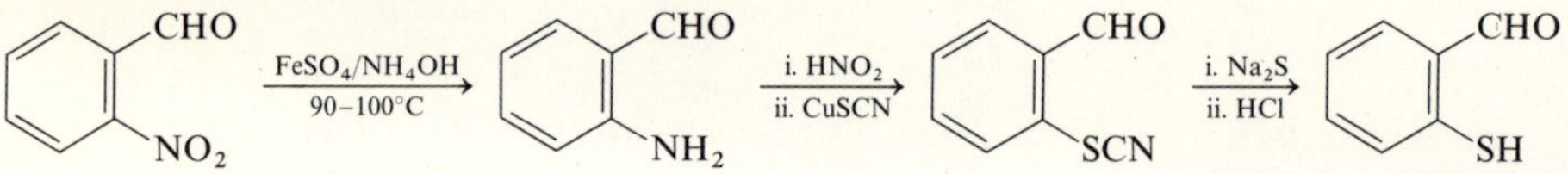

Scheme 5

An alternative but equally cumbersome synthesis utilizes the more accessible bis(2-carboxyphenyl)disulfide, easily prepared from anthranilic acid. By conversion to the corresponding anilide, then by successive reduction with zinc dust and acetic acid (yielding 2-mercaptobenzoyl anilide) and lithium aluminum hydride, the aldehyde (6) is obtained in fair overall yield.[21] Although 2-mercaptobenzaldehyde gives negligible yields of the parent thiocoumarin, under Perkin conditions (acetic anhydride and potassium acetate),[22] it may be effectively utilized to make 3-substituted derivatives by condensation of the aldehyde group with an active methylene derivative, sometimes followed by acid-catalyzed cyclization (Scheme 6). Thus, with ethyl benzoylacetate,

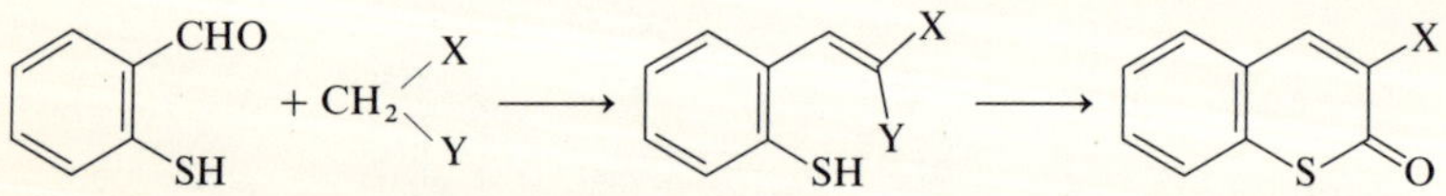

Scheme 6

3-benzoylthiocoumarin is isolated (40%),[23] while condensation with cyanoacetamide yields the 3-carboxamide by way of the 2-imine.[24] Using butyro-

---

[19] A. I. Tolmachev, L. M. Shulezhko, and A. A. Kisilenko, *J. Gen. Chem. USSR (Engl. Transl.)* **37**, 342 (1967).
[20] P. Friedländer and E. Lenk, *Ber. Dtsch. Chem. Ges.* **45**, 2083 (1912).
[21] D. Leaver, J. Smolicz, and W. H. Stafford, *J. Chem. Soc.*, 740 (1962).
[22] W. D. Cotterill, C. J. France, R. Livingstone, and J. R. Atkinson, *J. C. S., Perkin 1*, 817 (1972).
[23] G. Herbertz, H. Wamhoff, and F. Korte, *Z. Naturforsch., Teil B* **23**, 312 (1968).
[24] Merck and Co. Inc., British Patent 977,073 (1961) [*CA* **62**, 9141 (1965)].

lactone the 3-(2-hydroxyethyl-) or 3-(2-bromethyl)thiocoumarin may be isolated by cyclization of the initial condensation product with acetic acid containing either sulfuric or hydrobromic acid.[25]

The instability and inaccessibility of 2-mercaptobenzaldehyde has led other groups to explore more useful variants, usually with a suitable S-protection. Thus, the readily made 2-*tert*-butylthiobenzaldehyde (7) is stable and produced in high yield.[26] Condensation with acids, esters, or cyano compounds containing an adjacent active methylene group proceeds readily as does the subsequent PPA-catalyzed cyclization, giving thiocoumarin (70%) and its 3-substituted derivatives (e.g., X = $CONH_2$, COOH, CO_2Et, COPh, NHCOPh, Ph, etc) in good yield, especially when Y = CN.[26] (Scheme 7). An alternative approach has been to make 2-methylmercaptobenzaldehydes (8) from 2-methylmercaptobenzoic acid or its derivatives by reduction to the alcohol ($LiAlH_4$) followed by a two-step oxidation to the

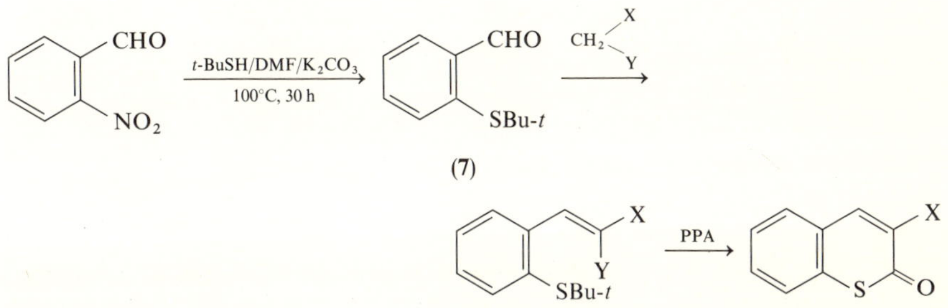

SCHEME 7

aldehyde.[27] The same workers prepared the corresponding methyl (9) and phenyl ketones (10) from the same 2-methylmercaptobenzoic acids by classical methods[28,29] and subjected them to the Reformatsky reaction followed by dehydration to give the corresponding cinnamic esters (11).[27]

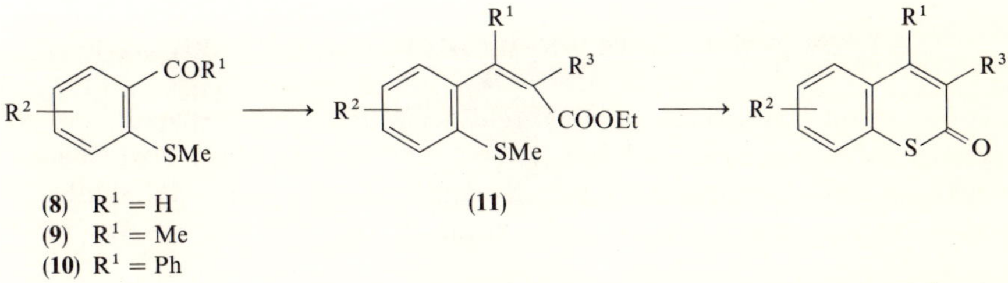

(8) R^1 = H
(9) R^1 = Me
(10) R^1 = Ph

(11)

SCHEME 8

[25] H. W. Zimmer and J. M. Holbert, U.S. Patent 3,287,459 (1966) [*CA* 66, 46,331 (1967)].
[26] O. Meth-Cohn and B. Tarnowski, *Synthesis*, 56 (1978).
[27] A. Ruwet and M. Renson, *Bull. Soc. Chim. Belg.* 79, 61 (1970).
[28] A. Ruwet and M. Renson, *Bull. Soc. Chim. Belg.* 78, 571 (1969).
[29] A. Ruwet and M. Renson, *Bull. Soc. Chim. Belg.* 75, 157 (1966).

Cyclization (Scheme 8) was accomplished using either phosphoryl chloride or aluminum chloride.[30,31] A range of substituents were introduced into the benzene ring (alkyl, aryl, or CN) or the 3- or 4-position (alkyl or aryl) and selenium successfully replaced sulfur in these reactions.

Another ready source of a suitable 2-mercaptobenzaldehyde derivative (**12**) stems from the Rieche formylation of di-*p*-tolyl disulfide[32] (Scheme 9).

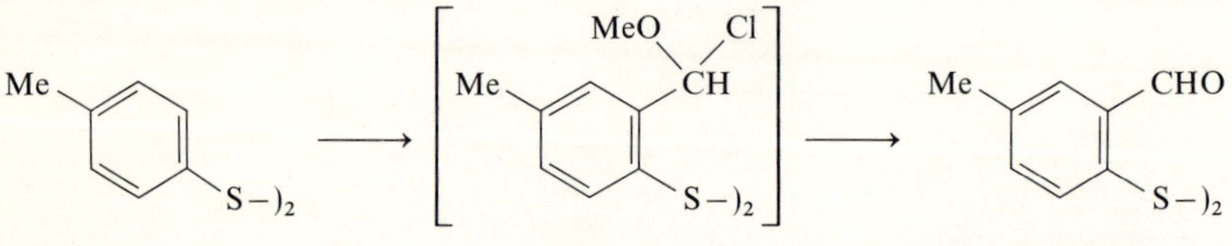

SCHEME 9

Perkin reaction of this aldehyde with phenylacetic acid gives a mixture of the benzothiophene **13** (57%) and the thiocoumarin **14** (40%).[32]

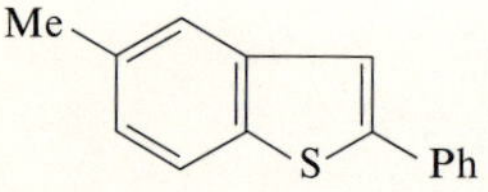

(13) 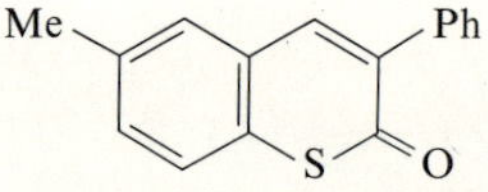   (14)

The interaction of 2-(acetylmercapto)benzoyl chloride with sodio diethyl malonate gives 3-carbethoxy-4-hydroxythiocoumarin (Eq. 1).[33,34]

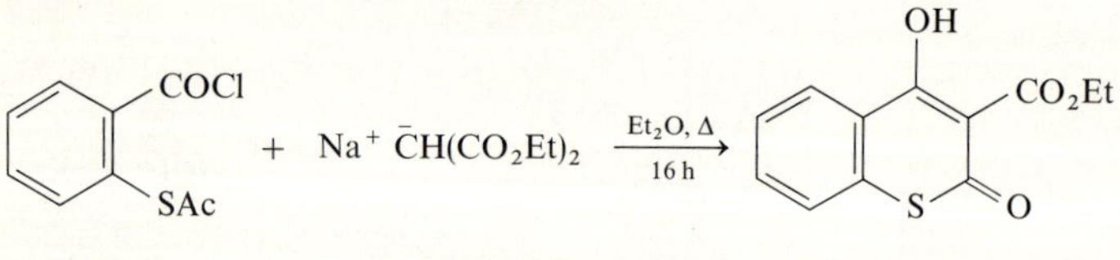

$$\tag{1}$$

C. From 2-Mercaptocinnamic Acid and Its Derivatives

In Section II,B several cinnamic acid derivatives were prepared from 2-mercaptobenzaldehydes and the corresponding ketones. These were cyclized either *in situ* or separately to give thiocoumarins. An alternative synthetic approach introduces the mercapto group as the last stage by way of a

---

[30] A. Ruwet and M. Renson, *Bull. Soc. Chim. Belg.* **77**, 465 (1968).
[31] A. Ruwet and M. Renson, *Bull. Soc. Chim. Belg.* **78**, 449 (1969).
[32] K. J. Brown and O. Meth-Cohn, *Tetrahedron Lett.*, 4069 (1974).
[33] C. Mentzner, P. Meunier, J. Lecocq, and D. Billet, *Bull. Soc. Chim. Fr.* **12**, 430 (1945).
[34] E. Boschetti, D. Molho, and L. Fontaine, French Patent 1,521,366 (1968) [*CA* **71**, 13022 (1969)].

Sec. II.D] THIOCOUMARINS 121

diazonium salt, another route pioneered by Friedländer (Scheme 10).[35] Cyclization to thiocoumarin was accomplished with acetic anhydride, or by PPA and phosphorus pentoxide, which gave poorer results.[21]

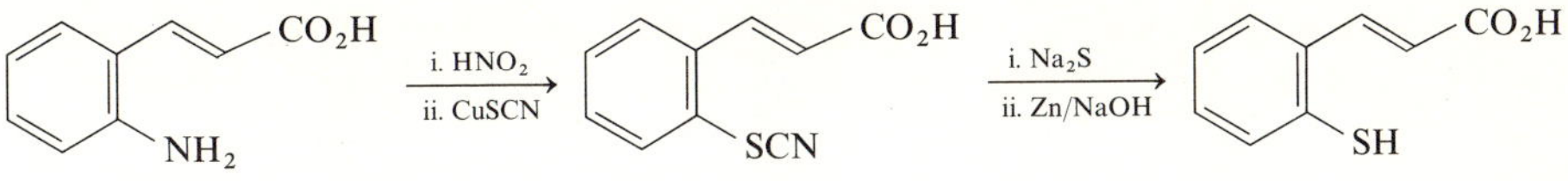

SCHEME 10

4- And 5-acetylamino-2-mercaptocinnamic acid has been efficiently cyclized yielding 7- and 6-aminothiocoumarin, respectively, by this approach.[36,37] These have been extensively utilized for the synthesis of other thiocoumarins (see p. 125).

D. FROM OTHER HETEROCYCLES

Thiochromenes[1,38,39] (e.g., **15**) are converted to thiocoumarins by oxidation with chromium trioxide in pyridine.[40] In a similar way thiochromylium salts (e.g., **16**) are converted to thiocoumarins by oxidation with either active

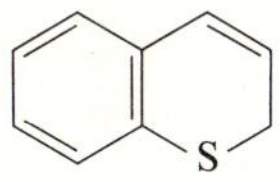

(15) 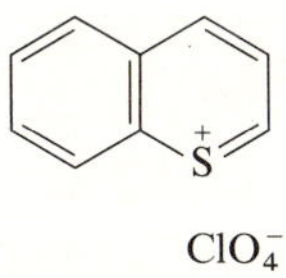   (16) 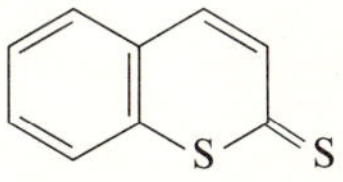   (17)

manganese dioxide in hot chloroform,[41,42] by air through heating in acetonitrile solution[41] or by the action of aqueous alkali,[43] which also gives the corresponding thiochromone. Some thiochromones are believed to undergo thermal,[17] and possibly Lewis acid-catalyzed[18,44] isomerization to give thiocoumarins (Eq. 2, see p. 117). However, this rearrangement has been disputed by other workers.[19]

[35] C. Chmelewsky and P. Friedländer, *Chem. Ber.* **46**, 1903 (1913).
[36] A. Ricci, *Ann. Chim.* (*Rome*) **48**, 985 (1958).
[37] A. Ricci and A. Martani, *Ann. Chim.* (*Rome*) **53**, 588 (1963).
[38] R. C. Elderfield, "Heterocyclic Compounds," Vol. 2, Chapter 6. Wiley, New York, 1950.
[39] S. W. Schneller, *Adv. Heterocycl. Chem.* **18**, 59 (1975).
[40] A. Ruwet, J. Meesen, and M. Renson, *Bull. Soc. Chim. Belg.* **78**, 459 (1969).
[41] I. Degani and R. Fochi, *Ann. Chim.* (*Rome*) **58**, 251 (1968).
[42] I. Degani, R. Fochi, and G. Spunta, *Boll. Sci. Fac. Chim. Ind. Bologna* **26**, 31 (1968).
[43] B. D. Tilak and G. T. Panse, *Indian J. Chem.* **7**, 315 (1969).
[44] G. Pfister-Guillouzo and N. Lozac'h, *Bull. Soc. Chim. Fr.*, 1624 (1962).

$$\text{(structure)} \xrightarrow[\text{or } \Delta]{\text{AlCl}_3} \text{(structure)} \quad (2)$$

Finally, dithiocoumarins (e.g., **17**, see p. 131) are converted into thiocoumarins by the action of hydrogen peroxide in acetic acid.[45]

III. Applications of Thiocoumarins

The uses of thiocoumarins as pharmaceuticals are numerous and include hemorrhagic agents,[46] anticoagulants,[11,25,47-49] antiallergic agents,[50] and agents with anti-vitamin K activity,[14] anthelmintic activity,[24,51-54] chloeretic activity,[34,55] bactericidal action,[36,56-58] and photodynamic activity.[36,59] Insecticidal,[60-62] plant[63] and animal growth arrest activity,[64] and mitochondrial swelling inhibition[65] have been observed as have useful fluorescent characteristics[36] and polymer swelling agent properties.[66] The 4-hydroxy

[45] J. L. Charlton, S. M. Loosmore, and D. M. McKinnon, *Can. J. Chem.* **52**, 3021 (1974).
[46] C. Mentzner and P. Meunier, *Bull. Soc. Chim. Biol.* **25**, 379 (1943).
[47] J. R. Geigy A.-G., British Patent 784,281 (1957) [*CA* **52**, 6413 (1958)].
[48] J. R. Geigy A.-G., British Patent 805,748 (1958) [*CA* **53**, 9252 (1959)].
[49] P. S. Jamkhandi, S. Purushottam, S. Rajagopal, and S. M. Srinivasan, *Indian J. Exp. Biol.* **6**, 224 (1968).
[50] D. R. Buckle, B. C. C. Cantello, and H. Smith, British Patent 1,433,637 (1976) [*CA* **85**, 159888 (1976)].
[51] H. D. Brown, U.S. Patent 3,278,547 (1966) [*CA* **65**, 18593 (1966)].
[52] Merck and Co., Netherlands Patent 6,517,255 (1966) [*CA* **66**, 2565 (1967)].
[53] H. D. Brown, U.S. Patent 3,338,784 (1967) [*CA* **68**, 78284 (1968)].
[54] J. Di Nelta and J. R. Egerton, U.S. Patent 3,549,754 (1970) [*CA* **74**, 146,378 (1971)].
[55] E. Boschetti, D. Molho, L. Fontaine, French Medicinal Patent 6846 (1969) [*CA* **74**, 91193 (1971)].
[56] W. L. Fowlks, D. G. Griffiths, and E. L. Oginsky, *Nature (London)* **181**, 571 (1958).
[57] L. De Poi, M. Pitzurra, and M. Negri, *Boll. Chim. Farm.* **101**, 367 (1962).
[58] M. Giannella and M. Pigini, *Farmaco, Ed. Sci.* **28**, 157 (1973).
[59] A. Ricci, G. Pino, and L. Santamaria, *Boll. Chim. Farm.* **97**, 655 (1958).
[60] L. E. Smith, E. H. Siegler, and F. Munger, *J. Econ. Entomol.* **29**, 1027 (1936).
[61] L. E. Smith, E. H. Siegler, and F. Munger, *U.S., Dep. Agric., Circ.*, 523 (1939) [*CA* **33**, 8895 (1939)].
[62] W. M. Hoskins and W. F. Chamberlain, *Calif., Agric. Exp. Stn., Circ.*, 365 (1946) [*CA* **41**, 2849 (1947)].
[63] R. H. Goodwin and C. Taves, *Am. J. Bot.* **37**, 224 (1950).
[64] E. Del Pianto and E. Pantano, *Atti Accad. Naz. Lincei, Cl. Sci. Fis., Mat Nat., Rend.* **10**, 230 (1951).
[65] P. Van Caneghem, *Biochem. Pharmacol.* **23**, 3491 (1974).
[66] H. Tacheci, M. Sander, and P. Pintaske, Netherlands Patent 67, 16,163 (1968) [*CA* **69**, 36806 (1968)].

derivatives have been used to generate a range of merocyanine dyes[67,68] with possible optical sensitizer properties.

IV. Spectroscopic and Physical Properties of Thiocoumarins

Ultraviolet[42] and near ultraviolet spectra[69] of a number of thiocoumarins have been recorded and discussed. Infrared spectra, in particular concerned with carbonyl absorptions, have been frequently collected,[19,44,70,71] and comparisons made with those of thiochromones.[18] All the usual spectra of thiocoumarin itself have been recorded and analyzed[41] as well as the ^{13}C-NMR spectrum[72] and the fluorescent characteristics.[73]

3-Substituted thiocoumarins have been thoroughly studied, and their infrared, NMR, and mass spectra critically examined.[26,74] Their NMR spectra show a characteristic low-field singlet for the 4-proton and, with a strongly electron-withdrawing 3-substituent, H-5 is observed downfield from the complex aromatic multiplet, as a double doublet.

The spectroscopic distinction between thiocoumarins and thiochromones is best made by observing the low-field 5-proton in the NMR spectrum of the latter and also the characteristic retro-Diels–Alder fragmentation always observed in the mass spectrum of thiochromones. Thiochromones also show distinctive, strong ultraviolet absorptions at 250–270 nm.[75]

Thiocoumarins characteristically lose carbon monoxide from their strong molecular ion in the mass spectrum, thus forming a benzo[b]thiophene cation, invariably accompanied by a large metastable peak.[26,74] Similar observations are recorded for selenocoumarins.[76]

MO calculations have been conducted on some thiocoumarins[77–79] and binding energies derived from photoelectron spectra have been reported.[80]

[67] P. C. Rath and K. Rajagopal, *Indian J. Chem.* **7**, 1273 (1969).
[68] P. C. Rath and K. Rajagopal, *Indian J. Chem.* **9**, 91 (1971).
[69] A. Mangini and D. Dal Monte Casoni, *Atti Accad. Sci. Ist Bologna, Cl. Sci. Fis., Rend.* **5**, 20 (1958) [*CA* **54**, 18065 (1960)].
[70] V. Prey, B. Kerres, and H. Berbalk, *Monatsh. Chem.* **91**, 774 (1960).
[71] A. Ruwet and M. Renson, *Bull. Soc. Chim. Belg.* **79**, 89 (1970).
[72] I. W. J. Still, N. Plavac, D. M. McKinnon, and M. S. Chauhan, *Can. J. Chem.* **54**, 280 (1976).
[73] R. H. Goodwin and F. Kavanagh, *Arch. Biochem.* **27**, 152 (1950).
[74] B. Tarnowski, M.Sc. Thesis, University of Salford, 1978.
[75] H. Nakazumi and T. Kitao, *Bull. Chem. Soc. Jpn.* **50**, 939 (1977).
[76] N. P. Buu-Hoi, M. Mangane, O. Perin-Roussel, M. Renson, A. Ruwet, and M. Marechal, *J. Heterocycl. Chem.* **6**, 825 (1969).
[77] R. Zahradnik and C. Parkanyi, *Collect. Czech. Chem. Commun.* **30**, 3016 (1965).
[78] A. I. Tolmachev, G. G. Dyadyusha, and L. M. Shulezhko, *Teor. Eksp. Khim.* **6**, 185 (1970) [*CA* **73**, 55511 (1970)].
[79] M. Kamiya and Y. Akahori, *Chem. Pharm. Bull.* **20**, 677 (1972).
[80] H. Nakazumi, T. Yoshida, S. Sawada, and T. Kitao, *Nippon Kagaku Kaishi* **5**, 849 (1976).

Thiocoumarin itself is a solid, mp 80–80.5°, which has the smell of coumarin.[35] Its basicity has been studied.[19,42]

V. Reactions of Thiocoumarins

The general chemistry of the thiocoumarin system has only been sparsely examined, leaving large uncharted areas of even the most basic kind. The known chemistry is collected below.

A. Reactions at the Carbonyl Group

Thiocoumarins form 1:1 complexes with Lewis acids such as $HgCl_2$, $SbCl_5$, and $SnCl_4$.[71]

The carbonyl group is easily converted to the thiocarbonyl, yielding 1,2-dithiocoumarin, by the action of phosphorus pentasulfide.[17,18,44] The chemistry of the dithiocoumarins is discussed separately (see p. 131). Interaction of thiocoumarin with phenylmagnesium bromide gives low yields of thioflavone (**19**) and the thiochromene **20**, probably by way of the thiobenzopyrylium salt (**18**, Scheme 11).[22]

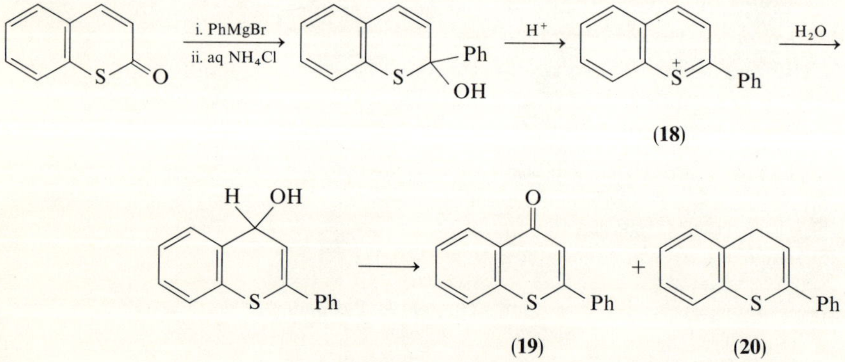

Scheme 11

B. Reactions at the 3,4-Double Bond

The 3,4-double bond of thiocoumarin behaves as a typical α,β-unsaturated carbonyl alkene group rather than as part of an aromatic ring system. Thus, with bromine in chloroform, a rapid addition occurs giving the 3,4-dibromo-3,4-dihydrothiocoumarin (**21**), which may easily be dehydrobrominated to yield 3-bromothiocoumarin (**22**, Scheme 12).[22]

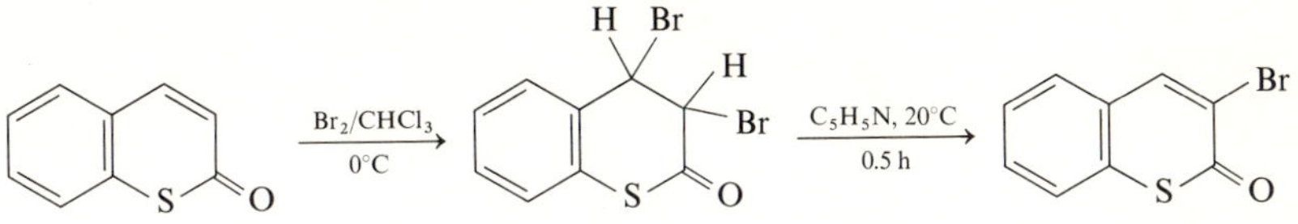

SCHEME 12

The 3,4-double bond is also easily reduced. Thus, for example, 3-benzoyl-thiocoumarin is transformed into its dihydro derivative without any adverse effect on the carbonyl groups (Eq. 3).[23]

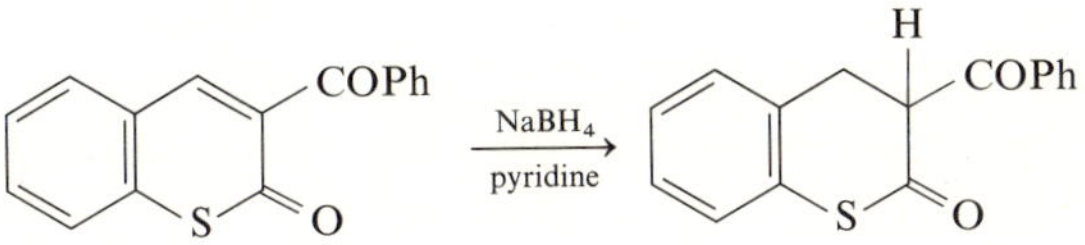

$$\text{(3)}$$

C. BENZENOID RING AND SUBSTITUENT REACTIVITY

Apart from the chemistry of the 4-hydroxy derivative (see p. 126) relatively little investigation in this area has been conducted. Thiocoumarin-3-carbonyl compounds have proved to be effective precursors for the synthesis of the pharmaceutically useful 3-(2-benzimidazolyl) derivatives. Thus, on interaction of the 3-carboxamide with *o*-phenylenediamine or the 3-aldehyde with *o*-nitroanilines, the above benzimidazoles or their *N*-oxides are formed.[51]

6-Amino- and 7-aminothiocoumarin have been thoroughly explored as useful precursors to a wide variety of thiocoumarins. Thus, substituents such as Cl, Br, I, SCN, CN, OH, and OMe have been introduced by way of the Sandmeyer reaction.[36] 6-Aminothiocoumarin is brominated in the 5-position, whereby 5-bromo- or 5,6-dibromothiocoumarin may be made by way of the corresponding diazonium salt, using hypophosphorous acid or cuprous bromide, respectively.[36] Alternatively, 6-acetylaminothiocoumarin may be nitrated in the 5-position and thus 5-nitro-, 5-amino-, 5,6-diamino-, or 6-chloro-5-nitrothiocoumarins can be prepared by classical syntheses.[81] These derivatives are useful intermediates for the synthesis of further fused thiocoumarins (e.g., **23**,[36] **24** and **25**[81]).

---

[81] A. Ricci, M. Negri, and C. Rossi, *Ann. Chim. (Rome)* **53**, 1507 (1963).

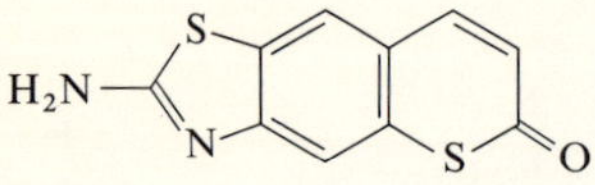

(23) 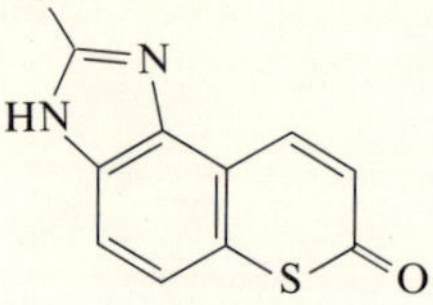   (24) 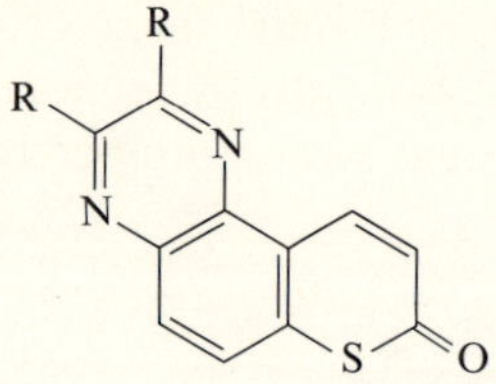   (25)

Acetoxymercuration of thiocoumarin occurs readily at the 6-position, offering a potential (but as yet unexploited) route to numerous substituted thiocoumarins.[82]

4-Chloro-3-nitrothiocoumarins readily undergo nucleophilic substitution with ammonia or amines giving useful diamines on further reduction (Scheme 13).[83]

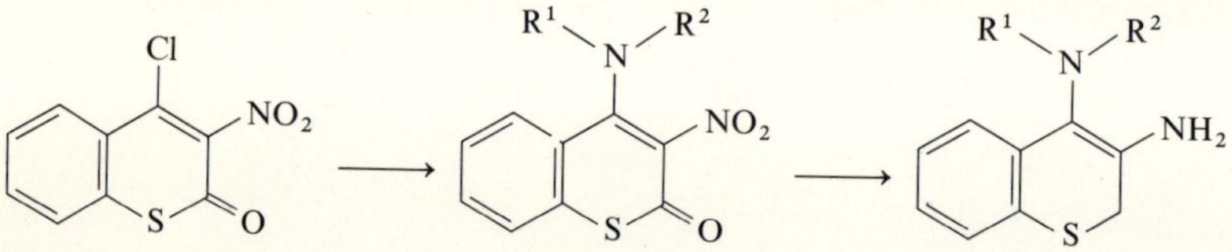

SCHEME 13

By far the most studied derivative of thiocoumarin is the 4-hydroxy compound (26), partly due to its ease of preparation and to some extent to its potential as a source of biologically significant derivatives. Thus, the thio analog (27) of the rodenticide Warfarin was examined, but showed only

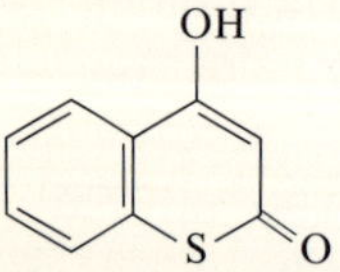

(26) 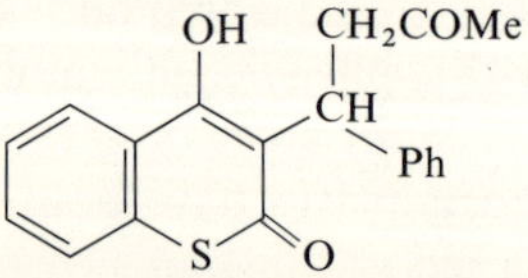   (27) 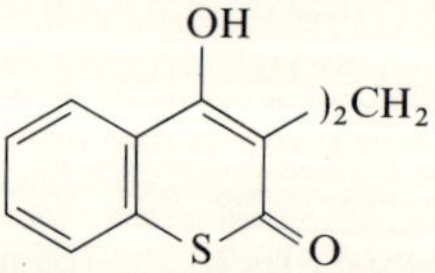   (28)

about one-tenth of the anticoagulant properties of its parent.[46] The large volume of chemistry exhibited by this system (26) stems from its β-keto ester-like structure and consequent "active" 3-position. Thus, condensation with formaldehyde yields 3,3'-methylene bis(4-hydroxythiocoumarin)s (28),[6,10,11,46,49,84] a reaction also applicable to the 4-methoxy derivative.[49] Similarly, the Mannich (Scheme 14) reaction takes place readily at room

[82] K. G. Naik and A. D. Patel, *J. Chem. Soc.*, 1043 (1934).
[83] V. L. Savel'ev, T. G. Afanas'eva, and V. A. Zagorevskii, *Chem. Heterocycl. Compd. (Engl. Transl.)* 420 (1976).
[84] P. S. Jamkhandi and S. Rajagopal, *Monatsh. Chem.* **97**, 1732 (1966).

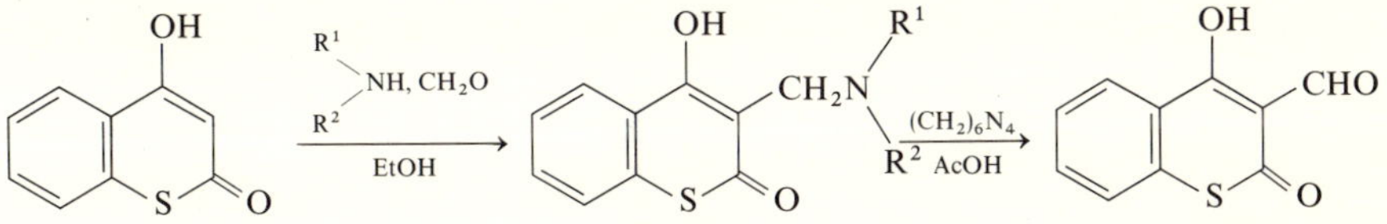

Sec. V.C] THIOCOUMARINS

SCHEME 14

temperature for primary amines, but requires reflux for secondary bases, the products giving the bis(4-hydroxythiocoumarinyl)methane (28) on treatment with mineral acid.[85] Treatment of the Mannich bases (29) with hexamine offers a convenient synthesis of the 4-hydroxy-3-aldehyde (30).[86] Knoevenagel and Perkin condensations of the aldehyde proceed with further ring annelation (Scheme 15), while with ethyl acetoacetate and aluminum chloride at 180°C in nitrobenzene, the methylpyranothiocoumarin (31; R = Me) is isolated in 54% yield.[86]

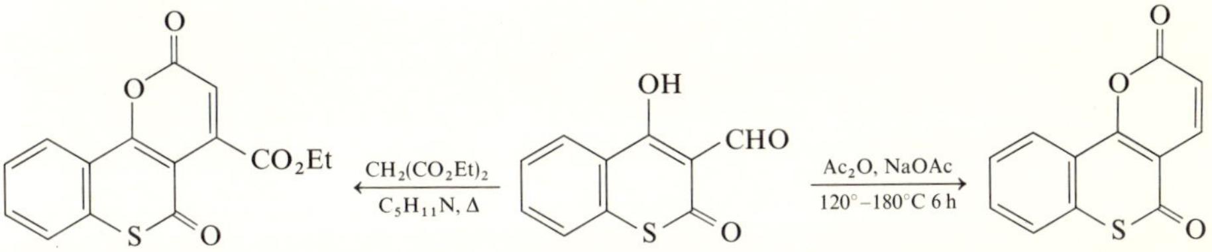

SCHEME 15

With aromatic aldehydes instead of formaldehyde, analogous bis-thiocoumarinyl derivatives (32) are formed, which undergo dehydration with acetic anhydride to give the pentacyclic systems 33.[49,84] Some aldehydes behave anomalously: p-dimethylaminobenzaldehyde yields the ylidene derivative 34,[68] and salicylaldehyde dehydrates after condensation to give compound 35,[84] by interaction of a 4-hydroxy and a salicylhydroxy group.

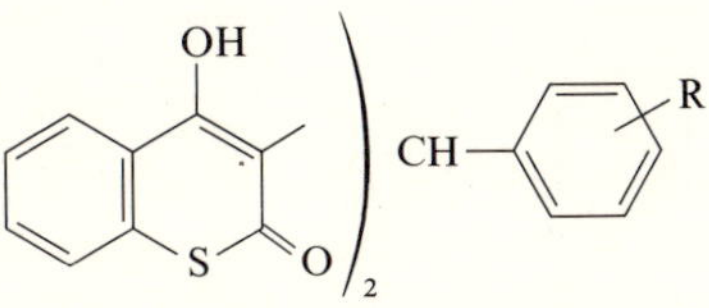

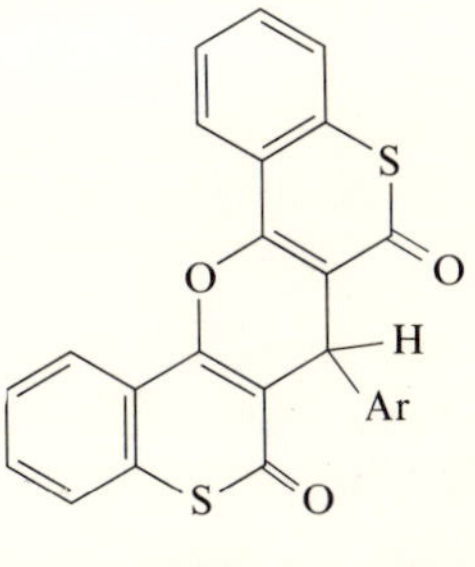

[85] P. S. Jamkhandi, S. Pirushottam, and S. Rajagopal, *J. Indian Chem. Soc.* **45**, 273 (1968).
[86] P. S. Jamkhandi and S. Rajagopal, *Indian J. Chem.* **11**, 708 (1973).

Alkylation at the 3-position of 4-hydroxythiocoumarin has been accomplished in moderate yields,[48,87] using secondary alcohols under acid catalysis. Another useful alkylation method involves Michael addition to an α,β-unsaturated ketone, as exemplified in Scheme 16.[47,88]

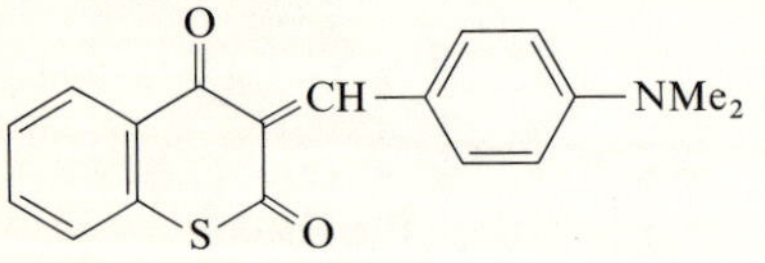

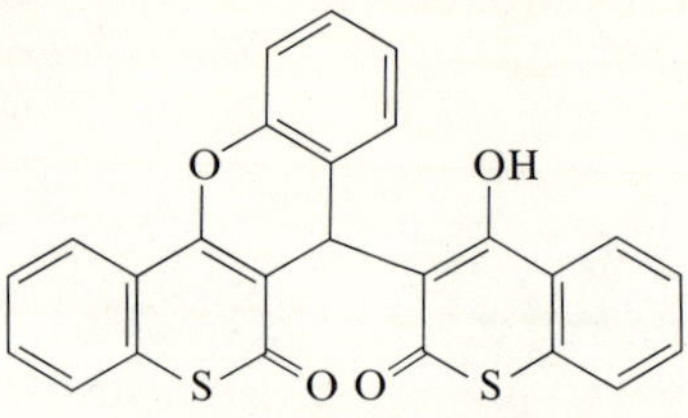

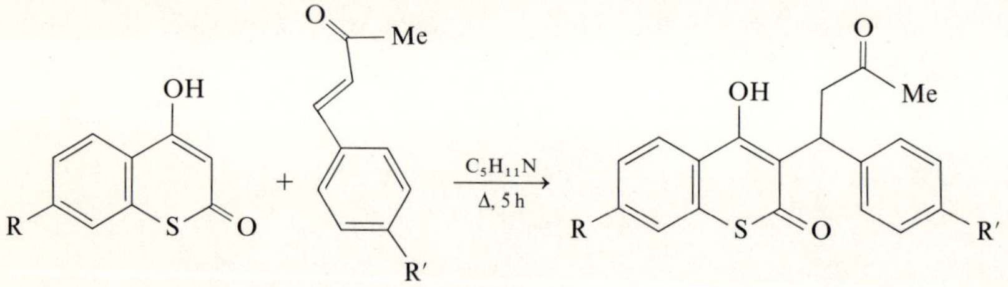

SCHEME 16

A variety of merocyanine dyes (**36**) have been generated (Scheme 17) by interaction of suitable oxazolium or thiazolium salts with 4-hydroxythiocoumarin.[67,68] Similar use of quinolinium salts is also reported.

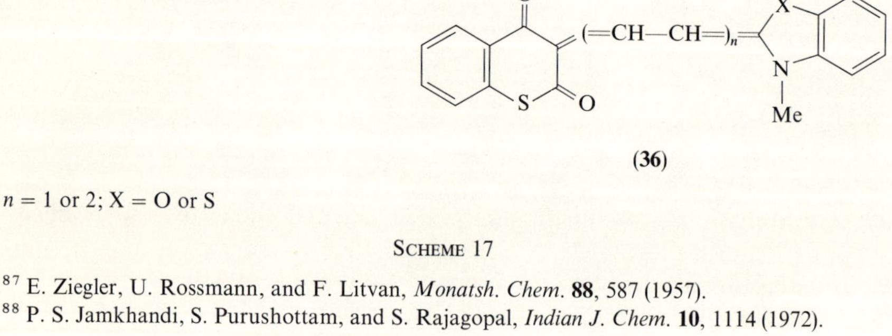

$n = 1$ or 2; $X = O$ or S

SCHEME 17

[87] E. Ziegler, U. Rossmann, and F. Litvan, *Monatsh. Chem.* **88**, 587 (1957).
[88] P. S. Jamkhandi, S. Purushottam, and S. Rajagopal, *Indian J. Chem.* **10**, 1114 (1972).

Acylation at the 3-position of 4-hydroxythiocoumarin can be achieved by numerous methods. Thus, aliphatic acids in phosphoryl chloride effect ready introduction of a 3-acyl group (Eq. 4).[89]

Schotten-Baumann benzoylation[90] or use of an acid chloride in pyridine[89,90] rapidly causes *O*-acylation giving 4-acyloxythiocoumarins (**37**) in high yield. These derivatives undergo the Fries rearrangement to give the 3-acyl derivatives (**38**), either by the action of hot pyridine[90] (quantitatively), aluminum chloride,[89] polyphosphoric acid,[90] or even partially by sublimation.[90] Cross-over experiments indicate that a mixture of intra- and intermolecular mechanisms is involved,[90] the products **38** also being available directly from 4-hydroxythiocoumarin.

Both the acetyl and benzoyl derivatives (**37**; R = Me and Ph, respectively) undergo photodimerization yielding the cyclobutanes **39** in 53 and 65% yields, respectively.[90]

3-Acetyl-4-hydroxythiocoumarin reacts in the expected ketonic manner at the side chain, in condensing with benzaldehyde and in forming anils and

---

[89] P. S. Jamkhandi and S. Purushottam, *Monatsh. Chem.* **99**, 1390 (1968).
[90] J. Lehmann and H. Wamhoff, *Justus Liebigs Ann. Chem.*, 1287 (1974).

phenylhydrazones. These last derivatives (40) may be cyclized to the pyrazolothiocoumarins (41) under acid catalysis. Interaction of the 3-hydroxy-4-acetyl derivative with diazonium salts results in elimination of the acetyl group, giving the azo compound 42.[91]

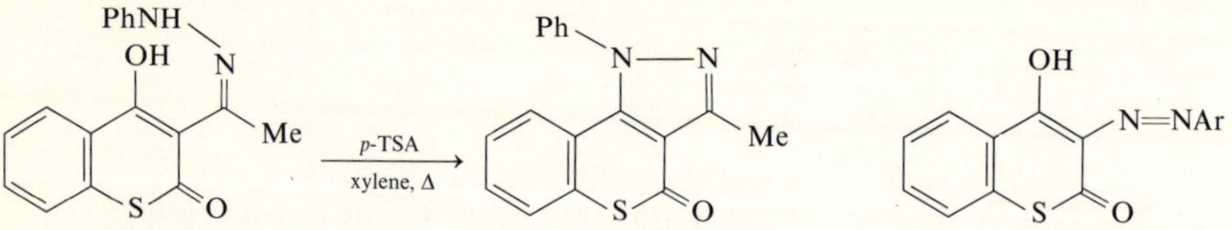

4-Hydroxythiocoumarin-3-carboxamides may be generated by prolonged heating of a 4-hydroxythiocoumarin with urea.[92]

The 4-hydroxythiocoumarins may be readily nitrated at the 3-position, using nitric acid in either acetic acid or chloroform solution.[50,93] The 3-nitro derivatives may be reduced to amines using standard conditions[93] and these aminohydroxy compounds (43) used as a source of further fused thiocou-

marins such as the oxazolo systems 44.[93] Acylation of the aminohydroxy derivatives takes place preferably at N.[93] These aminohydroxy compounds are also conveniently derived in high overall yield by the Japp-Klingemann reaction of 4-hydroxythiocoumarin with a diazonium salt followed by hydrogenation of the derived azo compound (Scheme 18).[93,94]

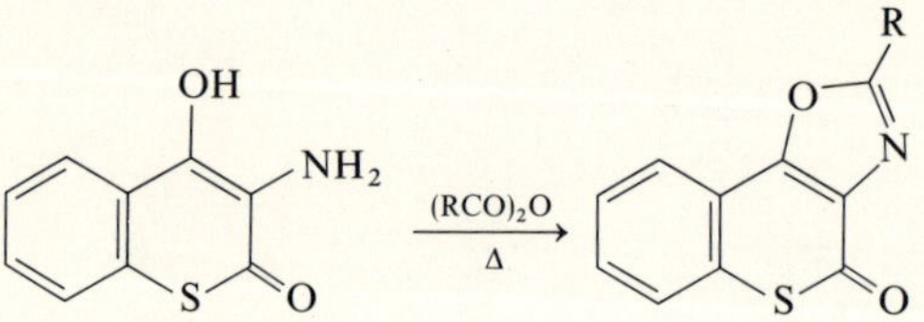

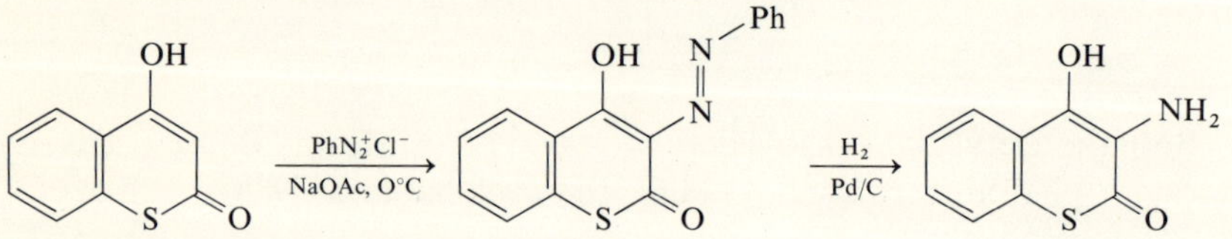

SCHEME 18

[91] W. Asker, M. H. Elnagdi, and S. M. Fahmy, *J. Prakt. Chem.* **313**, 715 (1971).
[92] H. C. Scarborough and W. A. Gould, *J. Org. Chem.* **26**, 3720 (1961).
[93] G. Peinhardt and L. Reppel, *Pharmazie* **28**, 729 (1973).
[94] E. Ziegler and E. Nolken, *Monatsh. Chem.* **91**, 850 (1960).

Nitrosation of 4-hydroxythiocoumarin leads to the 3-oximino-4-keto derivative (45).[58] Replacement of the 4-hydroxy group by a chloro (46), arylamino (47), or phenylhydrazino group (48) is readily accomplished by heating the compound with phosphorus pentachloride in chloroform, aniline, or phenylhydrazine, respectively.[91]

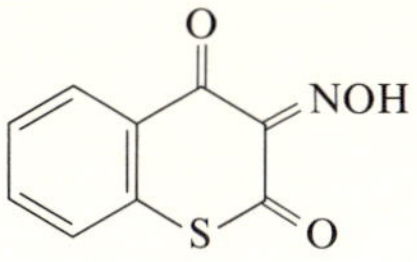

(45)

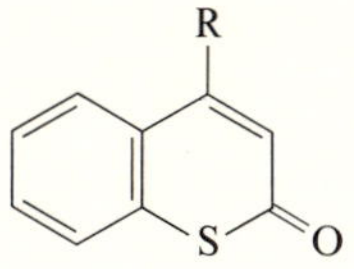

(46) R = Cl
(47) R = NHAr
(48) R = NHNHPh

VI. Dithiocoumarins

Dithiocoumarins (50) are generally prepared by treatment of thiocoumarins with phosphorus pentasulfide, isolation being best effected by formation of a mercury complex by addition of mercurous chloride, followed by decomposition of this complex with sodium sulfide.[7,18,44] An alternative ring-closure approach involves heating o-allylphenols (49) with sulfur in

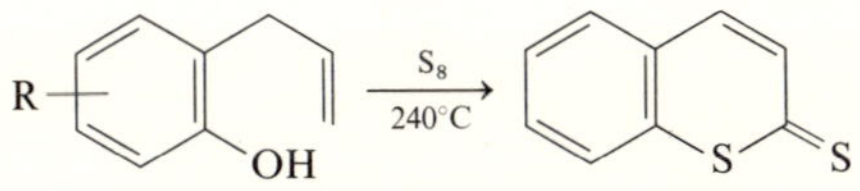

(49)   (50)

diethyl phthalate solution, the product being contaminated with other related systems.[95] The corresponding phenol ethers are also useful in some cases but also give rise to 1,2-dithiole-3-thiones. A similar, low-yield pathway from o-methoxyphenyl ethyl ketones, exemplified in Scheme 19, has been

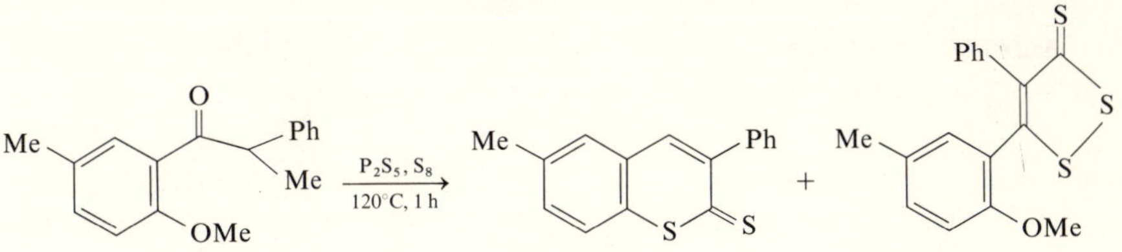

SCHEME 19

[95] Y. Mollier and N. Lozac'h, Bull. Soc. Chim. Fr., 651 (1958).

reported.[17] Just as thiochromones may be isomerized to thiocoumarins, they may also be converted to dithiocoumarins by heating with phosphorus pentasulfide and sulfur.[17,18] The tetrahydrodithiocoumarin (**52**) is readily made by condensation of dithiophenylacetic acid with the reactive chloroaldehyde (**51**) (Eq. 5).[96]

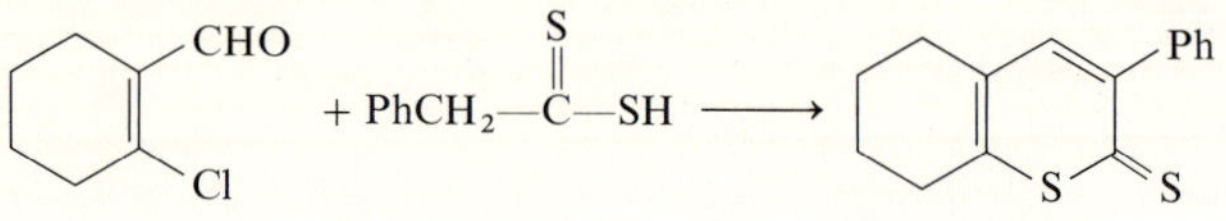

$$\text{(51)} + \text{PhCH}_2\text{—C(=S)—SH} \longrightarrow \text{(52)} \quad (5)$$

VII. 3,4-Dihydrothiocoumarins

The ready addition of bromine and hydrogen to the 3,4-double bond of thiocoumarins has already been referred to (see p. 125). Relatively little has been published on the chemistry of the 3,4-dihydrothiocoumarins. The parent system has been made by reduction of 2-mercaptocinnamic acid with sodium amalgam and alkali, followed by acid-catalyzed (or thermal) cyclization (Scheme 20).[22,35] Similar methods have been employed for substituted analogs.

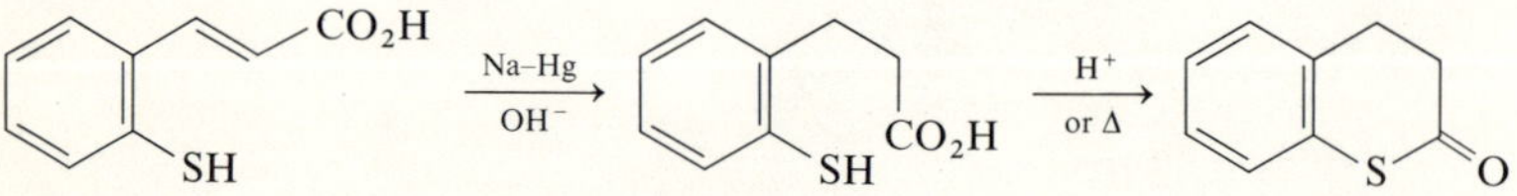

SCHEME 20

An unusual approach to the system is outlined in Scheme 21.[97]

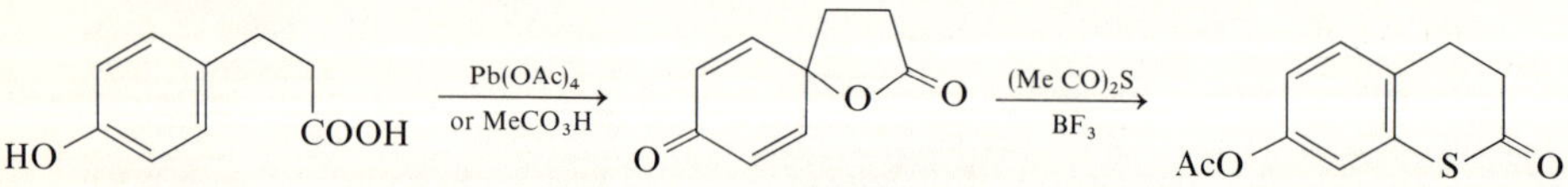

SCHEME 21

Another unexpected example which proceeds quantitatively is shown below (Eq. 6), the intermediate being of uncertain structure.[74]

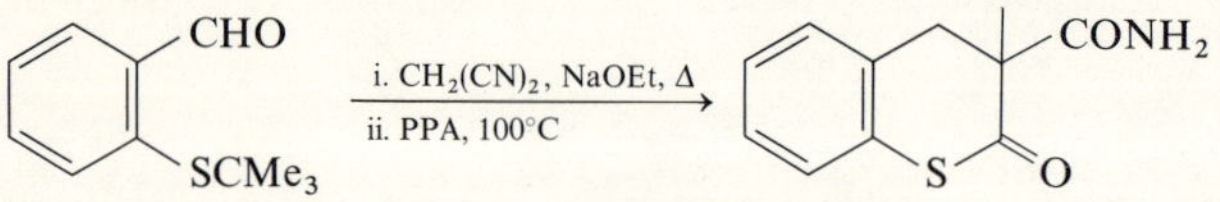

$$\quad (6)$$

[96] S. Scheithauer and R. Mayer, *Z. Chem.* **9**, 59 (1969).
[97] J. S. Davies, C. H. Hassall, and J. A. Schofield, *Chem. Ind.* (*London*), 740 (1963).

Sec. VII] THIOCOUMARINS

The dihydro compounds behave as typical thiolactones and are easily ring-opened with base. The 3-benzoyl-3,4-dihydrothiocoumarin (**53**, see p. 125) becomes, on ring opening, a useful intermediate to other systems (Scheme 22).[23] Under acid-catalyzed conditions, this same benzoyl derivative rearranges, presumably via a ring-opened intermediate, to a thiochromene (Eq. 7).[23] Thiochromenes are also derived from the action of a Grignard reagent on a dihydrothiocoumarin (Scheme 23).

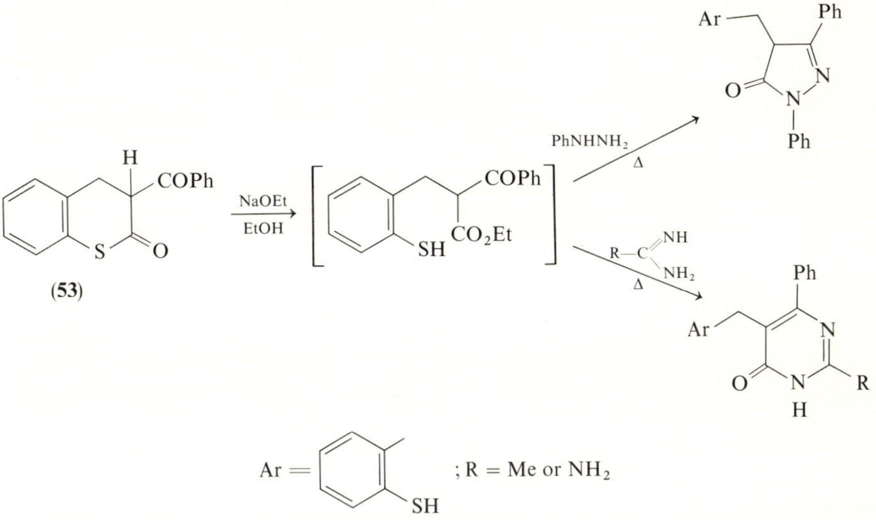

SCHEME 22

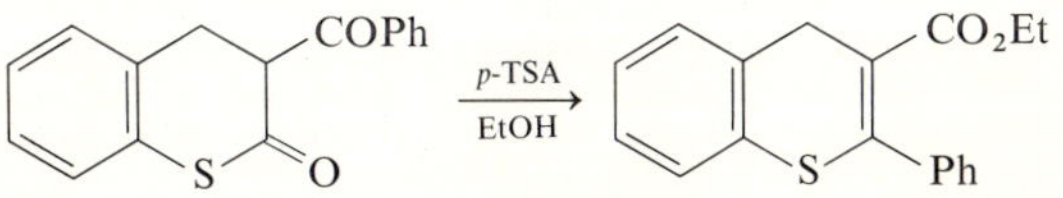

(7)

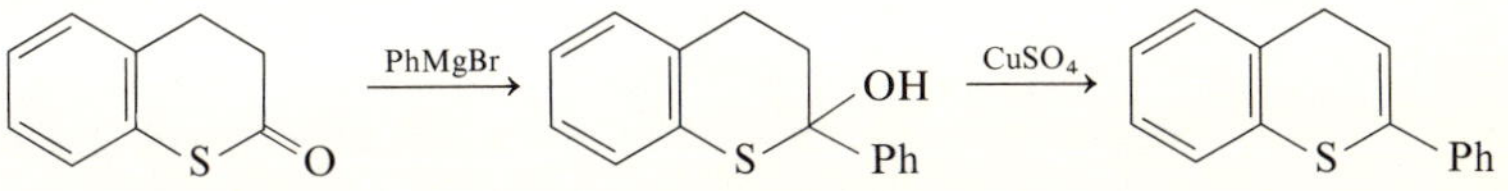

SCHEME 23

Benzo[c]furans

WILLY FRIEDRICHSEN

Institut für Organische Chemie der Universität Kiel, Kiel, West Germany

I. Introduction	135
II. Theoretical Aspects	137
III. Benzo[c]furan and Its Alkyl- and Monoaryl-Substituted Derivatives	142
A. Trapping Experiments	142
B. Syntheses	151
IV. 1,3-Diarylbenzo[c]furans	161
A. Syntheses	161
1. Reaction of Arylphthalides with Arylmagnesium Halides	162
2. Reduction of Diaroylbenzenes	163
3. Ring Closure of Diaroylcyclohexadienes	166
4. Rearrangement of Diarylphthalins	167
5. Further Syntheses	172
6. Compilation of Known 1,3-Diarylbenzo[c]furans	182
B. Diels–Alder Reactions	182
C. Higher Cycloaddition Reactions	191
D. Reactions with Singlet Oxygen	194
1. Preparative Aspects	194
2. Mechanistic Aspects	197
E. Photochemical Reactions	204
F. Other Reactions	209
G. Luminescence, Electroluminescence, and Lasing Properties	211
V. Spectroscopic Properties	215
A. UV Spectra	215
B. ^{1}H-NMR Spectra	217
C. Mass Spectra	217
D. Photoelectron Spectra	218
VI. Cyclobuta[c]furans and Cyclobutabenzo[c]furans	218
VII. Benz-Annelated and Hetero-Substituted Benzo[c]furans and Larger Ring[c]-Fused Furans	219
VIII. Benzo[c]furan-4,7-diones	234

I. Introduction

The chemistry of the *o*-quinonoidal heterocycles benzo[c]indole (isoindole, **1**) and benzo[c]thiophene (**2**) has been previously reviewed in this series.[1,2] The analogous selenium compound, benzo[c]selenophene (3), has only recently been synthesized.[3] In this article the chemistry of benzo[c]furan (4),[4-6] benz-annelated and hetero-substituted derivatives, and benzo[c]-furan-4,7-diones (5 and derivatives) will be presented. The literature is covered up to mid-1978.

The currently accepted name for **4** in *Chemical Abstracts* (*CA*) is isobenzofuran; the systematic name is benzo[c]furan. It should be mentioned that L. F. Fieser strongly advocated[7] and used[8-10] the spelling "isobenzofurane." The numbering of the system is as shown in **4**; in the older literature, numbering as in **6** has been used.[11]

The first synthesis of a stable benzo[c]furan, namely 1,3-diphenylbenzo[c]furan (**7a**), was reported in 1905.[12] As pointed out by Blicke *et al.*,[13-15]

---

[1] J. D. White and M. E. Mann, *Adv. Heterocycl. Chem.* **10**, 113 (1969).
[2] B. Iddon, *Adv. Heterocycl. Chem.* **14**, 331 (1972).
[3] L. E. Saris and M. P. Cava, *J. Am. Chem. Soc.* **98**, 867 (1976).
[4] E. Bergmann, *Bull. Soc. Chim. Fr.*, 19 (1948).
[5] R. C. Elderfield, *in* "Heterocyclic Compounds" (R. C. Elderfield, ed.), Vol. 2 Wiley, New York, 1951.
[6] M. J. Haddadin, *Heterocycles* **9**, 865 (1978).
[7] L. F. Fieser and M. Fieser, "Style Guide for Chemists." Reinhold, New York, 1960.
[8] L. F. Fieser and M. J. Haddadin, *J. Am. Chem. Soc.* **86**, 2081 (1964).
[9] L. F. Fieser and M. J. Haddadin, *Can. J. Chem.* **43**, 1599 (1965).
[10] L. F. Fieser and M. Fieser, "Reagents for Organic Synthesis," Vols. 1, 2, 3, 4. Wiley, New York, 1967, 1969, 1972, 1974, resp.
[11] M. M. Richter, "Lexikon der Kohlenstoffverbindungen," Part III, p. 3840. Voss, Leipzig, 1911.
[12] A. Guyot and J. Catel, *C. R. Hebd. Seances Acad. Sci., Ser. C* **140**, 1348 (1905); *Bull. Soc. Chim. Fr.* [3] **35**, 1124 (1906).
[13] F. F. Blicke and O. J. Weinkauff, *J. Am. Chem. Soc.* **54**, 1454 (1932).
[14] F. F. Blicke and R. A. Patelski, *J. Am. Chem. Soc.* **58**, 273 (1936).
[15] F. F. Blicke and R. A. Patelski, *J. Am. Chem. Soc.* **58**, 276 (1936).

von Baeyer[16] worked with 1,3-diaryl-substituted benzo[c]furans **7b–e**, evidently without knowing it.

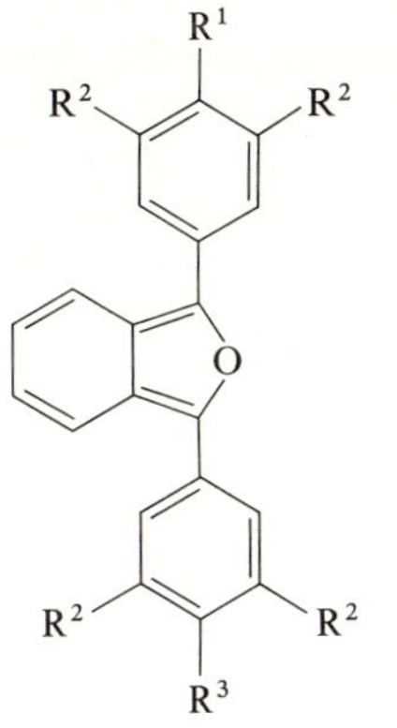

a: $R^1 = R^2 = R^3 = H$
b: $R^1 = R^3 = Cl; R^2 = H$
c: $R^1 = R^3 = OH; R^2 = H$
d: $R^1 = R^3 = OH; R^2 = Br$
e: $R^1 = R^3 = OAc; R^2 = Br$
f: $R^1 = OH; R^2 = R^3 = H$

(7)

They were formulated by him as anthranols. Also, von Pechmann described a "monooxyphenylanthranol"[17] which presumably[18] was a mixture of **7f** and 9-(*p*-hydroxyphenyl)anthrone (see Section IV,A,4). Now many 1,3-diarylbenzo[c]furans are known (Section IV,A,6); the preparation of the parent compound[19,20] (**4**) and of simple derivatives has also been described (Section III,B).

II. Theoretical Aspects

The parent compound (**4**), a derivative of *o*-quinodimethane, is unstable; in solution it reacts almost instantaneously with typical dienophiles like maleic anhydride and methyl vinyl ketone to give the corresponding Diels–Alder adducts (Section III). Simple HMO calculations show that the total π-electron energy of **4** is smaller than that of benzo[*b*]furan (**8**). It can be

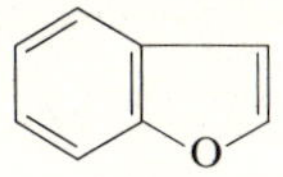

(8) 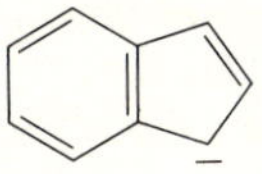   (9)

---

[16] A. von Baeyer, *Justus Liebigs Ann. Chem.* **202**, 36 (1880).
[17] H. von Pechmann, *Ber. Dtsch. Chem. Ges.* **13**, 1608 (1880).
[18] F. F. Blicke and R. J. Warzynski, *J. Am. Chem. Soc.* **62**, 3191 (1940).
[19] R. N. Warrener, *J. Am. Chem. Soc.* **93**, 2346 (1971).
[20] D. Wege, *Tetrahedron Lett.*, 2337 (1971).

shown easily by perturbation theory that the difference between these energies (ΔX_π) depends on the heteroatom parameters employed. Substitution of a methine group in the isoconjugate benzocyclopentadienyl anion (**9**) by an oxygen atom in position 1 or 2 leads to an expression for ΔX_π as in Eq. (1), where h and k have their usual meanings.

$$\Delta X_\pi = (q_1 - q_2)h + 2(p_{1,7a} - p_{1,2})(k - 1) + (\pi_{1,1} - \pi_{2,2})h^2/2$$
$$+ (\pi_{1,7a;1,7a} - \pi_{1,2;1,2})(k - 1)^2 + 2(\pi_{1,7a;1} + \pi_{1,2;1} - 2\pi_{1,2;2})h(k - 1) \quad (1)$$

From tabulated values of π-electron densities (q_i), bond orders ($p_{i,j}$), atom–atom ($\pi_{i,i}$), bond–bond ($\pi_{i,j;i,j}$), and bond–atom polarizabilities ($\pi_{i,j;k}$), ΔX_π has been calculated. These values (Table I) resemble those obtained by direct solution of the Hückel matrices.

TABLE I
ΔX_π VALUES[a] (IN β) AS A FUNCTION OF HETEROATOM PARAMETERS

h	2.0	1.5	2.0	2.0	2.0[b]
k	1.2	0.8	0.8	0.7	0.34
$\Delta X_\pi^{(1)c}$	0.2428	0.2472	0.3172	0.3358	0.4028
$\Delta X_\pi^{(2)d}$	−0.0714	0.0564	0.0726	0.1094	0.2446
ΔX_π	0.1714	0.3018	0.3898	0.4452	0.6474
ΔX_π^e	0.1646[f]	0.2523	0.2912	0.3247	0.4299

[a] As defined in Eq. (1).
[b] See Hess et al.[23]
[c] First-order perturbation.
[d] Second-order perturbation.
[e] Calculated from Hückel matrices.
[f] See V. Horak, C. Parkanyi, J. Pecka, and R. Zahradnik, *Collect. Czech. Chem. Commun.* **32**, 2272 (1967).

HMO calculations with $\alpha_O = \alpha_C + 2.50\beta$ and a bond-distance–bond-order relationship as in Eq. (2) (n_i, n_j: row number of element i, j; Z_i, Z_j: effective

$$d_{i,j} = 0.371 + 0.3181(n_i + n_j) - 0.1477(Z_i/n_i + Z_j/n_j)$$
$$- [0.02 + 0.0523(Z_i/n_i + Z_j/n_j)]p_{i,j} \quad (2)$$

nuclear charges, calculated after Slater's rules) together with Eq. (3) also showed that **8** is more stable than **4**.[21]

$$\beta = \beta_0 \exp[-5.075(d - 1.396)] \quad (3)$$

[21] A. Julg and R. Sabbah, *C. R. Hebd. Seances Acad. Sci., Ser.* **285**, 421 (1977); A. Julg and O. Julg, *Theoret. Chim. Acta* **22**, 353 (1971).

For a number of heterocycles, resonance energies per π-electron (REPE) have been calculated by a modified HMO method[22]; Coulomb and resonance integrals were estimated from heats of atomization.[23] Whereas benzo[b]furan (8) has a substantial REPE (0.036β), benzo[c]furan (4) shows a very small value (0.002β). The resonance energies obtained in this way are—as in other cases—in good agreement with the experimental properties of these compounds. Similar conclusions concerning the stability and reactivity of 4 and 8 have been reached through a graph theoretical concept; in most cases the topological resonance energy (per electron) [TRE(PE)] parallels REPE.[24-26] Theoretical studies using Dewar's variant of the MO procedure of Pople have shown[27] that 4, with a resonance energy (E_R) of 2.4 kcal mol^{-1}, is almost nonaromatic, and that the addition of reagents to the 1 and 3 positions should lead to a great increase (about 18 kcal mol^{-1}) in E_R. The bond lengths of 4 have been calculated to be close to polyene values. Klasinc et al., using the semiempirical SCF–MO method of Pariser and Parr, made the qualitative predictions[28] that (1) positions 1 and 3 are most reactive to the attack of radical and electrophilic reagents; (2) positions 4 and 7 are most reactive to nucleophilic reagents, and (3) that addition reactions will most probably occur at bonds with the highest bond-order values (4,5). The authors also calculated the singlet–singlet transition energies and oscillator strengths with a configuration interaction procedure for 4 and some other molecules; a value of 3.75 eV ($f = 0.37$) was obtained for the longest wavelength transition (exp 3.62).[20] Recently, two nonempirical calculations for 4[29,30] and 8[29,31] (and hetero analogs) were published. When the differences in the total energies between 8 and 4 are compared, the correct trends (0.025454 au[29] and 0.13753 au,[30] respectively) are obtained, although these figures differ considerably. In the latter work,[30] a value of 30.3 kcal mol^{-1} for the resonance energy (E_R) of 4 was obtained, compared with 56.0 kcal mol^{-1} for 8; it should be noted, however, that there was a change of definition of E_R. The instability of 4 is attributed to a small HOMO–LUMO separation (compared, e.g., with 8),[29] to the low E_R and to a low-lying triplet state.

---

[22] B. A. Hess and L. J. Schaad, J. Am. Chem. Soc. 93, 905 (1972).
[23] B. A. Hess, L. J. Schaad, and C. W. Holyoke, Tetrahedron 28, 3657 (1972).
[24] J. V. Knop, N. Trinajstić, I. Gutman, and L. Klasinc, Naturwissenschaften 60, 475 (1973).
[25] I. Gutman, M. Milun, and N. Trinajstić, J. Am. Chem. Soc. 99, 1692 (1977).
[26] A. Graovac, I. Gutman, and N. Trinajstić, "Topological Approach to the Chemistry of Conjugated Molecules." Springer-Verlag, Berlin and New York, 1977.
[27] M. J. S. Dewar, A. J. Harget, N. Trinajstić, and S. D. Worley, Tetrahedron 26, 4505 (1970).
[28] L. Klasinc, E. Pop, N. Trinajstić, and J. V. Knop, Tetrahedron 28, 3465 (1972).
[29] J. Koller, A. Azman, and N. Trinajstić, Z. Naturforsch., Teil A 29, 624 (1974).
[30] M. H. Palmer and S. M. F. Kennedy, J. C. S., Perkin 2, 81 (1976).
[31] M. H. Palmer and S. M. F. Kennedy, J. C. S., Perkin 2, 1893 (1974).

TABLE II
Charge Densities in 4

				Method						
							Nonempirical[29]		Nonempirical[30]	
Atom	HMO[a,b]	Dewar[27,b]	PP[28]	CNDO/2[b,c]	CNDO/2[c,d]	INDO[b,c]	INDO[c,d]	b	d	b	d
2	1.7603	1.9852	1.7585	1.7298	8.1338	1.7095	8.1643	1.6728	8.1898	1.7079	8.4795
3	1.0320	0.9660	1.0724	1.0932	5.9249	1.0902	5.8765	1.1096	5.9610	1.0090	5.8986
3a	1.0562	1.0380	1.0333	1.0312	6.0104	1.0421	6.0198	1.0434	6.0331	1.0530	6.0603
4	1.0062	1.0005	1.0048	1.0048	6.0014	1.0047	5.9801	1.0042	6.0599	1.0009	6.1351
5	1.0254	1.0029	1.0102	1.0058	5.9951	1.0083	5.9790	1.0065	6.0646	1.0022	6.1488

[a] W. Friedrichsen ($h = 2.0$; $k = 0.8$).
[b] π-Electron density.
[c] W. Friedrichsen and I. Schwarz, unpublished.
[d] Total electron density.

In Table II, charge densities are given and compared with those values which have been obtained by the aforementioned methods. As can be seen, CNDO/2, INDO, and nonempirical calculations show an enhancement of charge density on the oxygen atom. This phenomenon represents a sum of two effects: a gain from the σ-orbitals and a loss from the π-orbitals.

Table III compares the bond orders calculated both by the HMO and by the semiempirical methods. The geometry of a molecule may be crucial to quantum chemical calculations[32]; it is not known for **4**. Whereas Dewar et al.[27] used an iterative procedure (correlation between bond lengths and bond orders), other workers[29] preferred standard geometry[33] or a combination of naphthalene and furan values.[30]

TABLE III
BOND ORDERS AND BOND LENGTHS CALCULATED FOR **4**

Bond	HMO[a]	HMO[b,c]	Dewar[27,b]	PP[28]	CNDO/2[d]
1–2	0.4171	1.34	1.378	0.4123	
3–3a	0.6920	1.37	1.355	0.7102	0.8012
3a–7a	0.4437	1.45	1.459	0.4562	0.3896
3a–4	0.5044	1.47	1.462	0.4659	0.3653
4–5	0.7576	1.35	1.353	0.7889	0.8699
5–6	0.5621	1.46	1.454	0.5289	

[a] W. Friedrichsen ($h = 2.0$; $k = 0.8$).
[b] Bond lengths (Å).
[c] See Julg et al.[21]
[d] W. Friedrichsen and I. Schwarz, unpublished.

In a rotational and vibrational analysis of the first singlet transition of benzo[c]furan, the CNDO/S method was used to calculate the bond orders in the ground and excited state[34]; these values were transformed into changes in bond length using an empirical relation described by Coulson.[35] Also, the set of angle changes was calculated in a manner that minimizes the strain energy in the excited state.

The electronic structure of **4** (and the other o-quinonoid heterocycles **1**, **2**, and **3**) has been discussed in the light of NMR data.[36] For a series of hydrocarbons containing formally cisoid butadiene fragments (**10**; R^1 and R^2 not explicitly stated), the ratio of the vicinal coupling constants $J_{ratio} (= J_{4,5}/J_{5,6})$

[32] H. M. Niemeyer, *Tetrahedron* **33**, 1369 (1977).
[33] J. A. Pople and M. Gordon, *J. Am. Chem. Soc.* **89**, 4523 (1967).
[34] M. J. Robey and I. G. Ross, *Can. J. Phys.* **53**, 1814 (1975).
[35] C. A. Coulson, *Proc. R. Soc. London, Ser. A* **169**, 413 (1939).
[36] E. Chacko, J. Bornstein, and D. J. Sardella, *J. Am. Chem. Soc.* **99**, 8248 (1977).

(10)   (11)   (12)

was plotted versus the ratio of bond orders P_{ratio} ($= p_{4,5}/p_{5,6}$); from the linear correlation obtained ($J_{ratio} = 0.954 \cdot P_{ratio} + 0.037$; $r = 0.992$) a value of J_{ratio} for the model compound **11** (P_{ratio} from HMO data[37]) can be calculated ($J_{ratio} = 0.703$; 0.71^{36}). This value is very close to the observed J_{ratio} for **4** (0.73; 0.70^{36}). In the opinion of the authors therefore, representation **12** with bond fixation in the carbocyclic ring best depicts the structures of the o-quinonoid heterocycles. However, at least in the case of isoindole (**1**) the correlation between REPE and J_{ratio} seems to indicate delocalization throughout both rings.[38]

Recent work reports that the stability of the sulfur heterocycles increases from **2** to **14**[39]; analogous investigations in the benzo[c]furan series would be especially interesting.

(2)   >   (13)   >   (14)

III. Benzo[c]furan and Its Alkyl- and Monoaryl-Substituted Derivatives

A. Trapping Experiments

In 1956 the first report appeared of the possible generation of **4** in a retro Diels–Alder reaction.[40] When **16** (from **15**[41] and diazomethane) was heated in the presence of Cu powder to 180–200°C, in addition to pyrazole (**17**) a

---

[37] C. A. Coulson and A. Streitwieser, "Dictionary of π-Electron Calculations." Freeman, San Francisco, California, 1965.
[38] B. A. Hess and L. J. Schaad, *Tetrahedron Lett.*, 535 (1977).
[39] H. Hart and M. Sasaoka, *J. Am. Chem. Soc.* **100**, 4326 (1978).
[40] G. Wittig and L. Pohmer, *Chem. Ber.* **89**, 1334 (1956).
[41] An improved procedure for the preparation of **15** has been described: H. Wynberg, J. de Wit, and H. J. M. Sinnige, *J. Org. Chem.* **35**, 711 (1970).

Sec. III.A] BENZO[c]FURANS 143

resinous residue was obtained, which was formulated as a polymer of **4** (**18**) (for further examples see the following).

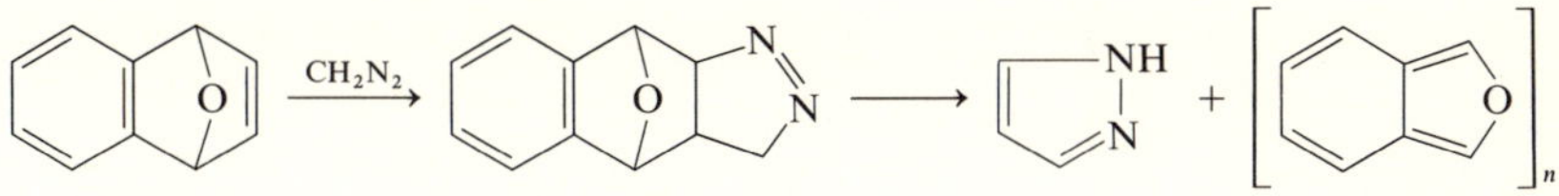

(15)   (16)   (17)   (18)

In 1964, the transient existence of **4** was proved unequivocally by Fieser and Haddadin.[8,9] The Diels–Alder adduct of **15** and tetracyclone (**19**; the stereochemistry of the carbonyl bridge is not known with certainty) when refluxed in diglyme in the presence of **15** delivered **4**, which was trapped as Diels–Alder adducts **21** (mp 265°C) and **22** (mp 175–176°C); these compounds were isolated in a ratio of 1:1.4 in nearly quantitative yield. The

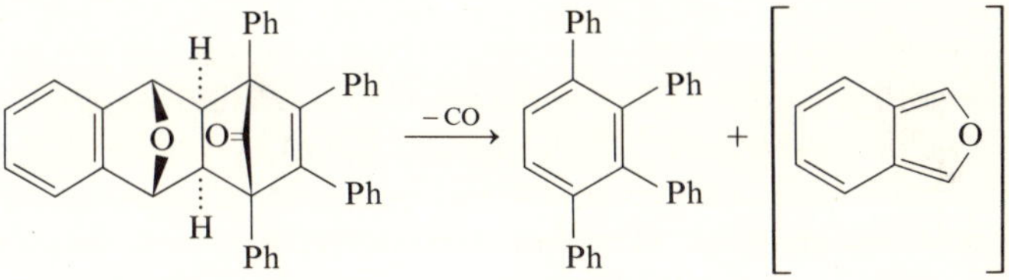

(19)   (20)   (4)

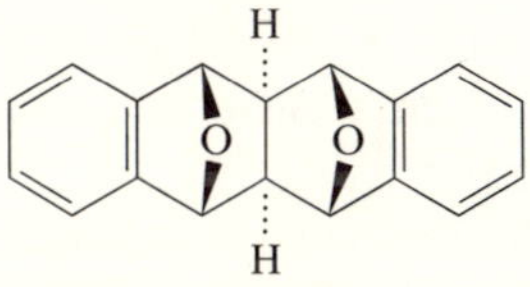

(21) 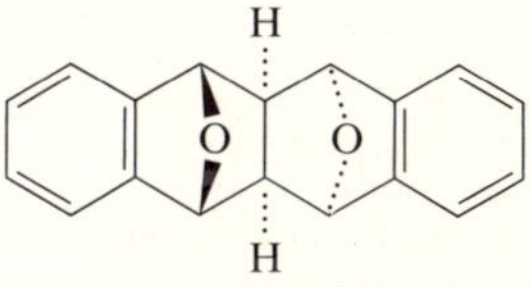   (22)

stereochemistry has been proved as in analogous cases by ^{1}H-NMR spectroscopy. In the same way **23**, which is available from α-pyrone and **15**, yielded **21** and **22** in a ratio of 1:1. When the decomposition was carried out in the presence of **24**, adducts **25** (mp 268–269°C) and **26** (mp 144–146°C) were formed (ratio 7.7:1). The intermediate **27** is highly reactive and could not be isolated.[42] This compound could be trapped, however, when **23** was

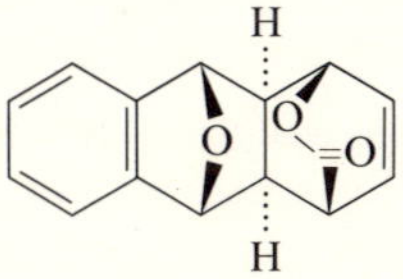

(23) 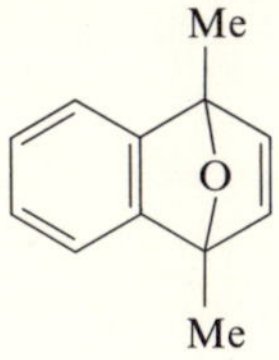   (24)

[42] R. McCulloch, A. R. Rye, and D. Wege, *Tetrahedron Lett.*, 5231 (1969).

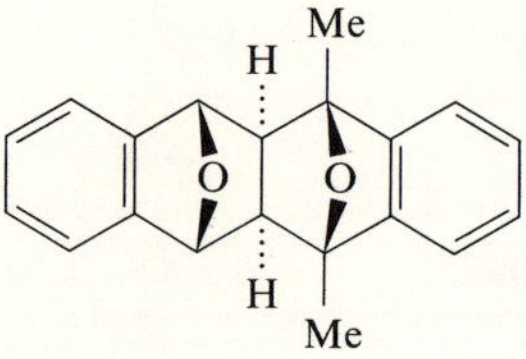

(25) 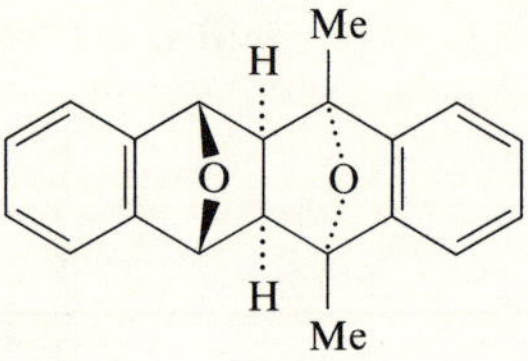   (26)

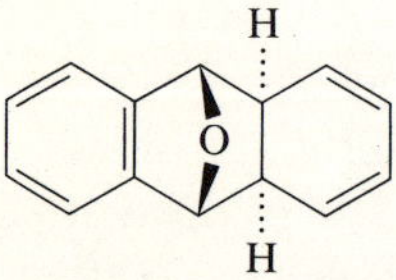

(27) 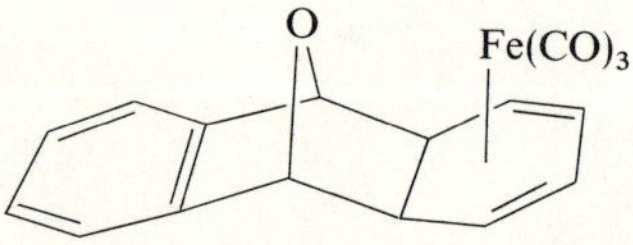   (28) 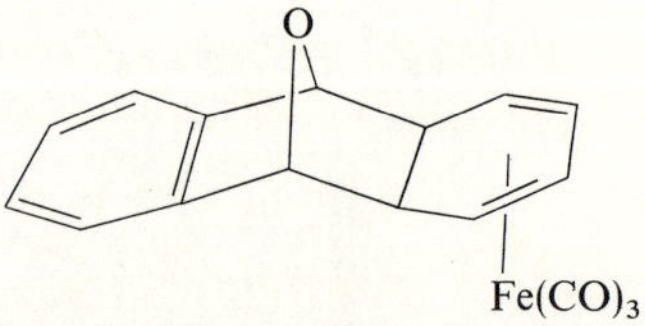   (29)

heated in toluene in the presence of $Fe_3(CO)_{12}$; two stereoisomeric carbonyl complexes with mp 174–175°C and 147–151°C (**28, 29**) were isolated in 6 and 1.2% yield, respectively.[43,44] Later, the isolation of **27** was reported[20]; when the Diels–Alder adduct of **15** and α-pyrone, which was formulated as in **30**, was absorbed on celite and sublimed at 10^{-2} mm through a tube heated to 130°C, diene **27** condensed as colorless crystals (6%; mp 93–96°C) on the cooler parts of the tube. It was identified from its spectral data and characterized as a Diels–Alder adduct with N-phenyltriazolinedione; **27** decomposes in dilute cyclohexane solution in a clean first-order reaction ($\Delta H^{\ddagger} = 25.9$ kcal mol^{-1}, $\Delta S^{\ddagger} = +3.7$ eu†). It was suggested that a symmetry-allowed, concerted cycloreversion to **4** and benzene takes place.[20]

In a manner analogous to that described for **19** and **23**, thermal decomposition of **31** and **32** yielded 1,3-dimethylbenzo[c]furan (**33**) as a transient species; it was trapped with **15** to give **25** and **34** (mp 159–160°C). When **31** was decomposed in the presence of **24**, only the exo–endo isomer **35** (mp

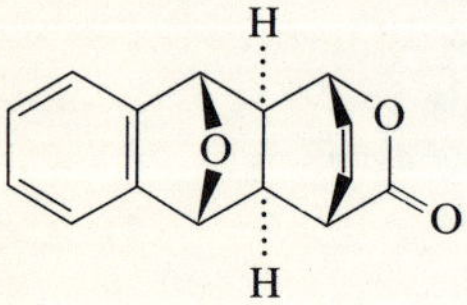

(30) 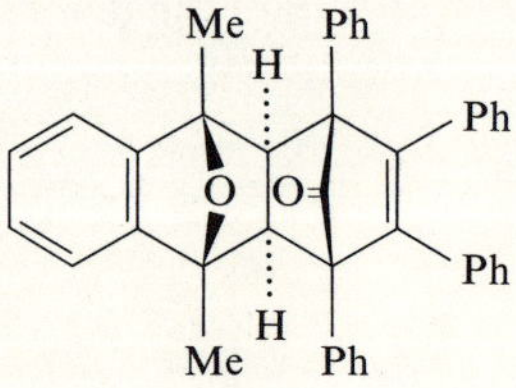   (31) 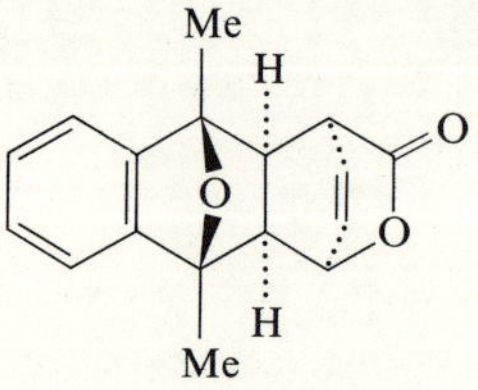   (32)

[43] W. Friedrichsen and E. Loock, unpublished.
[44] E. Loock, Diplomarbeit, Universität Kiel (1970).

† Recalculated from the data of Wege.[20]

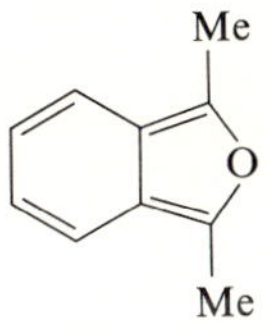

Sec. III.A] BENZO[c]FURANS 145

(33) 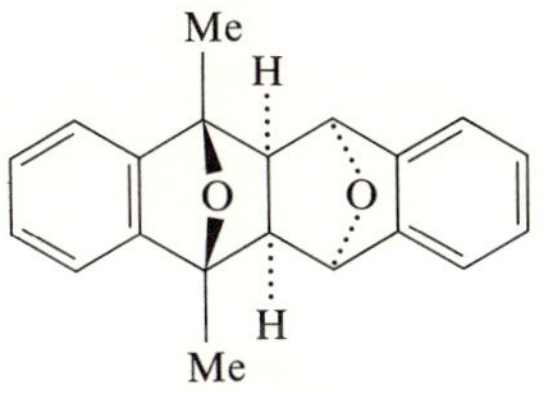 (34) 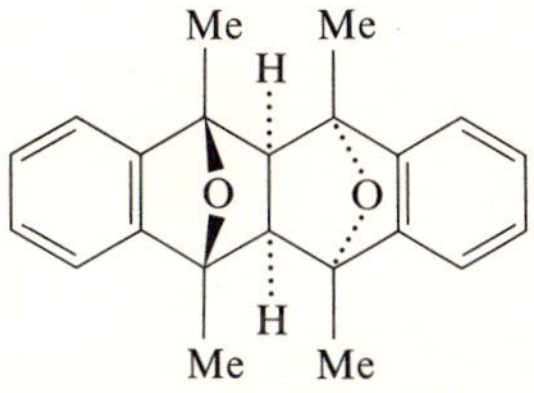 (35)

202–203°C) was formed; models show that the methyl groups exert a substantial steric hindrance to an exo–exo approach.

Since then, the transient generation of **4** through the retro Diels–Alder pathway has been used a number of times for synthetic purposes. Decomposition of **36** in the presence of dimethyl fumarate gave **37** (mp 66°C); with dimethyl maleate **38** (mp 105°C) and **39** (mp 146–147°C) were obtained (ratio, **38/39** = 2:1).[42] Decomposition of **19** in the presence of **40** (R = H)

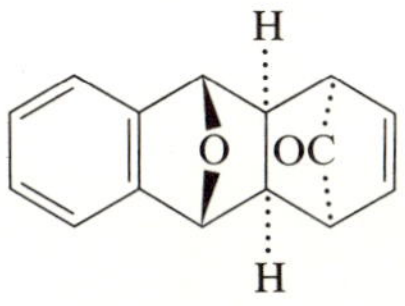

(36) 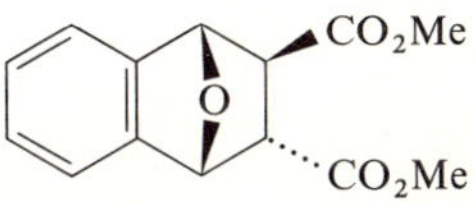 (37)

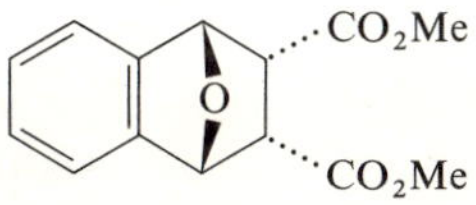

(38) 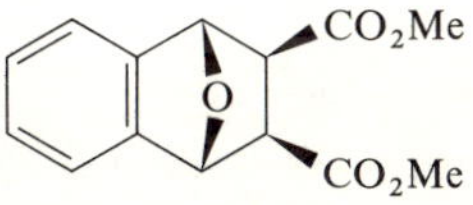 (39)

gave the exo and endo adducts (**41**, R = H, mp 218–219°C; **42**, mp 165–167°C) in a total yield of 90.7%[45]; with **40** (R = CH_3) only the exo adduct (**41**, R = CH_3, mp 207–208°C) was obtained.[46]

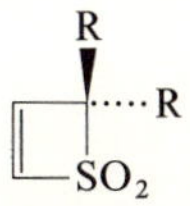

(40) 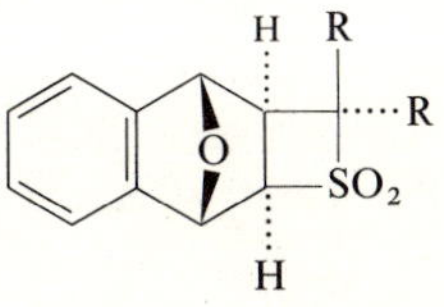 (41) 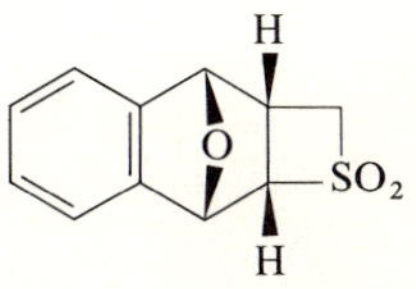 (42)

While attempts at acid-catalyzed dehydration with methanol/hydrogen chloride and acetic acid/hydrogen bromide left **41** (R = CH_3) unchanged,

---

[45] L. A. Paquette and T. R. Phillips, *J. Org. Chem.* **30**, 3883 (1965).
[46] L. A. Paquette, *J. Org. Chem.* **30**, 629 (1965).

treatment with polyphosphoric acid (2 days, room temperature) gave **43**; **41** (R = H) with polyphosphoric acid (100°C, 4–5 hr) yielded **44**.

In a manner analogous to the synthesis of **19**, the reaction of **15** with 2,5-dimethyl-3,4-diphenylcyclopentadienone gave **45**; photolysis at −45° yielded 1,4-dimethyl-2,3-diphenylbenzene, the diene **46**, and another photoproduct identified as **4**. When the photolysis was conducted in the presence of dimethyl fumarate, compound **37** was isolated in poor yield.[47]

Thermal decomposition of **23** in the presence of cyclopentene and cycloheptene gave the endo adducts **47** (mp 80–82°C, 52%) and **48** (49–50°C, 22%; TLC of the crude product indicated the possible presence of a very

[47] W. S. Wilson and R. N. Warrener, *Tetrahedron Lett.*, 5203 (1970).

Sec. III.A]   BENZO[c]FURANS   147

small quantity of the exo adduct)[48]; cycloheptatriene yielded both an endo (oily, 29%) and an exo isomer (82–83°C, 17%) as did norbornadiene (exo: 82°C, 46%; endo: 72–73°C, 26%; possibly **49** and **50** as indicated by the ^{1}H-NMR spectra; for comparison, see Cava and Scheel[49]).[48] Tropone gave an adduct with mp 206–208°C (35%)[48] (see the following).

In an attempt to realize $[\pi_4 + \pi_6]$-additions, the reaction of **4** with N-substituted azepines was investigated. Thermal decomposition of **19** in the presence of **51** (R = COOMe, COOEt, COMe, and SO$_2$Me) resulted in the formation of the Diels–Alder adducts **52**.[50] In xylene with tetracyanoethyl-

(51)   (52)   (53)

R = CO$_2$Me, CO$_2$Et, COMe, SO$_2$Me

ene, **19** yields **53** (mp 180°C, 204°C[19]).[51] Prolonged heating of **54**, which is available from **15** and furan, under drastic conditions (o-dichlorobenzene, 160°C) gave the isomer **55**.[52] Whether this isomerization occurs via (a) or (b) remains unclear; it seems improbable, however, that **4** and oxanorbornadiene (**56**) would resist such reaction conditions.

[48] D. W. Jones and G. Kneen, *J. C. S., Perkin 1*, 1647 (1976).
[49] M. P. Cava and F. M. Scheel, *J. Org. Chem.* **32**, 1304 (1967).
[50] L. A. Paquette, D. E. Kuhla, J. H. Barrett, and L. M. Leichter, *J. Org. Chem.* **34**, 2888 (1969).
[51] J. B. Bremner and Y. Hwa, *Aust. J. Chem.* **24**, 1307 (1971).
[52] T. Sasaki, K. Kanematsu, K. Hayakawa, and M. Uchide, *J. C. S., Perkin 1*, 2750 (1972).

Possible generation of **4** also occurs on heating of adducts **57** and **59**; 3-benzoylisoxazole (**58**) and 1-phenyltriazole (**60**) were isolated from the reaction mixtures.[52]

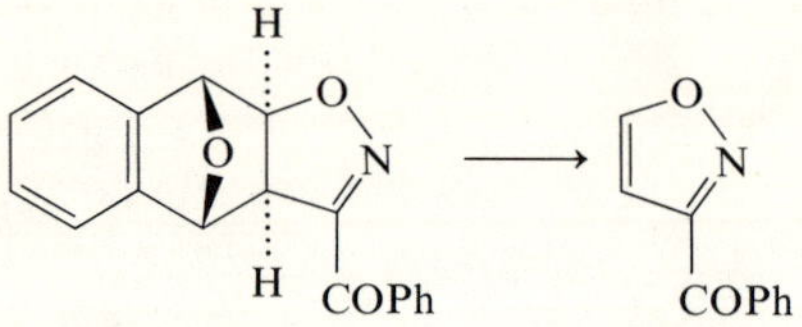

(57) (58)

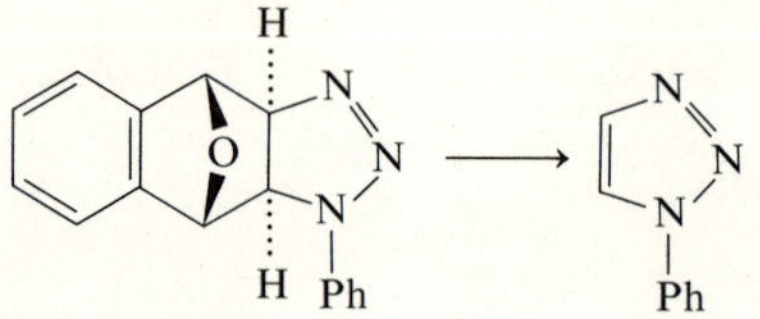

(59) (60)

Heating of **19** (xylene, 170°C) with **61** yielded isomers **62** (mp 294–295°C) and **63** (mp 300°C) in a total yield of 54% (ratio ~1:1)[53]; the oxanorbornadiene seems to react exclusively in an exo manner. In his paper on the synthesis of benzo[c]furan, Warrener[19] described the reaction of **15** with 3,6-dipyridyl-1,2,4,5-tetrazine (**64**). In a Diels–Alder reaction with inverse

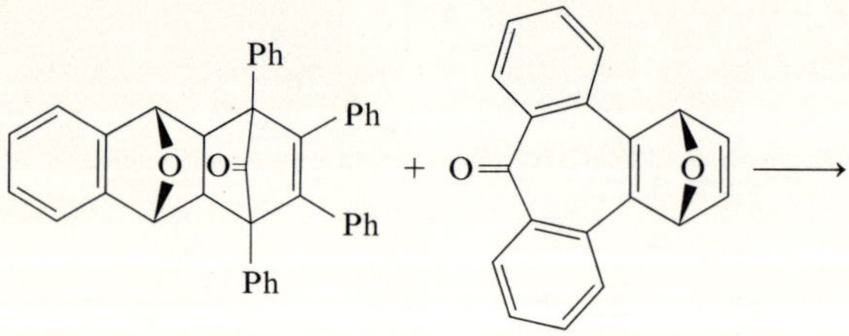

(19) (61)

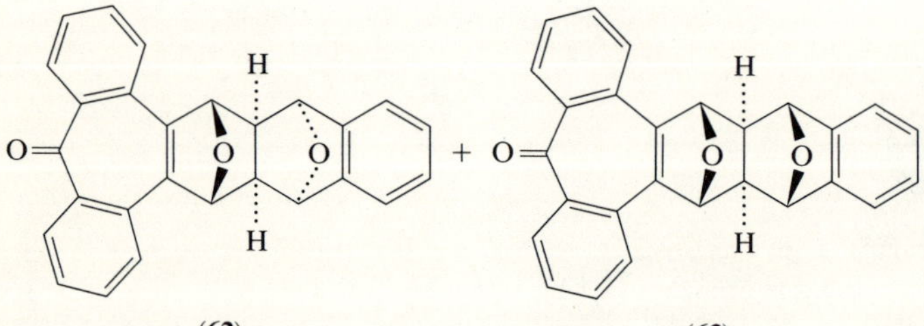

(62) (63)

[53] T. Sasaki. K. Kanematsu, K. Iizuka, and I. Ando, *J. Org. Chem.* **41**, 1425 (1976).

Sec. III.A] BENZO[c]FURANS 149

electron demand,[54] **65** is presumably formed first; loss of nitrogen leads to **66**, an unstable but isolable intermediate that decomposes to **4** and 3,6-dipyridylpyridazine (**67**). This pathway has been used by a number of workers to generate **4** in one step from **15** and **64** (method A) and to add it

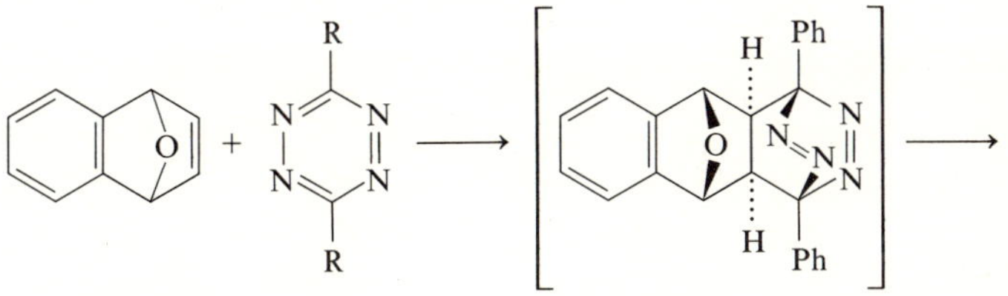

(15)   (64)   (65)

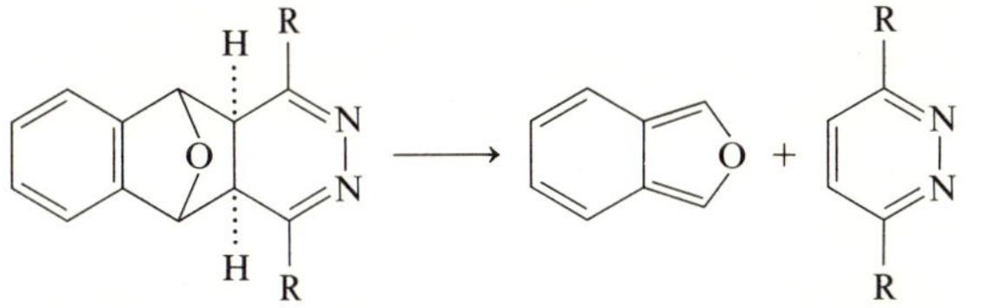

(66)   (4)   (67)

R = 2-pyridyl

to dienophilic compounds. In this way[55] the reaction of **15** and **64** in the presence of tropone gives the Diels–Alder products **68** (endo; mp 193–195°C, 4.5%) and **69** (exo-anti-endo; mp 179–180°C, 12%) and a ($\pi_4 + \pi_6$)-adduct (**70**, R = H; exo; mp 190–191°C, 12%). Whereas, when **4** is generated by thermal decomposition of **19** (method B), the products **68, 69**, and **70** (R = H) were isolated in 2, 1, and 36% yield, respectively. As mentioned earlier, under

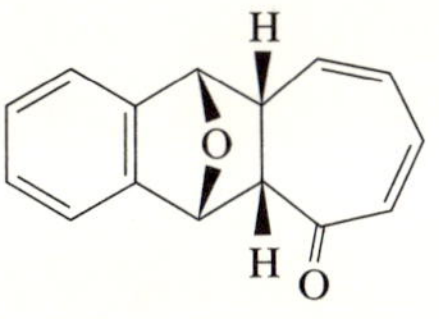

(68) 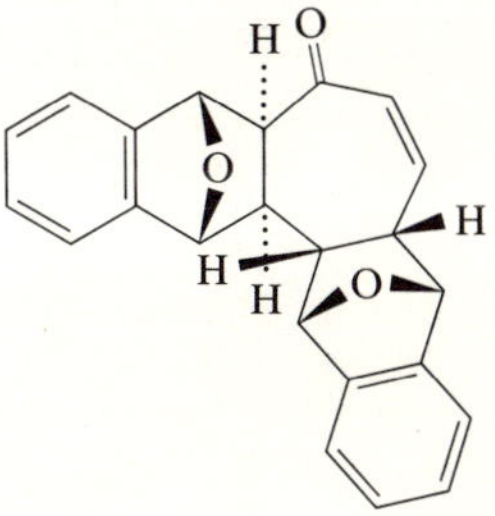   (69)

---

[54] A recent literature survey can be found in W. Friedrichsen and H. von Wallis, *Tetrahedron* **34**, 2509 (1978).
[55] H. Takeshita, Y. Wada, A. Mori, and T. Hatsui, *Chem. Lett.*, 335 (1973).

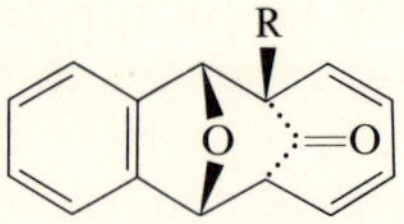

(70) 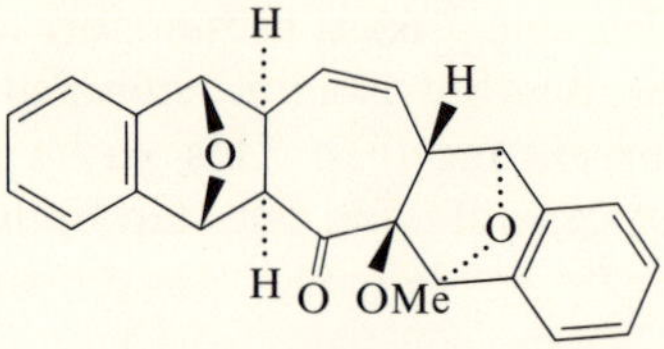   **(71)**

these conditions a product of mp 206–208°C was obtained,[48] whose structure has not been established. It seems that this compound is identical to **70**, although the melting point differs considerably from that obtained by Takeshita et al.[55] Since **68** is isomerized to **70** (R = H) (benzene, 160–165°C, 2 hr), the latter type of product is thermodynamically more stable than the former. When **4** is allowed to react with 2-chlorotropone, **70** (R = Cl, mp 173–174°C) was the only isolable product (method A: 22%; method B: 25%). 2-Methoxytropone reacted only under the conditions of method B; two adducts, **70** (R = OMe; mp 173–174°C, 17.5%) and **71** (exo-anti-exo; mp 264–265°C, 5.4%) were produced. When method A was employed, 2-methoxytropone was recovered quantitatively; 2-dimethylaminotropone did not react at all. Two Diels–Alder adducts, **72** (endo; mp 160–161°C, 48.1%) and **73** (endo-anti-endo; mp 250–251.5°C, 14.9%), were isolated when **19** was decomposed in the presence of 6,6-diphenylfulvene.[56] Compound **4**, generated by the tetrazine pathway, reacted with 6,6-dimethylfulvene to give a $[\pi_4 + \pi_6]$-adduct (**74**, colorless liquid), in addition to this, the already-known Diels–Alder adducts, **21** and **22** were isolated.

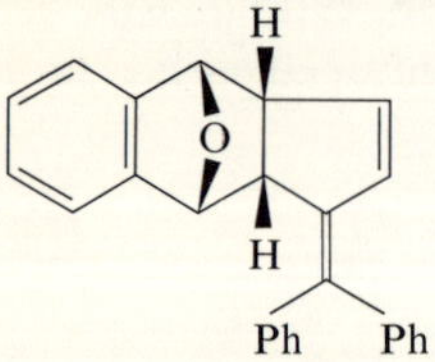

(72) 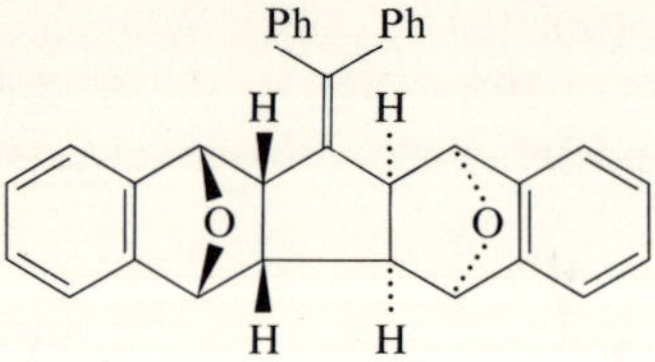   **(73)** 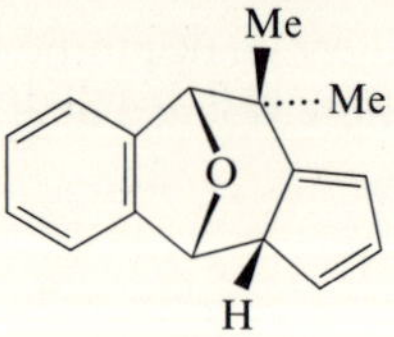   **(74)**

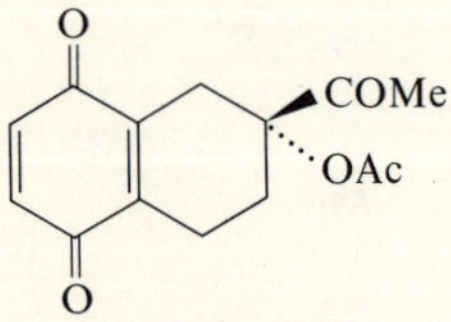

(75) 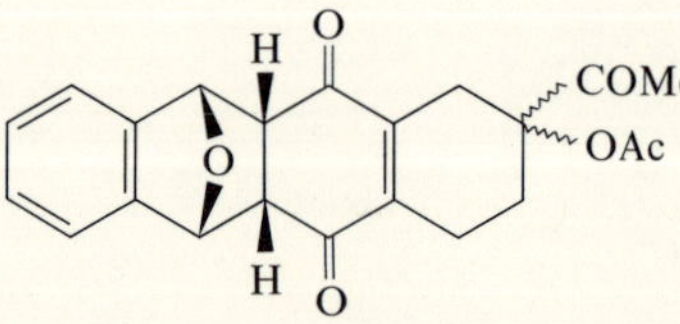   **(76)**

[56] H. Takeshita, A. Mori, S. Sano, and Y. Fujise, *Bull. Chem. Soc. Jpn.* **48**, 1661 (1975).

Sec. III.B] BENZO[c]FURANS 151

Recently, benzo[c]furan has been of value in the synthesis of anthracyclinones. As Kende et al.[57] found, **4**, generated from the precursor **23**, reacts with the quinone **75** (diglyme, 140°C) to give **76** and **77** in a total yield of 96% (ratio 3:1); dehydration of the mixture with acetic acid/sodium acetate (reflux, 16 hr) gives the desired product (**78**, 70%). Similar reaction sequences have been conducted with other benzo- and naphthoquinones.[58]

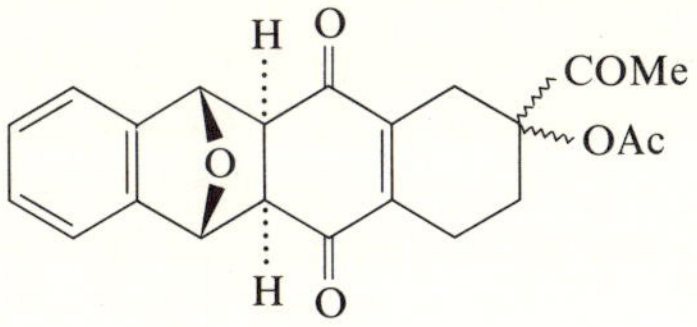

(77) 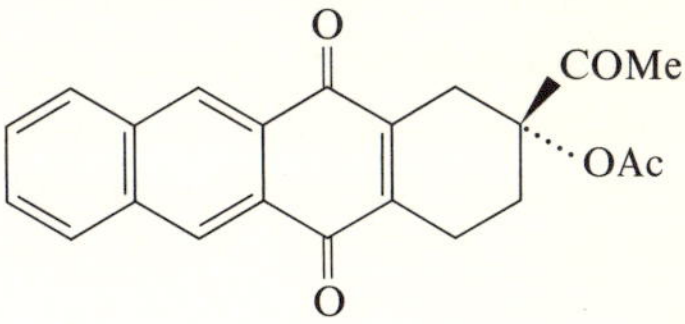   (78)

B. SYNTHESES

In 1971, two papers concerning the parent compound (**4**) appeared almost simultaneously. As was pointed out in Section III,A, Warrener[19] found that 3,6-dipyridyl-1,2,4,5-tetrazine (**64**) reacts with **15** to give **66**; a lack of coupling between the methine protons is consistent with the assigned stereochemistry. Compound **66** is quite stable in solution below −20°C but slowly decomposes on storage in the solid state. It reacted with N-methylmaleimide to give Diels–Alder adducts of **4**; namely **79** and **80** (not isolated; composition

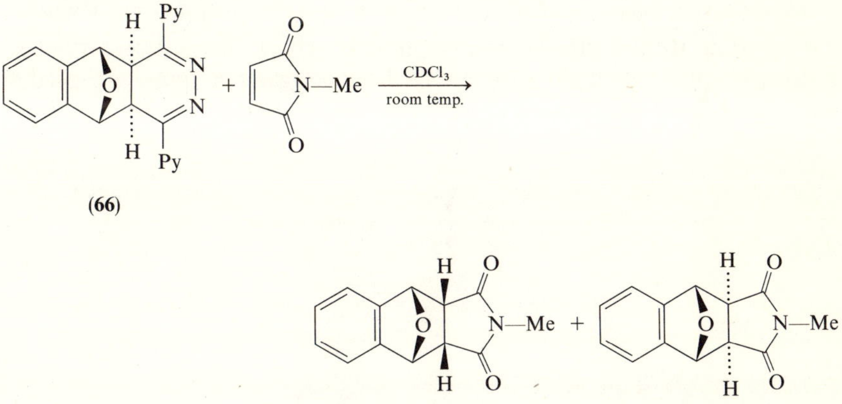

(66)

(79)   (80)

---

[57] A. S. Kende, D. P. Curran, Y. Tsay, and J. E. Mills, *Tetrahedron Lett.*, 3537 (1977).
[58] Y. Tsay, Ph.D. Thesis, University of Rochester (1977) (quoted in Kende et al.[57]).

determined by ^{1}H-NMR spectroscopy). When **66** was decomposed at 120°C and 0.1 torr, **4** could be isolated in a cold trap; the nonvolatile pyridazine (**67**) remained in the pot. Compound **4** was obtained as a colorless, crystalline solid with mp ~20°C; it was characterized as a Diels–Alder adduct with tetracyanoethylene. Also, Wege[20] succeeded in the preparation and isolation of **4**; as mentioned in Section III,A, the α-pyrone adduct **30** on heating gave **27** and **4**, which was isolated in 30% yield as a colorless solid melting on warming to room temperature. With dimethyl fumarate it reacts almost instantaneously to give **37**. In 1972, another synthesis of **4**[59] used 1,4-oxido-1,2,3,4-tetrahydronaphthalene (**81**) as starting material; pyrolysis at 650°C and 0.1 torr in an unpacked quartz tube resulted in quantitative formation of **4** as colorless crystals which melted at 20°C to a colorless or faintly yellow liquid with a typically disagreeable odor. This reaction sequence appears especially attractive for preparative purposes; larger quantities can be obtained in a relatively short time (10 g/hr). Whether this method is also superior with respect to the purity of **4** remains to be established: a product prepared by the tetrazine route was reported to be contaminated with the pyridazine **67**.[30]

$$(81) \xrightarrow{650°/0.1\ torr} (4) + C_2H_4$$

Wiersum and Mijs[59] found that in ice-cold ethereal solution **4** reacts instantaneously with maleic anhydride, *N*-phenylmaleimide, and methyl vinyl ketone to give **82**, **83**, and **84**, respectively, in quantitative yield as endo–exo mixtures (endo:exo = 3:1). Longer reaction times are needed with styrene and cyclohexene (2 hr, room temp.): compounds **85** and **86** are

(82)   (83)   (84)

(85)   (86)

[59] U. E. Wiersum and W. J. Mijs, *Chem. Commun.*, 347 (1972).

produced in 20–30% yield (homopolymerization occurs); with isobutene there is no reaction.

The retro Diels–Alder reaction has also been proved of value in a number of related cases; benzo[c]indoles,[60] isobenzofulvenes,[61,62] furans, and fulvenes[63] have been prepared by the tetrazine route, while the Wiersum–Mijs method has been used for the preparation of benzo[c]indole itself.[64]

Dilute oxygen-free solutions of **4** (10^{-4} *M*) are stable up to 70°C, but in concentrated solutions polymerization occurs at and below room temperature. An interesting phenomenon was observed with temperature change: whereas solutions are colorless at room temperature, they show a reversible yellow to pink coloration on cooling to dry ice/acetone (in $CHCl_3$ and CCl_4).[59]

The mass spectrum of **4** shows a strong parent peak ($m/e = 118$) as well as major peaks at $m/e = 90$ ($M^+ - CO$) and $m/e = 89$ ($90 - H$).[19] The ^{1}H-NMR spectrum is solvent-dependent; in DMSO-d_6 the furanoid protons appear as a singlet at 8.40 ppm.[19] In $CDCl_3$ (10°C) this signal is found at 8.00 ppm, whereas in CCl_4 it is shifted further upfield to 7.80 ppm.[59] The protons of the carbocyclic ring have been analyzed as an AA'BB' system (Section V).[30]

The UV spectrum has been measured in hexane[20] and cyclohexane[19]; other solvents also seem to have been used.[19] In the 305–343 nm region, the UV spectrum strongly resembles that of benzo[c]thiophene.[2,65] The lack of solvent dependence together with a mirror relationship to its fluorescence spectrum (in isopentane–methylcyclohexane 3:1 at 77°K) is indicative of the π,π^*-character of the lowest singlet transition.[19]

In the last few years, other simple benzo[c]furans have been prepared or at least detected or formulated as transients (see Table IV).[9,59,66–73]

---

[60] G. M. Priestley and R. N. Warrener, *Tetrahedron Lett.*, 4295 (1972).
[61] H. Tanida, T. Irie, and K. Tori, *Bull. Chem. Soc. Jpn.* **45**, 1999 (1972).
[62] P. L. Watson and R. N. Warrener, *Aust. J. Chem.* **26**, 1725 (1973).
[63] W. S. Wilson and R. N. Warrener, *Chem. Commun.*, 211 (1972).
[64] J. Bornstein, D. E. Remy, and J. E. Shields, *Chem. Commun.*, 1149 (1972).
[65] R. Mayer, H. Kleinert, S. Richter, and K. Gewald, *J. Prakt. Chem.* **20**, 244 (1963).
[66] E. Chacko, D. J. Sardella, and J. Bornstein, *Tetrahedron Lett.*, 2507 (1976).
[67] R. Faragher and T. L. Gilchrist, *J. C. S., Perkin 1*, 336 (1976).
[68] M. Hamaguchi and T. Ibata, *Chem. Lett.*, 287 (1976).
[69] L. Contreras, C. E. Slemon, and D. B. McLean, *Tetrahedron Lett.*, 4237 (1978).
[70] M. Avram, D. Constantinescu, I. G. Dinulescu, and C. D. Nenitzescu, *Tetrahedron Lett.*, 5215 (1969).
[71] I. G. Smith and R. T. Wikman, *J. Org. Chem.* **39**, 3648 (1974).
[72] F. J. Petracek, N. Sugisaka, M. W. Klohs, R. G. Parker, J. Bordner, and J. D. Roberts, *Tetrahedron Lett.*, 707 (1970).
[73] J. Rigaudy, M. C. Perlat, D. Simon, and N. K. Cuong, *Bull. Soc. Chim. Fr.*, 493 (1976); J. Rigaudy, A. Defoin, and J. Baranne-Lefort, *Angew. Chem.* **91**, 443 (1979); *Angew. Chem., Int. Ed. Engl.* **18**, 413 (1979).

TABLE IV
SIMPLE BENZO[c]FURANS

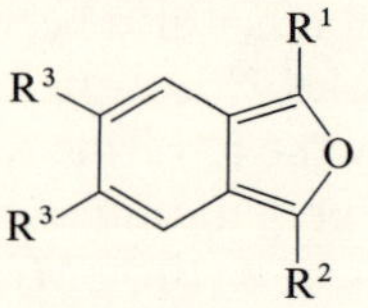

(87)

R^1	R^2	R^3	mp (°C)	References
H	Me	H	Crystalline	59, 66
H	PhCH$_2$	H	—	66
H	Ph	H	—	67
H	OMe	H	—	68
H	OEt	H	—	69
H	OEt	OMe	—	69
Me	Me	H	Crystalline	9, 59, 66
t-Bu	t-Bu	H	—	70
Me	Ph	H	—	67
t-Bu	Ph	H	Oily	71
CN	Ph	H	63–64	72
COOH	Ph	H	—	72
Me	o-HOC$_6$H$_4$	H	—	73
CO$_2$Me	OMe	H	—	68

The flash-pyrolysis technique has been especially useful for the synthesis of unstable benzo[c]furans; the 1-methyl, 1,3-dimethyl, and 1-benzyl derivatives have been prepared in this way in quantitative yield.[66] The UV and ^{1}H-NMR spectra are in accord with their structures. These compounds are extremely reactive, resinifying rapidly on standing at room temperature. With N-phenylmaleimide, the Diels–Alder adducts **88** (endo, mp 201–201.6°C, 18%), **89** (exo, mp 153–153.5°C, 14%) and **90** [endo, mp 215–217°C, 60% (includes both endo and exo)] are obtained; **87** ($R^1 = R^3 = H$, $R^2 =$

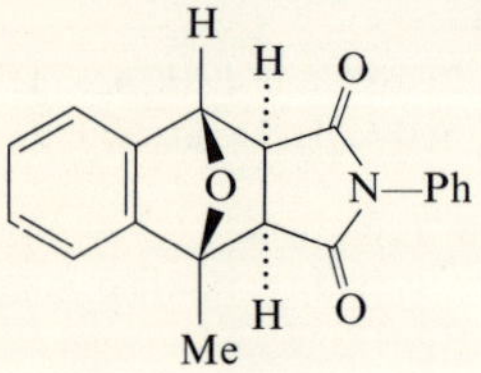

(88) 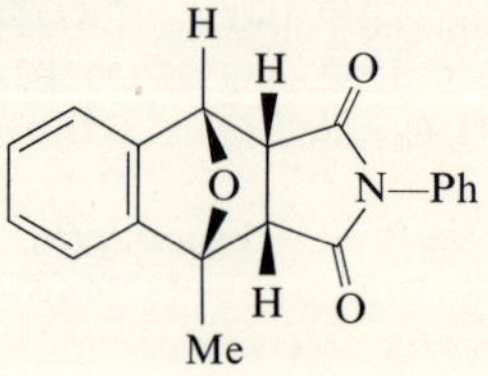   (89)

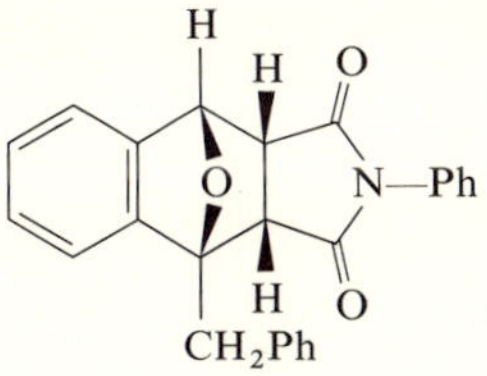

Sec. III.B] BENZO[c]FURANS 155

(90)

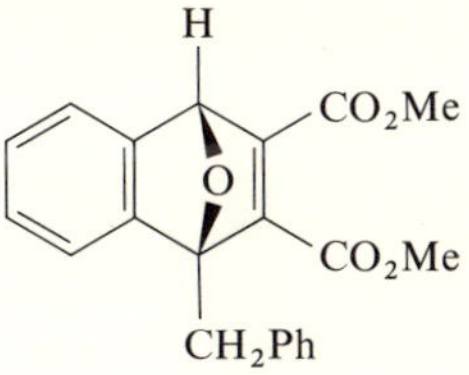

(91)

CH$_2$Ph) reacts with dimethyl acetylenedicarboxylate to give **91**.[71] In solution, an interesting equilibrium is established between **87** (R^1 = R^3 = H, R^2 = CH$_2$Ph) and its tautomer (Eq. 4). On addition of trifluoroacetic acid the equilibrium is reached instantaneously; triethylamine arrests the H-shift promptly.[66]

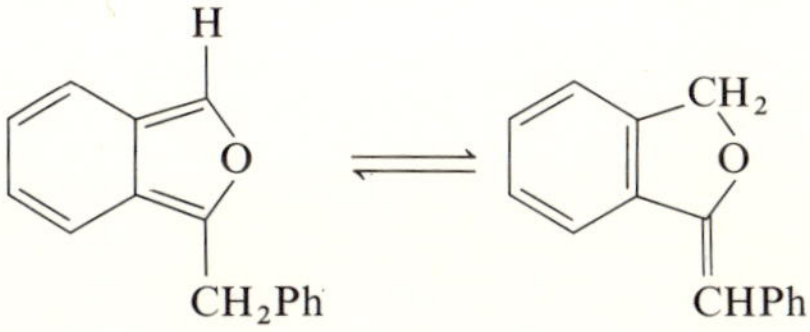

$$\text{(4)}$$

2-Methylbenzophenone gives, on photochemical bromination (refluxing CCl$_4$), 2-bromomethylbenzophenone in 80% yield. When the bromination was conducted at room temperature and the crude product treated in refluxing CHCl$_3$ with N-phenylmaleimide, dimethyl maleate, diethyl fumarate, methyl vinyl ketone, dimethyl acetylenedicarboxylate, and 1,4-naphthoquinone adducts **92** (mp 227–228°C, 75%), **93** (mp 120–121°C, 65%), **94** (mp 129–130°C, 51%), **95** (mp 81–82°C, 23%; an unidentified product was also isolated), **96** (mp 149–150°C, 62%) and **97** (mp 315–316°C, 19%) were obtained. 1-Phenylbenzo[c]furan is assumed to be an intermediate in these reactions.[67] Whereas the formation of **92–96** can be simply

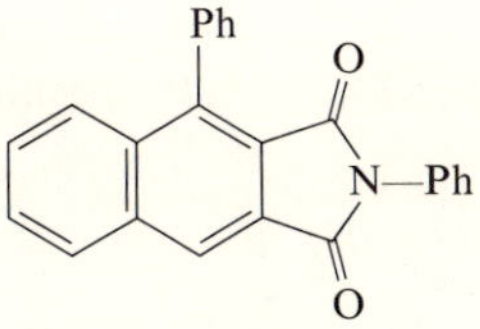

(92)

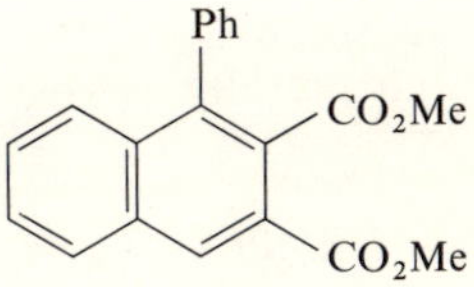

(93)

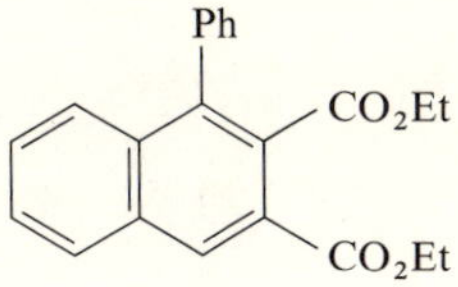

(94)

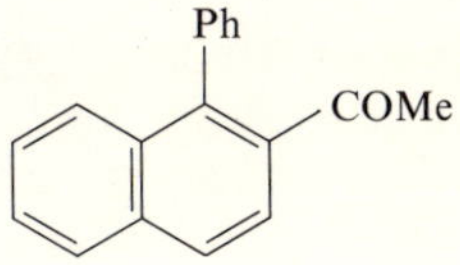

(95)

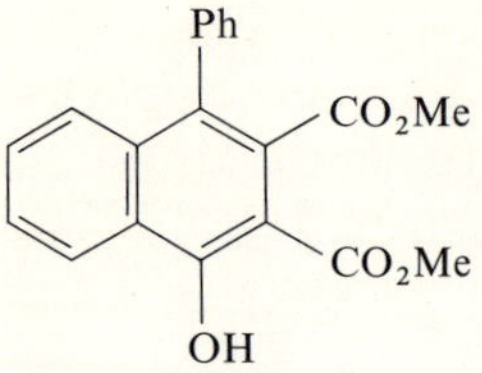

(96) 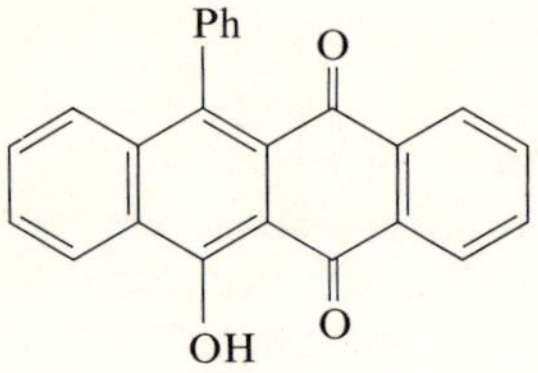 **(97)**

formulated as Diels–Alder reactions with consecutive acid-catalyzed dehydration, the origin of **97** remains unclear; ring opening, deprotonation, and subsequent aerial dehydrogenation is possible (Scheme 1).

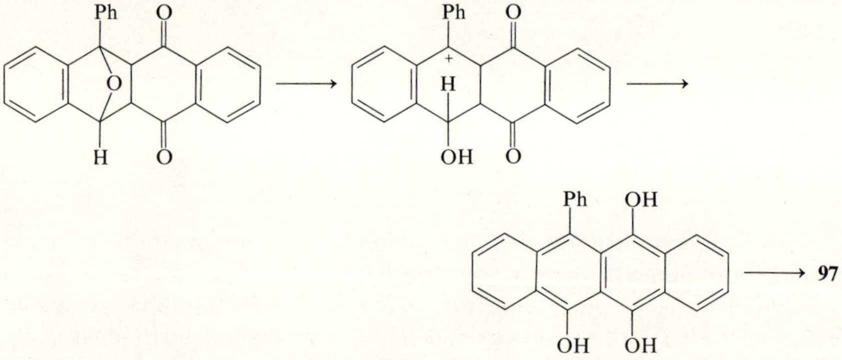

$\longrightarrow$ 97

SCHEME 1

When 2-ethylbenzophenone is brominated as described earlier and the crude product treated with N-phenylmaleimide (refluxing CCl_4) and dimethyl acetylenedicarboxylate (refluxing $CHCl_3$), the Diels–Alder adducts **98** (mp 183–184°C, 40%; presumably endo, when the ^{1}H-NMR data are compared with those of **88** and **89**) and **99** (mp 126.5–127°C, 57%) are isolated; again a benzo[c]furan (**87**; R^1 = Me, R^2 = Ph, R^3 = H) is assumed to be an intermediate.[67]

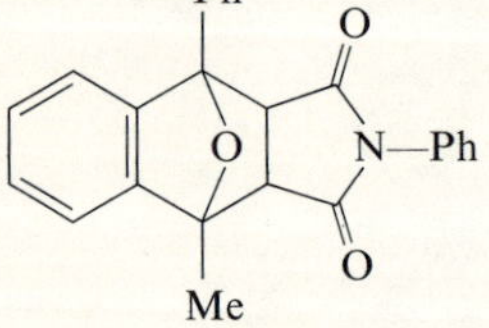

(98) 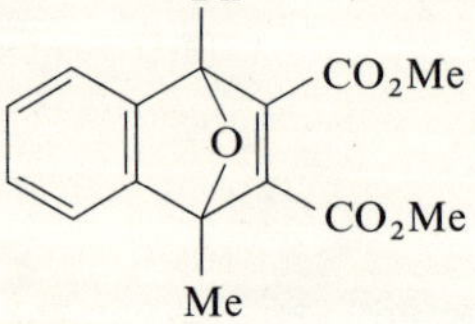 **(99)**

In an attempt to prepare di-*tert*-butylbenzocyclobutadiene by base-catalyzed dehydrobromination of **100**, a mixture of oxygen-containing substances was obtained; in some experiments a compound could be isolated

Sec. III.B] BENZO[c]FURANS 157

which was formulated as 1,3-di-*tert*-butylbenzo[c]furan.[70] On standing in air, diketone **101** was formed. The ^{1}H-NMR spectrum of the compound, believed to be **87** ($R^1 = R^2 = t$-Bu; $R^3 = H$), showed signals at 1.16 (s,18H) and 5.60–6.11 ppm (4H); as this spectrum is significantly different from that of **87** ($R^1 = t$-Bu, $R^2 = $ Ph, $R^3 = H$)[71] the structural assignment remains doubtful.

(100) (101)

(102) (103)

Compound **87** ($R^1 = t$-Bu, $R^2 = $ Ph, $R^3 = H$) has been prepared by addition of *tert*-butylmagnesium chloride to 1-phenylphthalide (**102**) and subsequent dehydration with *p*-toluenesulfonic acid. It is described as a yellow oil with brilliant fluorescence under UV light, which on oxidation with sodium dichromate yields diketone **103**.[71] Compound **87** (R = *t*-Bu, $R^2 = $ Ph, $R^3 = H$) with dimethyl acetylenedicarboxylate gives **104** (mp 128–129°C, 80%); on reduction **105** is obtained; **105** is also accessible from **87** ($R^1 = t$-Bu, $R^2 = $ Ph, $R^3 = H$) and dimethyl maleate (mp 151–152°C, endo); a maleic anhydride adduct has also been described (mp 147.5–148°C, 83%). Diels–Alder adducts have been also prepared from the benzo[c]furan precursors **106** ($R^1 = CH_2$Ph, $R^2 = $ H, Ph) in the presence of catalytic amounts of acid; **107** (mp 141–142.5°C, 70%) and **108** (mp 140–141.5°C, 49%) were obtained, which on catalytic hydrogenation (Pd/C) gave the exo isomers **109** (110–111°C) and **110** (90–92°C, 125.5–127°C, dimorphism). The basis for these stereochemical assignments remains unclear, however. Compounds

(104) (105) (106)

(107) (108) (109) (110) (111) (112)

109 and **110** are converted to the corresponding naphthalenes on treatment with acid. In the former case, a by-product was obtained which was formulated as **111**; it is difficult to understand why such a compound does not aromatize under the conditions employed. The acid-catalyzed ring opening of **107** yields **112**.

Thermal decomposition of the photooxides **113** (R = Me, Ph) may result in the formation of benzo[c]furans (**114**); both compounds were trapped as Diels–Alder adducts (**115**, R = Me, Ph). It was possible, however, to prepare **114** (R = Ph) by another route (Section IV,A,5).

(113) (114) (115)

Thermal decomposition of the diazo compounds **116a,b** in methanol or ethanol, in the presence of Cu(acac)$_2$ as catalyst, resulted in an intramolecular carbene reaction; the benzo[c]furans **117a,b** were not isolated, but **117b** could be trapped with N-phenylmaleimide and dimethyl acetylenedicarboxylate, to give compounds **118** (endo:exo = 1:1, 90%) and **121** (92%). Dimethyl fumarate yielded adduct **119** (15%) and supposedly **120** (83%); concerning this structure the same objections can be raised as in the case of **111**. Without trapping agents, cyclic ortho esters were formed.

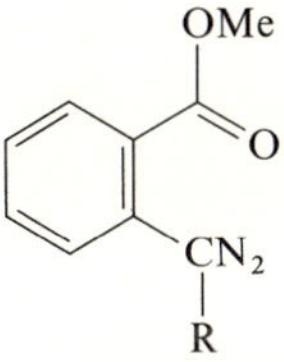

Sec. III.B] BENZO[c]FURANS 159

(116 a, b) 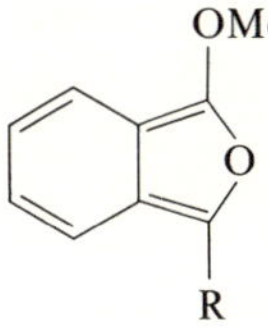 **(117 a, b)** 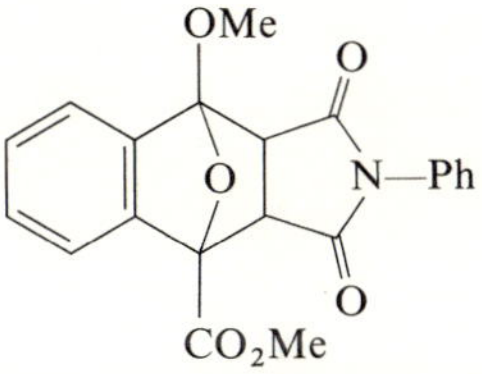 **(118)**

a: R = H; b: R = CO_2Me

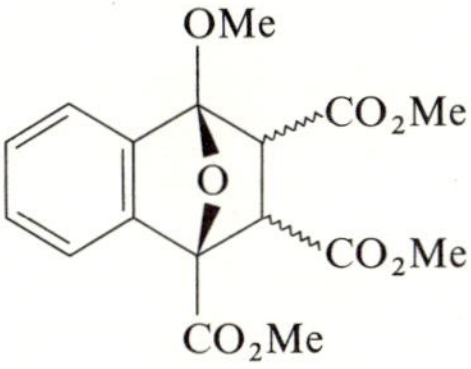

(119) 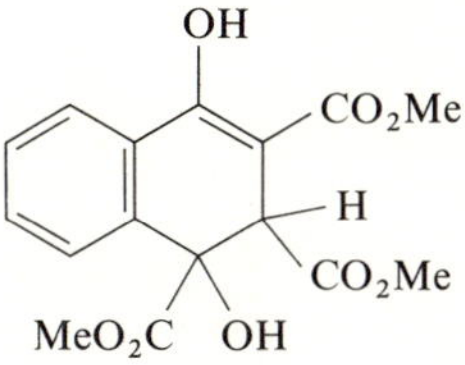 **(120)** 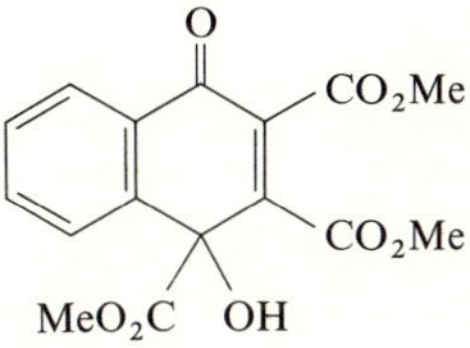 **(121)**

The phthalide ortho esters **122** (R = H, OMe) with dimethyl acetylenedicarboxylate at 140–150°C ($CHCl_3$; 6 hr) gave naphthalene derivatives (**123**, R = H, OMe); with **122** (R = H) and diethyl fumarate ($CHCl_3$, 170°C, 2 hr) a mixture of two compounds (**124**, R = H, ethyl) was obtained. An equilibrium as shown in Eq. (5) is possibly involved.

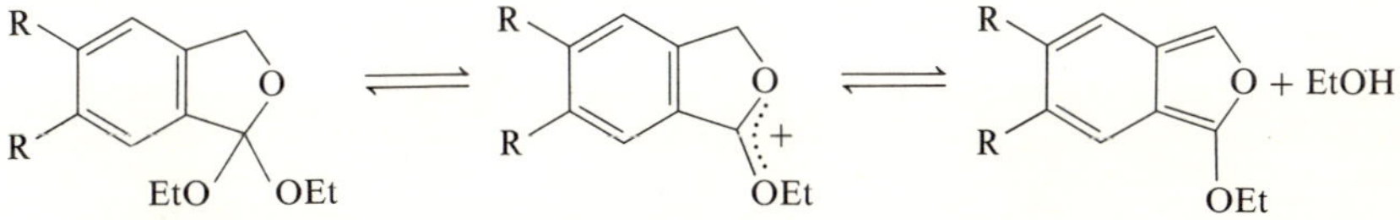

(122)

(5)

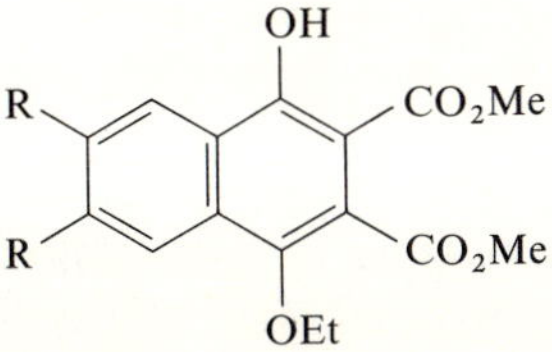

(123) 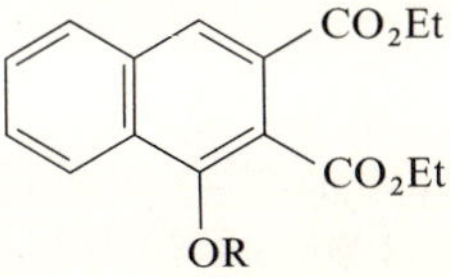 **(124)**

The benzo[c]furan (**126**) is probably formed on treatment of the tetralone derivative **125** with o-phthalaldehyde in the presence of acetic anhydride/sulfuric acid.[74]

---

[74] E. Aufderhaar, J. E. Baldwin, D. H. R. Barton, D. F. Faulkner, and M. Slaytor, *J. Chem. Soc. C.*, 2175 (1971).

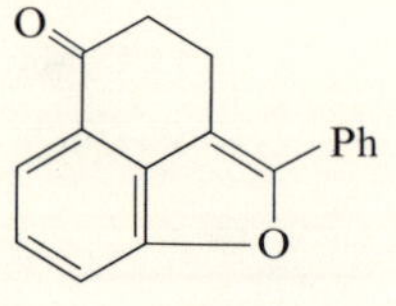

(125) 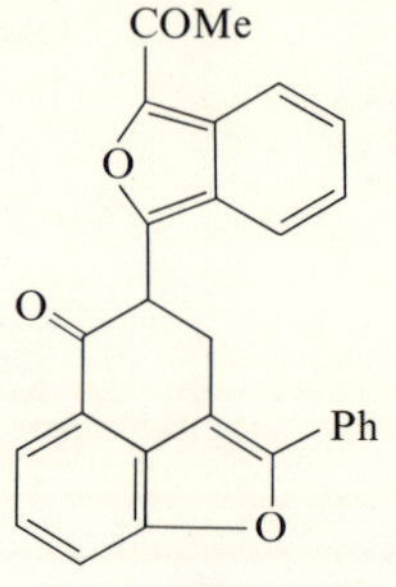 (126)

Electron-withdrawing substituents seem to stabilize the benzo[c]furan system as they do in the analogous benzo[c]thiophene and benzo[c]indole series. 1-Cyano-3-phenylbenzo[c]furan (128) was obtained from the aldehyde 127 as stable yellow needles; hydrolysis with base yields the carboxylic acid (129). Catalyzed hydrogenation of these compounds leads to 1,3-dihydrobenzo[c]furans (130, R^1 = CH_2NH_2, COOH; R^2 = Ph). Increased

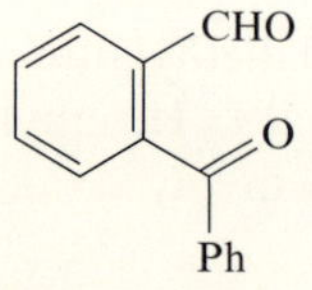

(127) 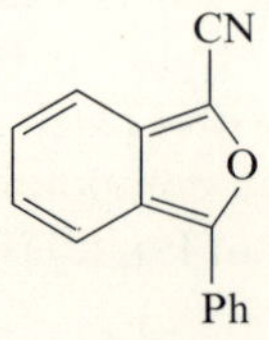 (128)

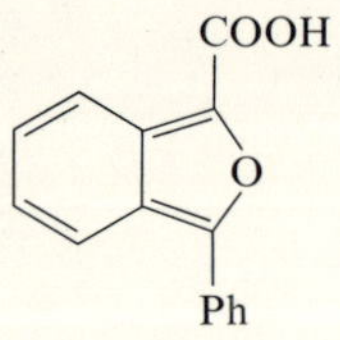

(129) 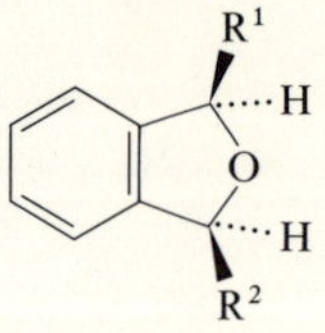 (130)

stability also results from substitution with halogens: 4,5,6,7-tetrahalogenobenzo[c]furans (132, R = F,Cl) are available[75,76] in quantitative yields through flow pyrolysis (600°C, 0.3 torr N_2) from 131. Whereas these compounds are stable at room temperature for a prolonged period ($t_{1/2}$ ≈ 58 hr in $CDCl_3$), they react immediately with N-phenylmaleimide to give the Diels–Alder adducts 133 and 134 (endo:exo = 2:1).

[75] H. Heaney, S. V. Ley, A. P. Price, and R. P. Sharma, *Tetrahedron Lett.*, 3067 (1972).
[76] An interesting paper concerning the influence of fluorine and chlorine substitution on the aromaticity of carbocyclic systems appeared quite recently: B. A. Hess and L. J. Schaad, *Isr. J. Chem.* **17**, 155 (1978).

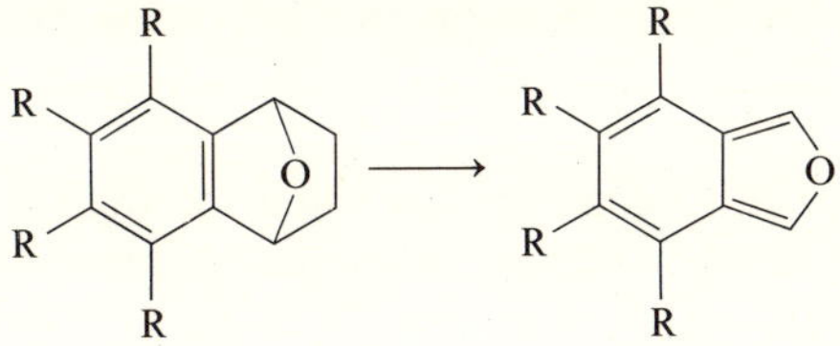

(131) (132)

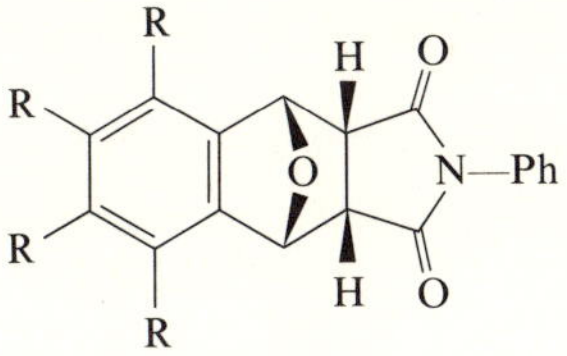

(133) 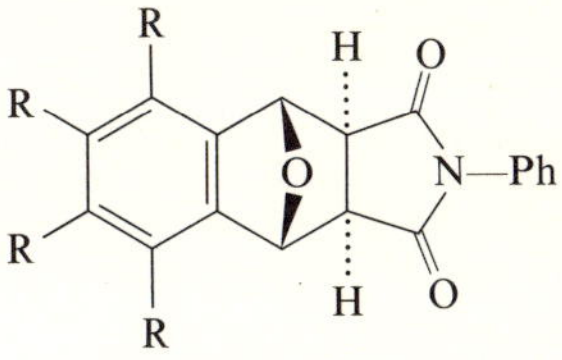 (134)

4,7-Dihydrobenzo[c]furan (136) has been prepared in a thermolysis reaction from 5,6-dimethylene-7-oxabicyclo[2.2.1]heptene (135) in benzene at 150°C. The mechanism of Scheme 2 accounts for this result.[77]

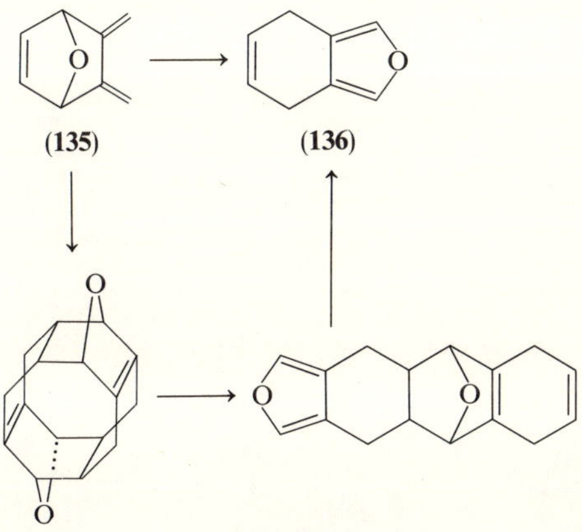

(135) (136)

SCHEME 2

IV. 1,3-Diarylbenzo[c]furans

A. SYNTHESES

On substitution at the 1- and 3-positions by aryl groups, the stability of the benzo[c]furan system increases considerably, so that these compounds

[77] W. R. Roth, H. Humbert, G. Wegener, G. Erker, and H. D. Exner, Chem. Ber. 108, 1655 (1975).

can generally be prepared quite easily in good yields. However, in solution 1,3-diarylbenzo[c]furans, like simple benzo[c]furans are susceptible to oxidation by air, especially in the presence of light (Section IV,D). It has been found valuable to conduct the work-up procedure in dim light; neglect of this precaution often results in substances contaminated by o-diaroylbenzenes.

Several methods are available for the preparation of 1,3-diarylbenzo[c]-furans; they will be discussed in the following subsections.

1. *Reaction of Arylphthalides with Arylmagnesium Halides*

One of the best procedures for the synthesis of 1,3-diphenylbenzo[c]furan (**138**) consists of the reaction of 3-phenylphthalide (**102**) with phenylmagnesium bromide,[12,78] especially when the reaction mixture is worked up in the presence of hydroquinone.[79] The primary product can be isolated as colorless crystals[12,80,81] with mp 145°C (decomposition above 100°C)[80,82]; in the presence of acid this unstable compound loses water very rapidly. The stereochemistry of the hydroxyphthalan is not known with certainty; presumably the cis isomer (**137a**) is formed first. In deuterated acetone equilibrium with the trans isomer (**137b**) is established.[80] For the synthesis of 1,3-diphenylbenzo[c]furan, the hydroxyphthalan need not be isolated.

(**102**)   (**137**)   (**138**)

(**137a**)   (**137b**)   (**139**)

[78] M. S. Newman, *J. Org. Chem.* **26**, 2630 (1961).
[79] A. Le Berre and R. Ratsimbazafy, *Bull. Soc. Chim. Fr.* [5], 229 (1963).
[80] P. Courtot and D. H. Sachs, *Bull. Soc. Chim. Fr.*, 2259 (1965).
[81] W. Baker, J. F. W. McOmie, G. A. Pope, and D. R. Preston, *J. Chem. Soc.*, 2965 (1961).
[82] A 1,3-diphenyl-1-hydroxyphthalan with mp. 109–113° (ethanol) has been obtained by reduction of 1,3-diphenyl-1-hydroperoxyphthalan: A. Rieche and M. Schulz, *Justus Liebigs Ann. Chem.* 653, 32 (1962).

The reaction sequence arylphthalide → 1,3-diarylbenzo[c]furan has been used often (cf. Table V). The Grignard reaction should be conducted in the inverse manner; normal addition results in formation of **139** as by-product.[12,83]

2. *Reduction of Diaroylbenzenes*

As has been known for a long time, reduction of *o*-dibenzoylbenzene (**140**) with zinc dust in acetic acid (2 days, occasional warming) yields **138** although the product could not be obtained in sufficient purity to be satisfactorily crystallized.[84,85] Under forcing conditions (3 hr, reflux) **141** was obtained.[81] The reduction is almost always better accomplished by the use of zinc dust and sodium hydroxide in ethanol.[86]

A number of 1,3-diarylbenzo[c]furans have been prepared by reduction of *o*-diaroylbenzenes with potassium[87,88] or sodium[89,90] borohydride. It

---

[83] A. Guyot and J. Catel, *C. R. Hebd. Seances Acad. Sci., Ser. C* **140**, 254 (1905).
[84] D. R. Boyd and D. E. Ladhams, *J. Chem. Soc.*, 2089 (1928).
[85] W. Steinkopf and W. Hanske, *Justus Liebigs Ann. Chem.* **541**, 238 (1939) (without experimental details).
[86] R. Adams and M. H. Gold, *J. Am. Chem. Soc.* **62**, 56 (1940).
[87] M. P. Cava, M. J. Mitchell, and A. A. Deana, *J. Org. Chem.* **25**, 1481 (1960).
[88] W. W. Zajac and D. E. Pickler, *Can. J. Chem.* **44**, 833 (1966).
[89] K. T. Potts and A. J. Elliott, *Org. Prep. Proced. Int.* **4**, 269 (1972).
[90] J. B. Miller, *J. Org. Chem.* **31**, 4082 (1966).

has been reported[88] that this method is superior to the older ones using zinc dust and acetic acid or alcoholic sodium hydroxide, although in some instances[87,89] the reduction–dehydration sequence has to be employed several times. Borohydride reduction under more drastic conditions leads to glycols which, when dehydrated, yield 9-arylanthracenes.[90] In the case of o-dibenzoylbenzene, both diastereoisomeric glycols have been isolated.[91] There are a number of reports concerning the reduction of o-dibenzoylbenzene with metals. Whereas with Na, Na/Hg, Mg/MgI$_2$, and Al/Hg **138** is obtained in various yields,[81] the one-electron reduction of **140** with sodium in tetrahydrofuran,[92] dimethoxyethane,[93] or potassium in dimethoxyethane[91] gives a bis(hydroxyphthalan) (**142**) which has been synthesized independently from **143**.[94] A radical-anion of the type shown may be an intermediate in the reduction process. Dimer **142** on treatment with acid

yields **138** and **140**; **138** has also been obtained by reduction of **140** with two equivalents of lithium or potassium in dimethoxyethane/*tert*-amyl alcohol.[91] The reduction of o-dibenzoylbenzene with alkali metals has been investigated in some detail[94,95]; in the case of o-dimesitoylbenzene the radical-anions have been studied by ESR.[96] The reduction of **140** to **138** has also been accomplished by prolonged boiling in triethyl phosphite.[97]

---

[91] J. A. Campbell, R. W. Koch, J. V. Hay, M. A. Ogliaruso, and J. F. Wolfe, *J. Org. Chem.* **39**, 146 (1974).
[92] B. J. Herold, *Tetrahedron Lett.*, 75 (1962).
[93] D. H. Eargle, *J. Org. Chem.* **39**, 1295 (1974).
[94] B. J. Herold, *Rev. Fac. Cienc., Univ. Lisboa, Ser. B* [2] **7**, 155 (1960).
[95] B. J. Herold, *Rev. Port. Quim.* **3**, 101 (1961).
[96] B. J. Herold, L. J. Alcazer, A. J. F. Carreia, A. J. P. Domingos, M. C. R. Lazana, and J. Santos Veiga, *Rev. Port. Quim.* **11**, 186 (1969).
[97] D. W. Jones, *J. C. S., Perkin 1*, 2728 (1972).

As expected, **140** can also be reduced photochemically; when a solution in isopropanol was exposed to sunlight, yellow crystals (mp ~ 160°C) separated which, on thermal treatment (200°C, 1 hr) in vacuum, yielded **138** (80%).[98] Evidently, a hydroxyphthalan is formed first.[80] When the photoreduction is conducted in the presence of acid, dimeric 1,3-diphenylbenzo[c]furan (**144**) is formed (Section IV,E).[80]

The syntheses described in this subsection rely upon the availability of o-diaroylbenzenes. A few methods seem to be of broader applicability. As Adams and co-workers have shown, the Diels–Alder reaction of dienes and diaroylethylenes gives 4,5-diaroylcyclohexenes, which on subsequent treatment with catalytic amounts of acid (sulfuric or phosphoric) form 4,7-dihydrobenzo[c]furans. Subjecting these compounds to the action of 2 moles of bromine in acetic acid in the presence of sodium acetate resulted in diaroylbenzenes (Eq. 6).[86,99–104]

---

[98] A. Schönberg, N. Latif, R. Moubasher, and W. I. Awad, *J. Chem. Soc.*, 374 (1950).
[99] R. Adams and T. A. Geissman, *J. Am. Chem. Soc.* **61**, 2083 (1939).
[100] R. Adams and R. B. Wearn, *J. Am. Chem. Soc.* **62**, 1233 (1940).
[101] R. Adams and M. H. Gold, *J. Am. Chem. Soc.* **62**, 2038 (1940).
[102] R. Adams (E. I. du Pont de Nemours Co.), U.S. Patent 2,325,727 (1943).
[103] R. Adams (E. I. du Pont de Nemours Co.), U.S. Patent 2,356,907 (1944).
[104] C. F. H. Allen and J. W. Gates, *J. Am. Chem. Soc.* **65**, 1283 (1943).

4,7-Dihydrobenzo[c]furans have been dehydrogenated directly to benzo[c]furans using p-chloranil in xylene.[105,106] The stereochemistry of the 4,5-diaroylcyclohexenes is generally not known. Sometimes two isomers have been isolated. As was reported by Adams and Geissman,[99] the reaction of 2,4-hexadiene with dibenzoylethylene gave a major product in 66% yield (mp 136–137°C); in addition, an isomeric compound (mp 86–88°C, 5%) was isolated. Two isomeric adducts (mp 120°C, 178–179°C) have also been obtained from the reaction of 1,4-diphenylbutadiene with dibenzoylethylene.[107] Both cis- and trans-dibenzoylethylene with 2,3-dimethylbutadiene gave the same stereoisomer.[86]

A further method for the synthesis of o-diaroylbenzenes has been described by Ried and Bönninghausen.[108] Tetracyclone (145), a potent diene for Diels–Alder reactions, together with dibenzoylacetylene (146) gives 147. This method has proved of value in a number of other cases[90,109] and is also applicable when the cyclopentadienone derivative is dimeric[90,109] as in the case of 2,5-dialkyl-3,4-diarylcyclopentadienones.[110]

3. Ring Closure of Diaroylcyclohexadienes

A useful synthesis of benzo[c]furans, based on experiments described by Ried,[108] has been published by White et al.[111] It consists of the Diels–Alder reaction of an acyclic diene with dibenzoylacetylene (146)[112] and subsequent ring closure of the resulting 1,2-diaroylcyclohexadiene with p-toluenesulfonic acid in benzene[111,113,114]; there has, however, been one report where this

---

[105] Y. Lepage and O. Pouchot, Bull. Soc. Chim. Fr., 2342 (1965).
[106] G. Freslon and Y. Lepage, C. R. Hebd. Seances Acad. Sci., Ser. C 280, 961 (1975).
[107] E. D. Bergmann, S. Blumberg, P. Bracha, and S. Epstein, Tetrahedron 20, 195 (1964).
[108] W. Ried and H. H. Bönninghausen, Justus Liebigs Ann. Chem. 639, 61 (1960).
[109] W. Friedrichsen, I. Kallweit, and R. Schmidt, Justus Liebigs Ann. Chem., 116 (1977).
[110] M. A. Ogliaruso, M. G. Romanelli, and E. J. Becker, Chem. Rev. 65, 261 (1965).
[111] J. D. White, M. E. Mann, H.-D. Kirshenbaum, and A. Mitra, J. Org. Chem. 36, 1048 (1971).
[112] G. Dupont and C. Paquot, C. R. Hebd. Seances Acad. Sci., Ser. C 205, 805 (1937).
[113] G. Doering, H. Straub, and E. Müller, Chem.-Ztg. 100, 291 (1976).
[114] J. M. Ruxer and G. Solladié, J. Chem. Res. M, 4946 (1978).

latter reaction has failed.[108] As expected, the cyclohexadienes on dehydrogenation with SeO_2 may give *o*-diaroylbenzenes[108]; in addition they

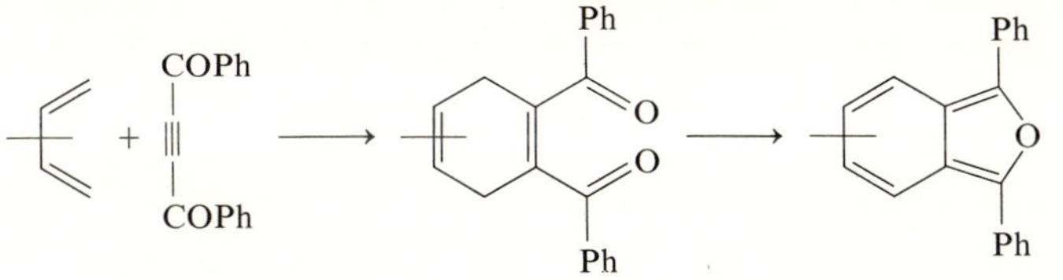

(146)

are useful starting materials for the synthesis of benzo[*c*]indoles and benzo[*c*]thiophenes.[111]

The Diels–Alder reaction of the acyclic diene and **146** may result in unexpected products, however. As reported by Ried,[108] the reaction of 1,4-diphenylbutadiene and **146** may, under drastic conditions, give the aromatized compound (**148**) and 1,3,4,7-tetraphenylbenzo[*c*]furan (**149**) in varying amounts, depending on the conditions. In methylglycol/propanol (6 hr reflux) a substance was isolated (mp 147–148°C, 59%) which has tentatively been formulated as **150**. One final goal of the sequence described seems to be unaffected by these anomalous reactions, however: Diels–Alder adducts of benzo[*c*]furans can be obtained in a one-step reaction from a diene, **146**, and an appropriate olefinic compound in acetic acid/acetic anhydride as shown in Eq. (7).[108]

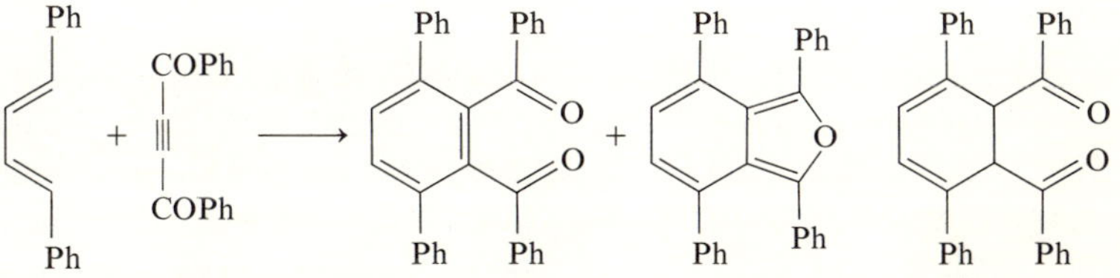

(146)   (148)   (149)   (150)

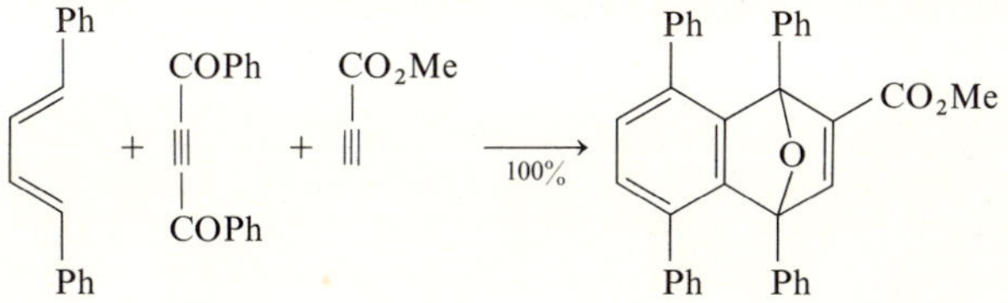

(7)

(146)

4. *Rearrangement of Diarylphthalins*

When diarylphthalides (phthaleins) are reduced, phthalins are formed (Eq. 8). When a diarylphthalin is treated with sulfuric acid, a rearrangement occurs; this reaction was studied intensively by von Baeyer.[16] Reinvestigation and extension of this work by Blicke et al.[115] has shown that the reaction is more complex than originally believed; it may take two courses, leading to formation of benzo[c]furans, anthrones, or to a mixture of both (Eq. 9).

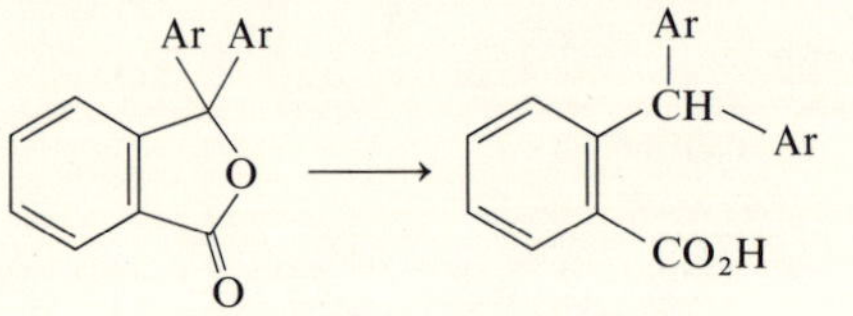

(8)

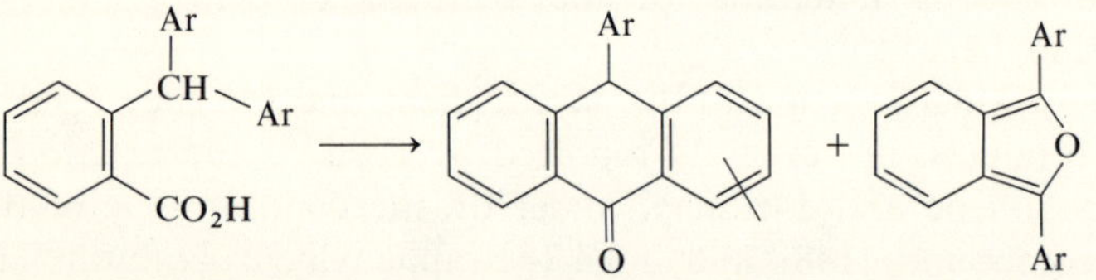

(9)

When 4′,4″-dihydroxyphthalin (**151a**), which is obtained from phenolphthalein with zinc and sodium hydroxide, is treated with sulfuric acid, a substance isolated as a green mass could not be purified; in ethereal solution it showed a strong green fluorescence. On standing in air, a compound was formed which von Baeyer[16] believed to be **153** ("phthalidein of phenol"); the green mass was formulated as an anthranol (**152**, "phthalidin of phenol"). Reinvestigation of this work has shown unequivocally that the oxidation product was an *o*-diaroylbenzene (**155a**) and that the primary product therefore must have been impure 1,3-bis(*p*-hydroxyphenyl)benzo[c]furan (**154a**).[13,15] In the same way, sulfuric acid treatment of **151b**,[13,18,115] **151c**,[15] and **151d**[15] gave the benzo[c]furans **154b–d**, whereas **151e** gave a dark green gum[15] (probably a benzo[c]furan or a mixture of isomers) which on oxidation with sodium dichromate formed **155e** in 30% yield. In the case

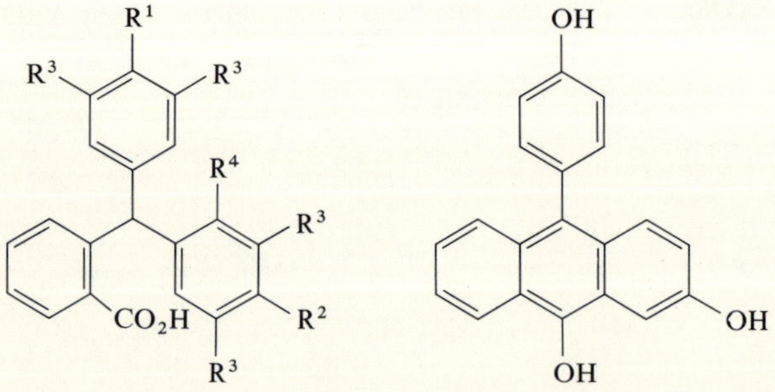

(151)   (152)

[115] F. F. Blicke and R. D. Swisher, *J. Am. Chem. Soc.* **56**, 1406 (1934).

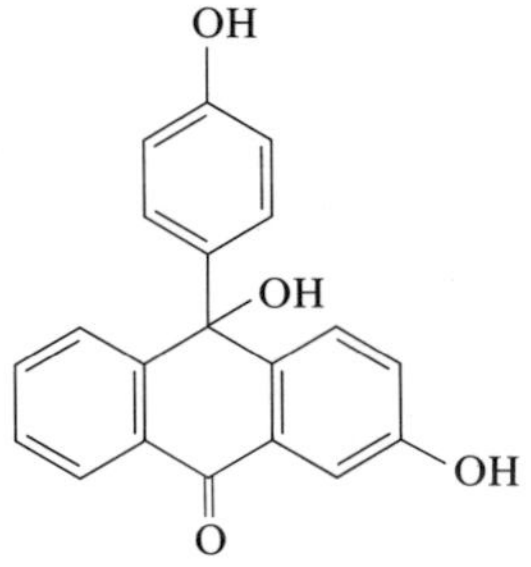

(153) 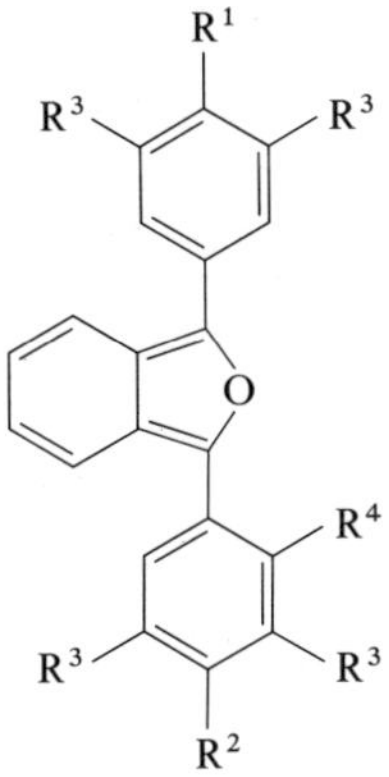   (154)

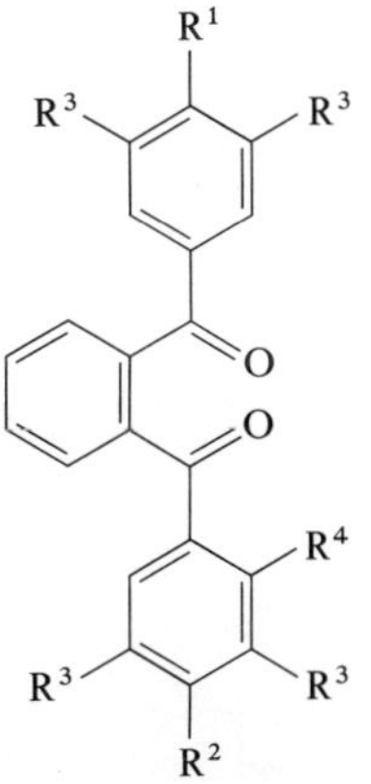

a: $R^1 = R^2 = OH; R^3 = R^4 = H$
b: $R^1 = R^2 = OMe; R^3 = R^4 = H$
c: $R^1 = R^2 = OH; R^3 = Br; R^4 = H$
d: $R^1 = R^2 = OAc; R^3 = Br; R^4 = H$
e: $R^1 = R^4 = OMe; R^2 = R^3 = H$
f: $R^1 = OMe; R^2 = R^3 = R^4 = H$
g: $R^1 = R^2 = Cl; R^3 = R^4 = H$
h: $R^1 = OH; R^2 = R^3 = R^4 = H$
i: $R^1 = R^2 = R^3 = R^4 = H$
j: $R^1 = R^2 = NMe_2; R^3 = R^4 = H$

(155)

of **151f**, a benzo[c]furan (probably **154f**) was isolated as a yellow amorphous product; on oxidation **155f** was obtained.[115] The isomeric 9-hydroxyanthrones **156** and **157** were also isolated. Compound **151g** on treatment with sulfuric acid gave an anthrone (**158a**) and trace amounts of a benzo[c]furan[14]; with zinc chloride only **158a** was obtained.[18] 4-Hydroxydiphenylphthalin (**151h**) under Friedel–Crafts conditions (zinc chloride, acetic anhydride) also gave a mixture[18,116] of an intramolecular acylation product (isolated as **159**) and a benzo[c]furan (probably **154h**).[117] Diphenylphthalin itself (**151i**) yielded an anthrone[18] (**158b**) as did 2-xanthylbenzoic acid (**160**); **161**

---

[116] H. von Pechmann, *Ber. Dtsch. Chem. Ges.* **13**, 1616 (1880).
[117] F. F. Blicke and R. A. Patelski, *J. Am. Chem. Soc.* **58**, 559 (1936).

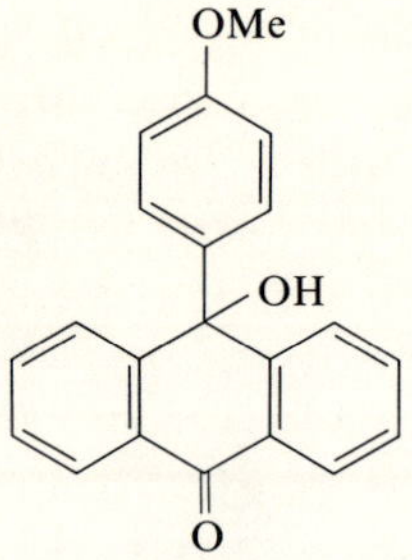

(156) 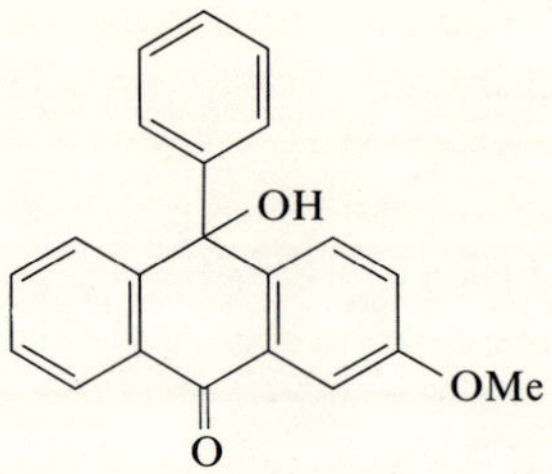 (157)

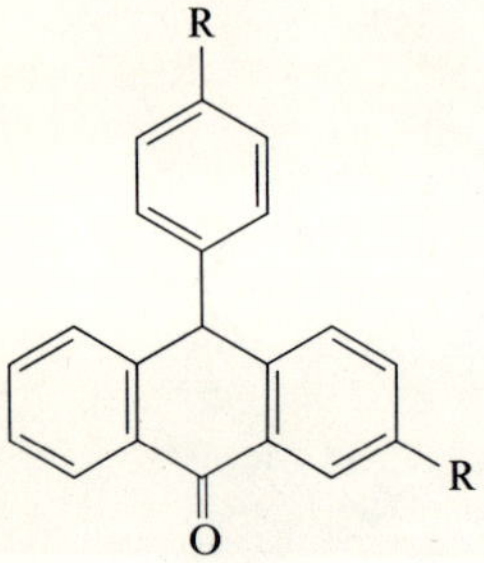

(158) 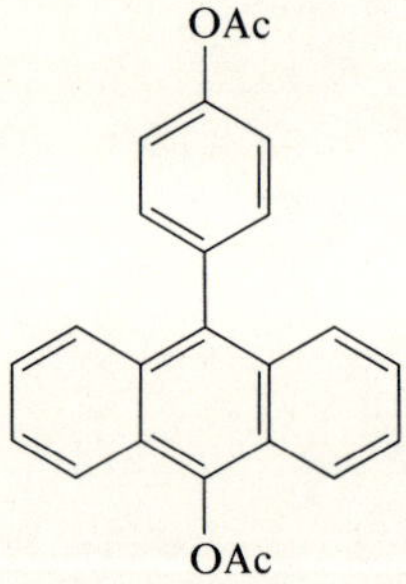 (159)

a: R = Cl; b: R = H

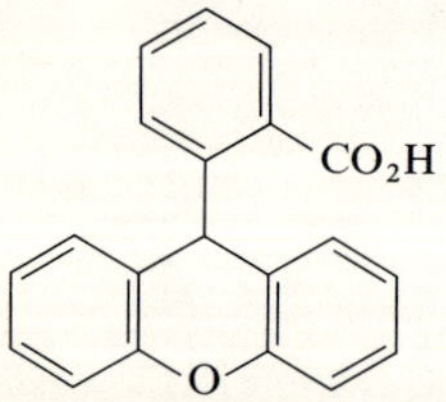

(160) 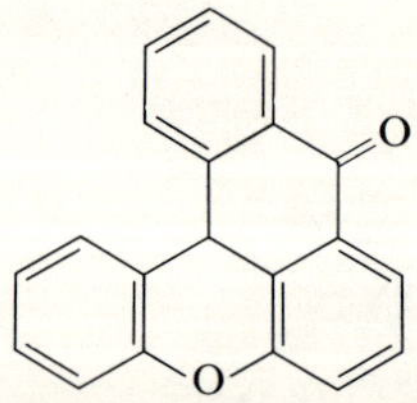 (161)

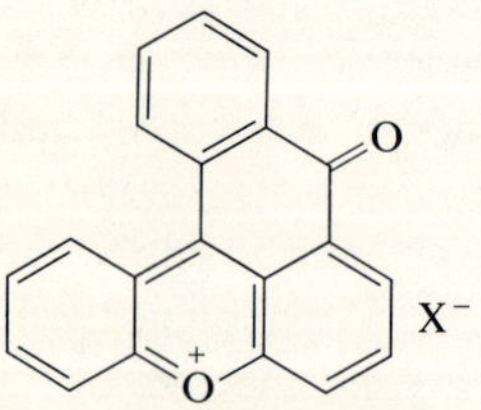

(162) 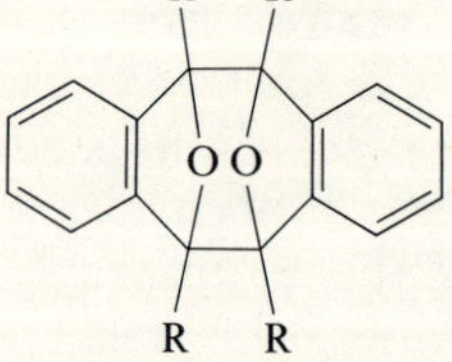 (163) R = (*p*-Me$_2$NC$_6$H$_4$)

("cöroxenol") and an oxonium salt (162, "cöroxonium salt") were also isolated.[118] The phthalin–anthrone conversion can also be brought about by thionyl chloride[115] or phosphorus pentachloride[119]; in one reported case (151c), the nature of the reaction product could be changed by the use of zinc chloride instead of sulfuric acid.[13] As reported by Guyot and Haller,[120] treatment of the phthalin 151j with acetic anhydride or phenyl isothiocyanate resulted in the formation of a dimeric benzo[c]furan (163) in 92% yield. The generation of the monomer (154j) could be effected by treatment of 163 with concentrated sulfuric acid at 100°C for about 15 minutes. In acetic acid dimerization to 163 takes place almost instantaneously. Whereas the anthrone formation is merely the result of an intramolecular acylation, the transformation to a benzo[c]furan has been suggested to occur intermolecularly in a manner illustrated for 151b.[117] In the first step, a retro Friedel–Crafts reaction to 164 (or the corresponding phthalide, 165) and anisole

[118] H. Decker and D. Ferrario, *Justus Liebigs Ann. Chem.* **348**, 225 (1906).
[119] A. Haller and A. Guyot, *Bull. Soc. Chim. Fr.* [3] **17**, 877 (1897).
[120] A. Guyot and H. Haller, *Ann. Chim. Phys.* [8] **19**, 297 (1910).

takes place; **165** then combines with anisole to yield either **166** or **167**, which then lose water to give **154b**. In support of this mechanism it has been found that a mixture of **165** and anisole when treated with sulfuric acid yields **154b**.[117] In three other instances it has been shown that benzo[c]furans can be obtained from a monoarylphthalide with phenol or anisole and sulfuric acid: compound **165** and phenol as well as 4′-hydroxyphenyl-phthalide and anisole yielded 1-(p-hydroxyphenyl)-3-(p-methoxyphenyl)-benzo[c]furan (after oxidation and methylation isolated as **155b**), whereas 4′-hydroxyphenylphthalide and phenol gave **154a**, which was converted to **155a** and isolated as such. If this mechanism applies to all cases, other isomers should also be present.

5. Further Syntheses

As was reported by Adams and Wearn,[100] 4,5-diaroylcyclohexenes react with bromine in chloroform to give dibromo compounds which, on treatment with acetyl chloride and catalytic amounts of sulfuric acid, give furan derivatives (stereochemistry probably as in Scheme 3)[121]; these latter compounds could also be obtained from 4,7-dihydrobenzo[c]furans with bromine.[100,101] A few examples have been described where these dibromo-furans were converted to benzo[c]furans with sodium acetate in acetic anhydride/acetic acid.[100,101] Another route which seems to be especially

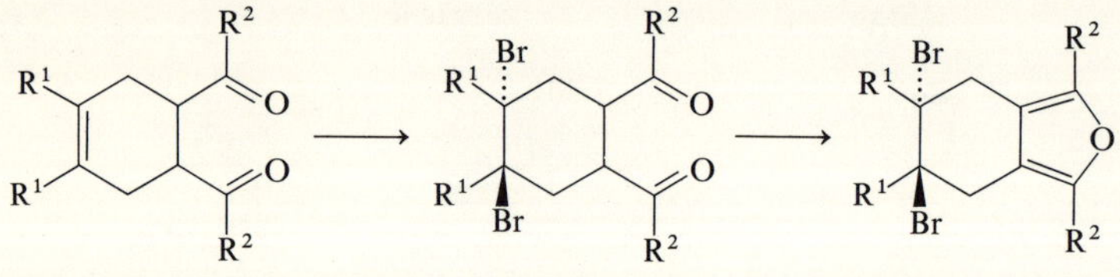

SCHEME 3

suited for the synthesis of 1-(2′,3′,5′,6′-tetramethylphenyl)-3-arylbenzo[c]-furans has been described by Fuson and Rife.[122] o-Duroylbromobenzene (**168**) is converted to the lithium derivative (**169**) which on treatment with aromatic aldehydes and subsequent acidic work-up gives 1-duryl-2-aryl-benzo[c]furans (**171**).

---

[121] "Beilsteins Handbuch der organischen Chemie," Vol. 17, p. 730 (3/4 Ergänzungswerk). Springer-Verlag, Berlin and New York, 1974.
[122] R. C. Fuson and W. C. Rife, *J. Org. Chem.* **25**, 2226 (1960).

Sec. IV.A] BENZO[c]FURANS 173

a: $R^1 = R^2 = H$; b: $R^1 = OMe$, $R^2 = H$; c: $R^1 = H$, $R^2 = OMe$
d: $R^1 = H$, $R^2 = Cl$; e: $R^1 = Cl$, $R^2 = H$; f: $R^1 = R^2 = OMe^{109}$

1,3-Diphenylbenzo[c]furan has been observed in a number of other reactions. When benzocyclobutadienequinone (172),[123] monothiophthalic anhydride (173),[124] chlorophthalide (the chloride of o-phthalaldehydic acid, 174),[125] and 2-cyanobenzaldehyde (175)[117] are treated with phenylmagnesium bromide, 138 could be obtained. This last reaction, which yields an imine in the primary stage, has also been conducted with p-methoxyphenylmagnesium bromide.[117]

1,4-Diphenylphthalazine N-oxide (176) on irradiation in acetone gave 138 in yields of up to 65%[126,127] (Scheme 4); the diazoketone 178 could be

---

[123] M. P. Cava, D. R. Napier, and R. J. Pohl, *J. Am. Chem. Soc.* **85**, 2076 (1963).
[124] R. H. Schlessinger and I. S. Ponticello, *Chem. Commun.*, 1013 (1968).
[125] M. Renson, *Bull. Soc. Chim. Belg.* **70**, 77 (1961).
[126] O. Buchardt, *Tetrahedron Lett.*, 1911 (1968).
[127] K. B. Tomer, N. Harrit, I. Rosenthal, O. Buchardt, P. L. Kumler, and D. Creed, *J. Am. Chem. Soc.* **95**, 7402 (1973).

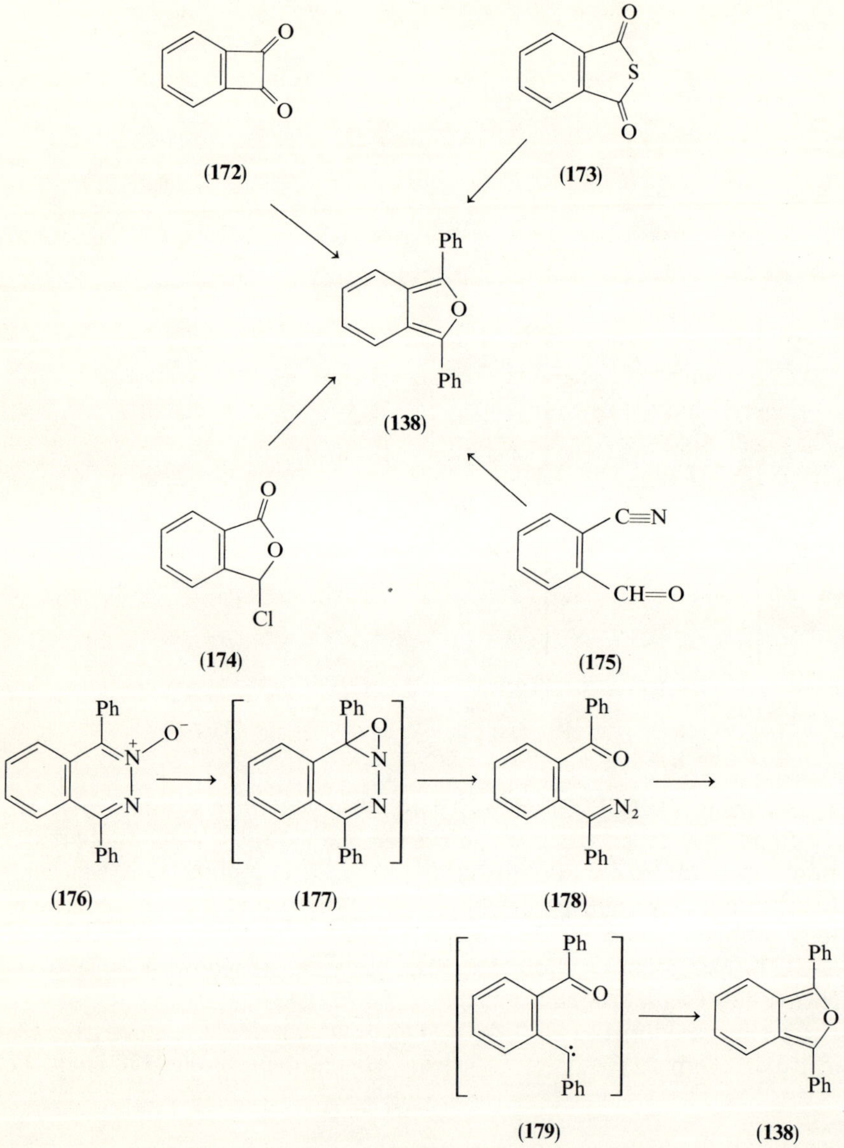

SCHEME 4

detected in the infrared spectrum by a sharp absorption band at 2040 cm^{-1}. The half-life was determined to be 115 minutes at 26°C. An analogous diazoketone of moderate stability (**180**) and its decomposition to **171a** has been described.[109]

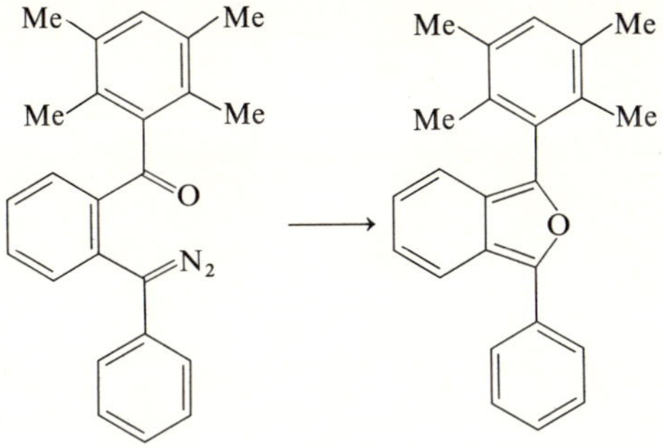

Sec. IV.A] BENZO[c]FURANS 175

(180) **(171a)**

Compound **138** has also been observed when **176** was treated with benzoyl or acetyl chloride in chloroform.[128] The following mechanism (Scheme 5) has been proposed to account for this result; a stable benzocyclobutadiene epoxide **182** (mp 287–288°C; cf. **181**) has recently been described.[129]

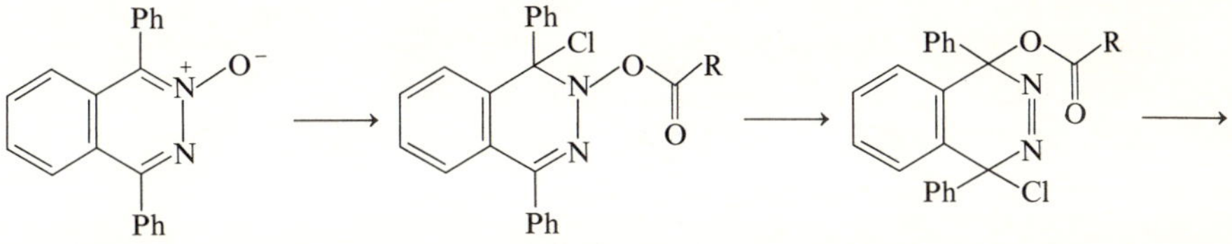

(176)

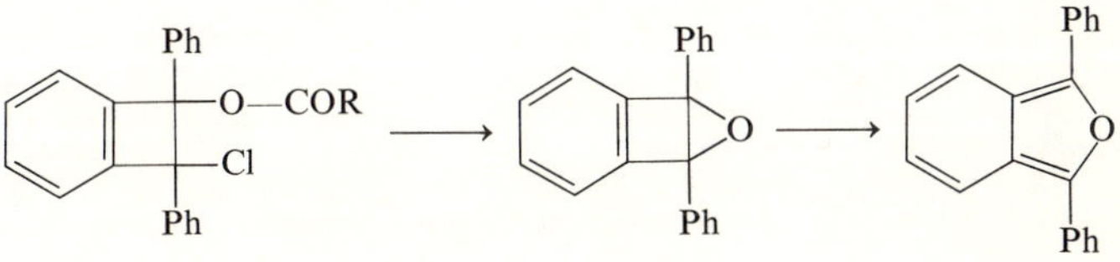

(181) **(138)**

SCHEME 5

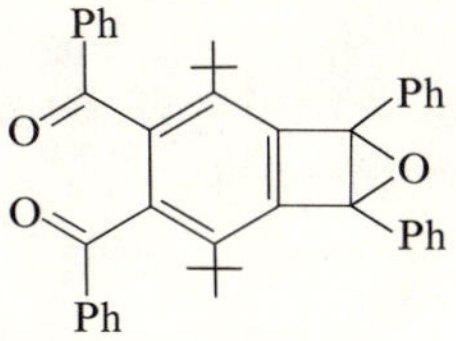

(182)

[128] E. Oishi, Y. Kawamura, D. Kojima, and E. Hayashi, *Yakugaku Zasshi* **97**, 1082 (1977).
[129] F. Toda, N. Dan, K. Tanaka, and Y. Takehira, *J. Am. Chem. Soc.* **99**, 4529 (1977).

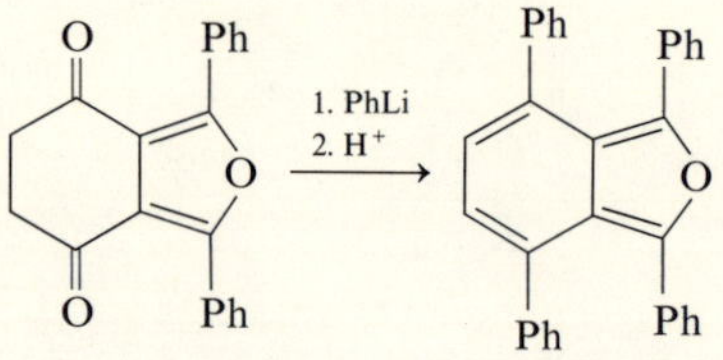

In another route to benzo[c]furan, Lepage and co-workers[130] treated diketone **183** with phenyllithium; subsequent acidic work-up gives 1,3,4,7-tetraphenylbenzo[c]furan (**184**).

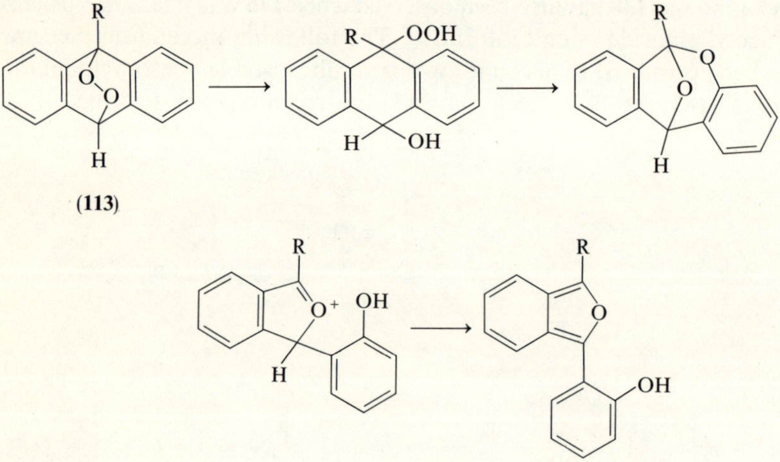

SCHEME 6

When the photooxide of 9-phenylanthracene (**113**, R = Ph)[131] is treated with aqueous acetic acid, 3-(o-hydroxyphenyl)-1-phenylbenzo[c]furan (**114**, R = Ph) is obtained in 22% yield.[132] The mechanism of Scheme 6 has been given for this rearrangement. Thermal decomposition of **113** (R = Me, Ph), which might occur as shown in Scheme 7, also gives **114** (R = Me, Ph); in these cases the benzo[c]furans have not been isolated. They could be trapped, however, with N-methylmaleimide.[73]

---

[130] A. Verine and Y. Lepage, *Bull. Soc. Chim. Fr.*, 1154 (1973).
[131] C. Dufraisse, L. Velluz, and L. Velluz, *Bull. Soc. Chim. Fr.*, 1260 (1937).
[132] J. Rigaudy, M. Perlat, and N. K. Cuong, *Bull. Soc. Chim. Fr.*, 2521 (1974).

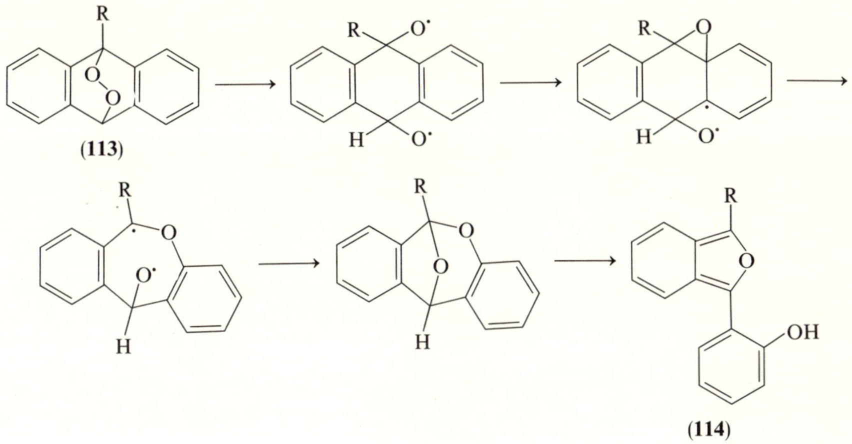

Sec. IV.A] BENZO[c]FURANS 177

(113)

(114)

SCHEME 7

Dufraisse[133] announced that **138** was formed when an alkaline solution of hydroxyindanone **185**[134,135] is treated with air or oxygen; up to 80% **138** could be obtained in this way.[135] This observation can be explained either by oxidation of an enolate ion to an oxiranol, with subsequent rearrangement, or by oxidative cleavage of the enolate ion to a benzilic acid (**186**), which after ring closure will lose CO_2 and H_2O to give **138** (Scheme 8).[136] Compound **138** is also formed when *o*-benzoylbenzil (**187**) is treated with potassium hydroxide in ethanol[136]; a benzilic acid rearrangement may be the first step of the sequence.

When 2,3-diphenylindenone (**188**) is treated with potassium hydroxide in alcohol, **138** is formed in various yields, depending on the alcohol employed (isopropanol 51%; ethanol 44%; methanol 2%)[136]; besides the β-alkoxyketone **189**, the *trans*-glycol **190** (mp 233–235°C,[135] 232–236°C[136]) and 2,3-diphenylindanone (**191**)[137] were also found (Scheme 9). A mechanism analogous to that in Scheme 8 can be envisaged. A similar rearrangement takes place when 2,3-diphenylindenone epoxide (**192**) is treated with base; with potassium hydroxide in ethanol **138** is obtained in 11% yield. The mode of rearrangement seems to depend on the concentration of base; when

---

[133] C. Dufraisse and S. Ecary, *C. R. Hebd. Seances Acad. Sci., Ser. C* **223**, 1143 (1946).
[134] C. F. Koelsch, *J. Am. Chem. Soc.* **58**, 1321 (1936).
[135] S. Ecary, *Ann. Chim. (Paris)* [12] **3**, 445 (1948).
[136] C. F. H. Allen and J. A. Van Allan, *J. Am. Chem. Soc.* **70**, 2069 (1948).
[137] A. Banchetti, *Gazz. Chim. Ital.* **81**, 419 (1951).

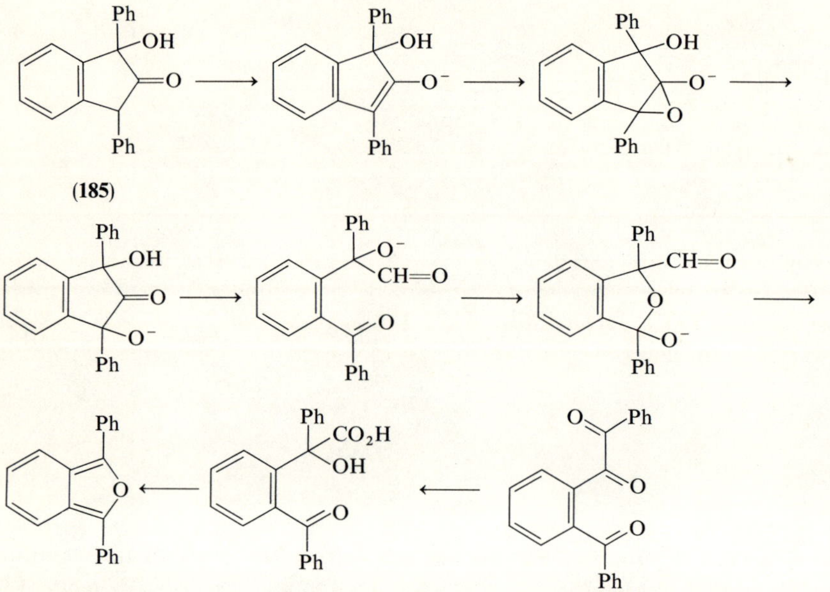

(185) (138) (186) (187)

SCHEME 8

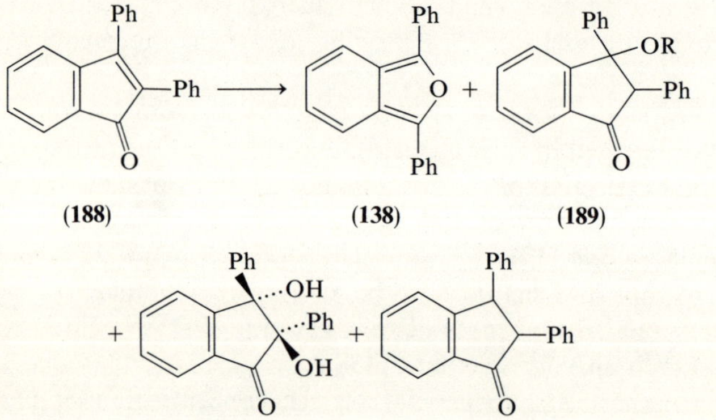

(188) (138) (189) (190) (191)

SCHEME 9

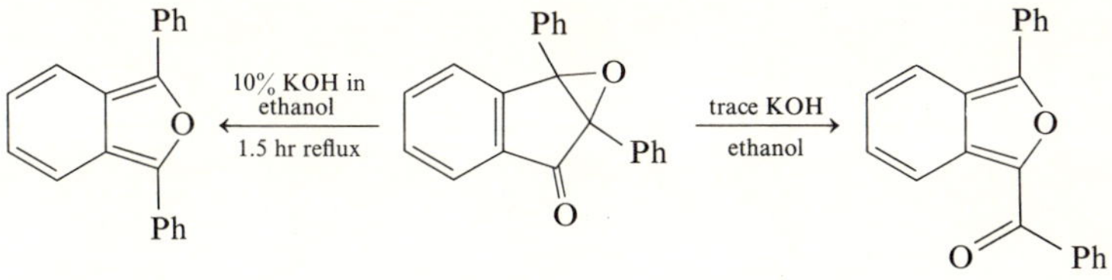

Sec. IV.A]   BENZO[c]FURANS   179

ethanolic **192** is heated in a flask pretreated with aqueous potassium hydroxide, 1-benzoyl-3-phenylbenzo[c]furan (**193**) is obtained in 73% yield.[138] The furan **193** is formed, too, when **192** is photolyzed in ethanol (Scheme 10)[139]; a mechanism has been suggested. 1-Benzoyl-3-phenylbenzo[c]furan

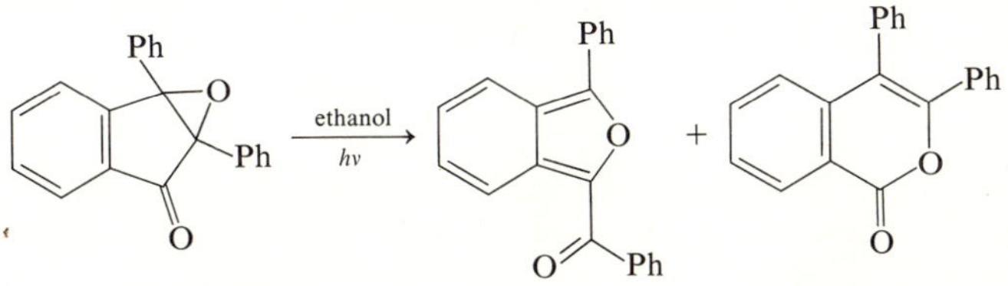

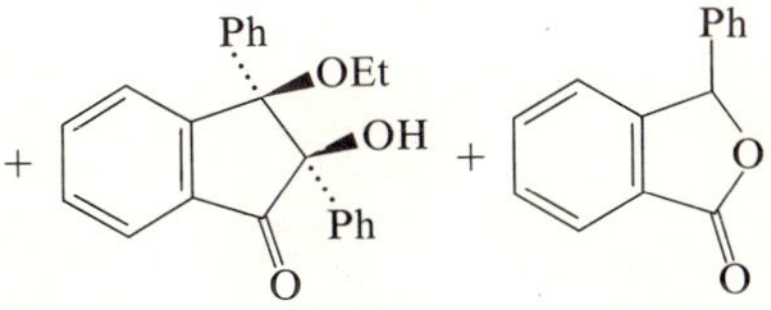

SCHEME 10

(**193**) was first described by Weitz and Scheffer[140] but not recognized as such. The hydrolysis of **192** with acetic acid/hydrogen chloride yielded a compound ("Hydrat")[140] which has been shown to be a glycol (**196**, probably cis)[141,142]; the glycol dissolves in alkali but cannot be recovered on acidification. Instead, a new brilliant yellow substance ("gelber Körper")[140] is formed in up to 90% yield,[136] which was proved to be **193**. The furan could be formed by a retroaldol condensation with subsequent ring closure as shown in Scheme 11. An analogous reaction has been performed with 5,6-dimethyl-2,3-diphenylindenone epoxide (Eq. 10).

[138] Unpublished work of R. Lopresti, quoted by Ullman and Henderson.[139]
[139] E. F. Ullman and W. A. Henderson, *J. Am. Chem. Soc.* **88**, 4942 (1966),
[140] E. Weitz and A. Scheffer, *Ber. Dtsch. Chem. Ges.* **54**, 2344 (1921).
[141] C. F. Koelsch and C. D. Le Claire, *J. Am. Chem. Soc.* **65**, 754 (1943).
[142] C. F. H. Allen and J. W. Gates, *J. Am. Chem. Soc.* **65**, 1230 (1943).

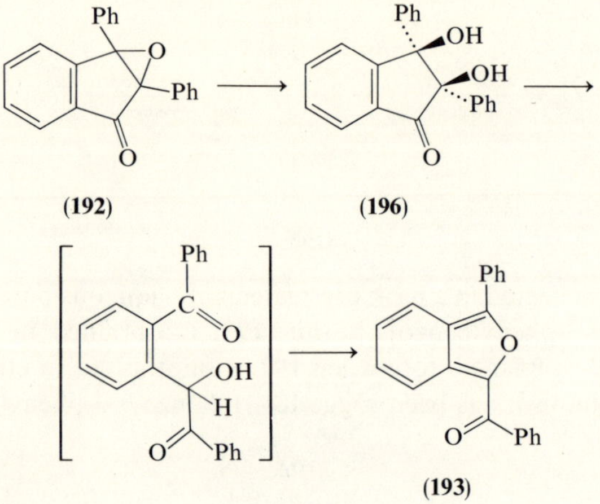

SCHEME 11

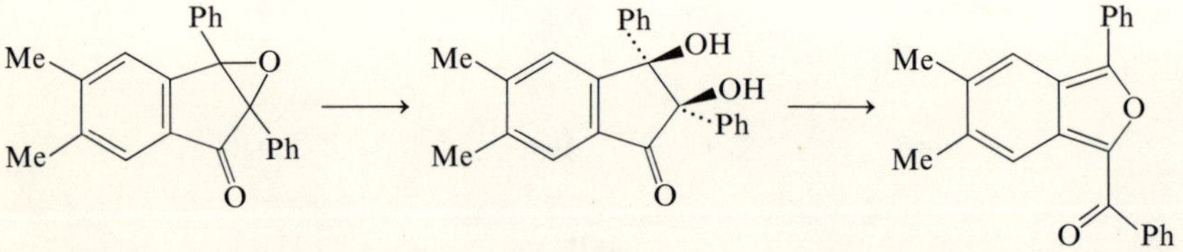

(10)

Compound **193** is not stable in alkaline solution, but is cleaved to give *o*-benzoylbenzoic acid and benzoic acid; cautious permanganate oxidation of **193** gives *o*-benzoylbenzil (**187**). It was not cleaved by hydrogen peroxide.

There was an early report of another benzo[*c*]furan, namely 1,3-dibenzoylbenzo[*c*]furan (**197**).[143] Whereas all benzo[*c*]furans are known to be susceptible to oxidative cleavage, this compound has been reported to resist further oxidation. A structure proof of the substance claimed to be **197** seems desirable.

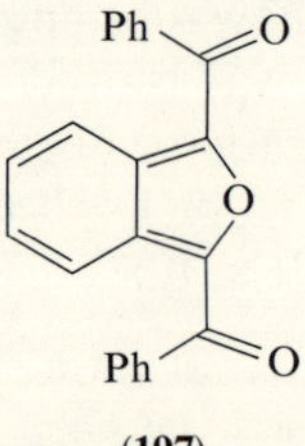

[143] R. Weiss and L. Sonnenschein, *Ber. Dtsch. Chem. Ges.* **58**, 1043 (1925).

TABLE V
1,3-DIARYLBENZO[c]FURANS

R^1	R^2	R^3	R^4	R^5	R^6	Selected references and remarks
Ph	Ph	H	H	H	H	12, 78, 79, 86, 87, 89, 135, 136, 144, 145; a
Ph	Ph	←————Various Me$_n$———→				86, 90, 99, 108, 111, 148, 149; b
Ph	Ph	Et	H	H	Et	151
Ph	Ph	H	CH=O	CH=O	H	152
Ph	Ph	CO$_2$Et	H	H	CO$_2$Et	153; c
Ph	Ph	OAc	H	H	OAc	130
Ph	Ph	←————Various Ph$_n$———→				104, 107, 108, 130, 154, 155
Ph	Ph	Ph	PhCH$_2$	PhCH$_2$	Ph	156
Ph	Ph	H	PhCO	PhCO	H	157, 157a
Ph	Ph	Ph	CH$_2$OH	CH$_2$OH	Ph	157b
Ar	Ph	H	H	H	H	78, 88, 109, 114, 115, 117, 122, 132, 158, 159
Ar	Ar'	H	H	H	H	13, 14, 15, 18, 78, 100, 101, 109, 117, 120, 122, 160
PhCO	Ph	H	H/Me	H/Me	H	140, 142
1-Nap[d]	Ph/1-Nap	H	H	H	H	161–163
Ar	Ar'	H	Me	Me	H	100, 101, 151
Ph	Ph	p-ClC$_6$H$_4$	H	H	Ph	164
Ar	Ar'	Ph	H	H	Ph	109, 165
ArCH=CH	ArCH=CH	H	H	H	H	151

[a] The substance can be purified by fractional sublimation under argon[146] or through zone melting.[147]

[b] The 4,7-dihydrobenzo[c]furan has also been described.[150]

[c] Not isolated; can be trapped.[153]

[d] 1-Naphthyl.

6. Compilation of Known 1,3-Diarylbenzo[c]furans

In Table V are listed the known benzo[c]furans which carry at least aryl substituents in the 1- and 3-positions.[144-165]

B. DIELS-ALDER REACTIONS

By far the most important property of benzo[c]furans is their capacity to act as 4π-components in cycloaddition reactions. Whereas the reactions described before 1969 were almost always of the Diels–Alder type, more recent investigations have shown that they can also participate in $[\pi_4 + \pi_4]$- and $[\pi_4 + \pi_6]$-addition (Section IV,C). In this chapter Diels–Alder reactions will be discussed. Benzo[c]furans have been used for two main purposes. First, Diels–Alder adducts with olefinic compounds can conveniently be dehydrated to naphthalene derivatives or higher condensed hydrocarbons not easily accessible by other methods; second, benzo[c]furans are excellent

---

[144] H. J. S. Winkler and G. Wittig, *J. Org. Chem.* **28**, 1733 (1963).
[145] A. Le Berre and G. Lonchambon, *Bull. Soc. Chim. Fr.*, 4328 (1967).
[146] J. Olmsted and G. Karal, *J. Am. Chem. Soc.* **94**, 3305 (1972).
[147] J. A. Howard and G. D. Mendenhall, *Can. J. Chem.* **53**, 2199 (1975).
[148] W. M. Horspool, J. M. Tedder, and Z. U. Din, *J. Chem. Soc. C*, 1694 (1969).
[149] M. E. Mann and J. D. White, *Chem. Commun.*, 420 (1969).
[150] A. Huth, H. Straub, and E. Müller, *Justus Liebigs Ann. Chem.*, 1893 (1973).
[151] C. J. Fox (Eastman Kodak Co.), U.S. Patent 3,784,376 (1974).
[152] D. Villesot and Y. Lepage, *Tetrahedron Lett.*, 1495 (1977).
[153] G. Freslon and Y. Lepage, *Bull. Soc. Chim. Fr.*, 2105 (1974).
[154] A. Zweig and J. B. Gallivan, *J. Am. Chem. Soc.* **91**, 261 (1969).
[155] Z. Zweig (American Cyanamid Co.), U.S. Patent 3,399,328 (1968).
[156] L. Lepage-Lomme and Y. Lepage, *C. R. Hebd. Seances Acad. Sci.* **272**, 2205 (1971); L. Lepage and Y. Lepage, *J. Heterocycl. Chem.* **15**, 1185 (1978).
[157] L. Lepage and Y. Lepage, *C. R. Hebd. Seances Acad. Sci.* **282**, 555 (1976).
[157a] L. Lepage and Y. Lepage, *J. Heterocycl. Chem.* **15**, 793 (1978).
[157b] M. Peyrot and Y. Lepage, *Bull. Soc. Chim. Fr.*, 2856 (1973).
[158] A. Guyot and F. Valette, *Ann. Chim. Phys.* [8] **23**, 363 (1911).
[159] N. Campbell and H. G. Heller, *J. Chem. Soc.*, 5473 (1965).
[160] E. Clar, F. John, and B. Hawran, *Ber. Dtsch. Chem. Ges.* **62**, 940 (1929).
[161] D. Bertin, *Ann. Chim. (Paris)* [12] **8**, 296 (1953).
[162] C. Seer and D. Dischendorfer, *Monatsh. Chem.* **34**, 1495 (1913).
[163] E. Buchta, H. Vates, and H. Knopp, *Chem. Ber.* **91**, 228 (1958).
[164] A. Zweig, A. K. Hoffmann, D. L. Maricle, and A. H. Maurer, *J. Am. Chem. Soc.* **90**, 261 (1968).
[165] A. Zweig, G. Metzler, A. Maurer, and B. G. Roberts, *J. Am. Chem. Soc.* **89**, 4091 (1967).

TABLE VI
Classes of Diels–Alder Adducts with 1,3-Diphenylbenzo[c]furan

Olefinic compound(s)	References and remarks
Monosubstituted ethylenes, styrenes	166–176
Nitroso compounds, Ph—N=S=O	177, 178
HNO_3	179
1,2-Disubstituted ethylenes	167, 173, 174, 180–183
Alkynes	184–189
1,4-Disubstituted butadienes	173
Cyclopropenes	190–200; a
Cyclopropenones	201–203
Azirines	204, 205
Cyclobutenes, benz-annelated and hetero-substituted analogs	46, 206–235, 285
Higher cycloalkenes, benz-annelated analogs; bicyclic and polycyclic compounds	236–271, 283, 284
Maleic anhydride, 3-sulfolene, maleimide and derivatives	161, 168, 175, 181, 218, 272, 274
Hetero-substituted cyclopentenes	181, 275–277
Quinones	135, 161, 273, 278–282
Azepines	50
Cycloalkynes	237, 286–301
Dehydrobenzenes	144, 302–309
Cyclic allenes	297, 298, 310–313

[a] The assignment of Geibel and Heindl[191] has to be revised.[190,192]

TABLE VII
Diels–Alder Reactions of Some 1,3-Diarylbenzo[c]furans

Benzo[c]furan	Class of olefinic compounds	References
5,6-Dimethyl-1,3-diphenyl	Maleic anhydride, substituted olefins, bicyclic compounds	86, 104, 174, 314, 315
4,7-Dicarbomethoxy-1,3-diphenyl	Dimethyl maleate, dibenzoyl-ethylene, maleic anhydride	316
5,6-Dimethyl-1,3-di-p-tolyl	Maleic anhydride, substituted maleimides	100, 175
1,3-Bis(3,5-dibromo-4-hydroxyphenyl)	Diethyl maleate, maleic anhydride	317
1,3-Di-α-naphthyl	Monosubstituted ethylenes, maleic anhydride	163, 318
1,3,4,7-Tetraphenyl	Monosubstituted ethylenes	107, 108, 319
Hexaphenyl	Monosubstituted ethylenes, disubstituted butadienes, maleic anhydride, maleimide, dehydrobenzene	320, 321

trapping agents for unstable olefins and acetylenes; 1,3-diphenylbenzo[c]-furan has often been used for this purpose. In Table VI, Diels–Alder reactions with 1,3-diphenylbenzo[c]furan are summarized.[166-313]

In Table VII, Diels–Alder reactions that have been carried out with other 1,3-diarylbenzo[c]furans are collected.[314-321]

In many cases the Diels–Alder reactions proceed quite easily and can be conducted even at room temperature. Frequently, the initial formation of a colored complex has been reported[86,178,249,273]; whether this complex

[166] A. Etienne, A. Le Berre, and G. Lonchambon, *C. R. Hebd. Seances Acad. Sci., Ser. C* **263**, 247 (1966).
[167] A. Etienne, A. Spire, and E. Toromanoff, *Bull. Soc. Chim. Fr.*, 750 (1952); A. Etienne and E. Toromanoff, *C. R. Hebd. Seances Acad. Sci.* **230**, 306 (1950).
[168] R. Weiss, A. Abeles, and E. Knapp, *Monatsh. Chem.* **61**, 162 (1932).
[169] C. Dufraisse and R. Priou, *Bull. Soc. Chim. Fr.* [5] **5**, 611 (1938).
[170] A. Etienne, *Ann. Chim. (Paris)* [12] **1**, 5 (1946); *C. R. Hebd. Seances Acad. Sci., Ser. C* **219**, 397 (1944).
[171] A. Etienne and A. Le Berre, *C. R. Hebd. Seances Acad. Sci., Ser. C* **236**, 1046 (1953).
[172] A. Etienne, *C. R. Hebd. Seances Acad. Sci., Ser. C* **217**, 694 (1944).
[173] G. Kaupp and E. Teufel, *J. Chem. Res. (M)*, 1301 (1978); G. Kaupp, personal communication.
[174] C. F. H. Allen, A. Bell, and J. W. Gates, *J. Org. Chem.* **8**, 373 (1943).
[175] A. I. Konovalov and B. N. Solomanov, *J. Org. Chem. USSR (Engl. Transl.)* **11**, 2178 (1975).
[176] L. A. Paquette, R. E. Moerck, B. Harirchian, and P. D. Magnus, *J. Am. Chem. Soc.* **100**, 1597 (1978).
[177] A. Mustafa, *J. Chem. Soc.*, 256 (1949).
[178] M. P. Cava and R. H. Schlessinger, *J. Org. Chem.* **28**, 2464 (1963).
[179] S. Ranganathan and S. K. Kar, *Tetrahedron* **31**, 1391 (1975).
[180] A. Etienne and A. Spire, *C. R. Hebd. Seances Acad. Sci., Ser. C* **230**, 2030 (1950).
[181] J. W. Lown and K. Matsumoto, *Can. J. Chem.* **49**, 3443 (1971).
[182] J. E. Baldwin and D. S. Johnson, *J. Org. Chem.* **38**, 2147 (1973).
[183] R. Weiss and E. Beller, *Monatsh. Chem.* **61**, 143 (1932).
[184] J. E. Berson, *J. Am. Chem. Soc.* **75**, 1240 (1953).
[185] R. A. F. Matheson, A. W. McCulloch, A. G. McInnes, and D. G. Smith, *Can. J. Chem.* **55**, 1422 (1977).
[186] C. G. Krespan, *J. Org. Chem.* **40**, 261 (1975).
[187] S. Lahiri, V. Dabral, M. P. Mahajan, and M. V. George, *Tetrahedron* **33**, 3247 (1977).
[188] R. S. Glass and D. L. Smith, *J. Org. Chem.* **39**, 3712 (1974).
[189] M. Hanack and F. Massa, *Tetrahedron Lett.*, 661 (1977).
[190] M. A. Battiste and C. T. Sprouse, *Tetrahedron Lett.*, 4661 (1970).
[191] K. Geibel and J. Heindl, *Tetrahedron Lett.*, 2133 (1970).
[192] M. P. Cava and K. Narasimhan, *J. Org. Chem.* **36**, 1419 (1971).
[193] T. H. Chan and D. Massuda, *Tetrahedron Lett.*, 3383 (1975).
[194] R. Breslow, G. Ryan, and J. T. Groves, *J. Am. Chem. Soc.* **92**, 988 (1970).
[195] I. G. Bolesov, L. G. Zaitseva, V. V. Plemenkov, I. B. Avezov, and L. S. Surmina, *J. Org. Chem. USSR (Engl. Transl.)* **14**, 64 (1978); I. G. Zaitseva, I. B. Avezov, V. V. Plemenkov, and I. G. Bolesov, *ibid.* **10**, 2227 (1974).
[196] K. B. Baucom and G. B. Butler, *J. Org. Chem.* **37**, 1730 (1970).

Footnotes continued from page 184

[197] D. T. Longone and D. M. Stehouwer, *Tetrahedron Lett.*, 1017 (1970).
[198] M. A. Battiste and C. T. Sprouse, *Tetrahedron Lett.*, 3165 (1969).
[199] R. Breslow, K. Ehrlich, T. Higgs, J. Pecoraro, and F. Zanker, *Tetrahedron Lett.*, 1123 (1974).
[200] M. Oda, Y. Ito, and Y. Kitahara, *Tetrahedron Lett.*, 977 (1978).
[201] R. Breslow and M. Oda, *J. Am. Chem. Soc.* **94**, 4787 (1972).
[202] M. Oda, R. Breslow, and J. Pecoraro, *Tetrahedron Lett.*, 4419 (1972).
[203] J. W. Lown, T. W. Maloney, and G. Dallas, *Can. J. Chem.* **48**, 584 (1970).
[204] A. Hassner and D. J. Anderson, *J. Org. Chem.* **39**, 2031 (1974).
[205] V. Nair, *J. Org. Chem.* **37**, 2508 (1972).
[206] C. D. Nenitzescu, M. Avram, I. G. Dinulescu, and G. Mateescu, *Justus Liebigs Ann. Chem.* **653**, 79 (1962).
[207] G. Wittig and E. R. Wilson, *Chem. Ber.* **98**, 451 (1965).
[208] M. Avram, I. Dinulescu, M. Elian, M. Farcasiu, E. Marica, G. Mateescu, and C. D. Nenitzescu, *Chem. Ber.* **97**, 372 (1964).
[209] T. R. Kelly and R. W. McNutt, *Tetrahedron Lett.*, 285 (1975).
[210] M. Avram, I. G. Dinulescu, E. Marica, G. Mateescu, E. Sliam, and C. D. Nenitzescu, *Chem. Ber.* **97**, 382 (1964).
[211] E. J. Corey and J. Streith, *J. Am. Chem. Soc.* **86**, 950 (1964).
[211a] G. Bianchi, C. De Micheli, A. Gamba, R. Gandolfi, and B. Rezzani, *J. C. S., Perkin 1*, 2222 (1977).
[212] G. Märkl and H. Schubert, *Tetrahedron Lett.*, 1273 (1970).
[213] M. G. Barlow, R. N. Haszeldine, and R. Hubbard, *J. Chem. Soc. C*, 90 (1971).
[214] D. C. Dittmer and N. Takeshina, *Tetrahedron Lett.*, 3809 (1964); D. C. Dittmer, N. Takeshina, and F. A. Davis, *Abstr., 148th Nat. Meet., Am. Chem. Soc., Chicago, Ill., 1964*, 69S.
[215] F. A. Kaplan and B. W. Roberts, *J. Am. Chem. Soc.* **99**, 518 (1977).
[216] B. W. Roberts and A. Wissner, *J. Am. Chem. Soc.* **94**, 7168 (1972).
[217] M. Avram, G. D. Mateescu, D. Dinu, I. G. Dinulescu, and C. D. Nenitzescu, *Stud. Cercet. Chim.* **9**, 435 (1961) [*CA* **57**, 4605 (1962)].
[218] M. P. Cava and J. P. Van Meter, *J. Org. Chem.* **34**, 538 (1969); *J. Am. Chem. Soc.* **84**, 2008 (1962); M. J. Haddadin, B. J. Agha, and R. F. Tabri, *J. Org. Chem.* **44**, 494 (1979).
[219] M. P. Cava and R. Pohlke, *J. Org. Chem.* **27**, 1564 (1962).
[220] M. P. Cava and A.-F. C. Hsu, *J. Am. Chem. Soc.* **94**, 6441 (1972).
[221] T. Bauch, A. Sanders, C. V. Magatti, P. Waterman, D. Judelson, and W. P. Giering, *J. Organomet. Chem.* **99**, 269 (1975).
[222] M. P. Cava and F. M. Scheel, *J. Org. Chem.* **32**, 1304 (1976).
[223] A. Sanders and W. P. Giering, *J. Organomet. Chem.* **104**, 49 (1976); *J. Am. Chem. Soc.* **97**, 919 (1975).
[224] A. Roedig, G. Bonse, and R. Helm, *Chem. Ber.* **106**, 2825 (1973).
[225] M. P. Cava and K. T. Buck, *J. Am. Chem. Soc.* **95**, 5805 (1973).
[226] M. P. Cava, H. Firouzabadi, and M. Krieger, *J. Org. Chem.* **39**, 480 (1974).
[227] M. P. Cava, B. Hwang, and J. P. Van Meter, *J. Am. Chem. Soc.* **85**, 4032 (1963).
[228] T. Miyamoto and Y. Odeira, *Tetrahedron Lett.*, 43 (1973).
[229] M. P. Cava and D. Mangold, *Tetrahedron Lett.*, 1751 (1964).
[230] R. Breslow, D. R. Murayama, S.-I. Murahashi, and R. Grubbs, *J. Am. Chem. Soc.* **95**, 6688 (1973).
[231] R. Breslow, J. Napiersky, and T. C. Clarke, *J. Am. Chem. Soc.* **97**, 6275 (1975).
[232] R. Breslow and P. L. Khanna, *Tetrahedron Lett.*, 3429 (1977).

Footnotes continued from page 184

[233] B. M. Adger, C. W. Rees, and R. C. Storr, *J. C. S., Perkin 1*, 45 (1975).
[234] B. M. Adger, M. Keating, C. W. Rees, and R. C. Storr, *Chem. Commun.*, 19 (1973).
[235] C. W. Rees, R. C. Storr, and P. J. Whittle, *Chem. Commun.*, 411 (1976).
[236] G. Wittig and T. F. Burger, *Justus Liebigs Ann. Chem.* **632**, 85 (1960).
[237] G. Wittig, J. Weinlich, and E. R. Wilson, *Chem. Ber.* **98**, 458 (1965).
[238] D. W. Jones and R. L. Wife, *J. C. S., Perkin 1*, 1654 (1976).
[239] E. J. Corey, F. A. Carey, and R. A. E. Winter, *J. Am. Chem. Soc.* **87**, 934 (1965).
[240] G. Wittig and A. Krebs, *Chem. Ber.* **94**, 3260 (1966).
[241] G. Wittig and R. Polster, *Justus Liebigs Ann. Chem.* **612**, 102 (1958).
[242] E. Vedejs, K. A. J. Snoble, and P. L. Fuchs, *J. Org. Chem.* **38**, 1178 (1973).
[243] A. J. Bridges and G. H. Whitham, *J. C. S., Perkin 1*, 2264 (1975).
[244] T. Sasaki, K. Kanematsu, and Y. Yukimoto, *Heterocycles* **2**, 1 (1974).
[245] A. G. Anastassiou and R. G. Griffiths, *J. Am. Chem. Soc.* **93**, 3083 (1971); J. E. Baldwin, A. H. Andrist, and R. K. Pinschmidt, *ibid.* **94**, 5845 (1972).
[246] T. Sasaki, K. Kanematsu, K. Hayakawa, and M. Sugiura, *J. Am. Chem. Soc.* **97**, 355 (1975).
[247] A. Padwa, W. Koehn, J. Masaracchia, C. L. Osborn, and D. J. Trecker, *J. Am. Chem. Soc.* **93**, 3633 (1971).
[248] E. Bergmann, *J. Am. Chem. Soc.* **74**, 1075 (1952).
[249] T. S. Cantrell and H. Shechter, *J. Org. Chem.* **33**, 114 (1968).
[250] G. I. Poos, H. Kleis, R. R. Wittekind, and J. D. Roseman, *J. Org. Chem.* **26**, 4898 (1961).
[251] J. M. Landesberg and J. Sieczkowski, *J. Am. Chem. Soc.* **93**, 974 (1971).
[252] B. M. Trost, *J. Am. Chem. Soc.* **91**, 918 (1969).
[253] B. M. Trost, G. M. Bright, C. Frihart, and D. Britelli, *J. Am. Chem. Soc.* **93**, 739 (1971).
[254] G. W. Nachtigall, Ph.D. Thesis, University of Colorado, Boulder (1968) (quoted in Cristol and Noven[255]).
[255] S. J. Cristol and A. L. Noreen, *J. Am. Chem. Soc.* **91**, 3969 (1969).
[256] G. R. Buske and W. T. Ford, *J. Org. Chem.* **41**, 1998 (1976).
[257] C. E. Dahl, Ph.D. Thesis, University of Texas at Austin (1971) (quoted in Carlton and Levin[259]).
[258] W. Washburn, Ph.D. Thesis, Columbia University, 1971 (quoted in Carlton and Levin[259]).
[259] J. B. Carlton and R. H. Levin, *Tetrahedron Lett.*, 3761 (1976).
[260] R. D. Miller, D. Kaufmann, and J. J. Mayerle, *J. Am. Chem. Soc.* **99**, 8511 (1977).
[261] P. Warner and S.-C. Chang, *Tetrahedron Lett.*, 3981 (1978).
[262] J. B. Carlton, R. H. Levin, and J. Clardy, *J. Am. Chem. Soc.* **98**, 6068 (1976).
[263] A. H. Alberts, J. Strating, and H. Wynberg, *Tetrahedron Lett.*, 3047 (1973).
[264] R. Greenhouse, W. T. Borden, K. Hirotsu, and J. Clardy, *J. Am. Chem. Soc.* **99**, 1664 (1977).
[264a] R. Greenhouse, W. T. Borden, T. Ravindranathan, K. Hirotsu, and J. Clardy, *J. Am. Chem. Soc.* **99**, 6955 (1977).
[265] W. G. Dauben and J. D. Robbins, *Tetrahedron Lett.*, 151 (1975).
[266] J. A. Chong and J. R. Wiseman, *J. Am. Chem. Soc.* **94**, 8627 (1972).
[267] K. B. Becker, *Helv. Chim. Acta* **60**, 81 (1977).
[268] J. R. Wiseman and W. A. Pletcher, *J. Am. Chem. Soc.* **92**, 956 (1970).
[269] W. G. Dauben and J. Ipaktschi, *J. Am. Chem. Soc.* **95**, 5088 (1973).
[270] J. A. Marshal and H. Faubl, *J. Am. Chem. Soc.* **92**, 948 (1970).
[271] K. J. Shea and S. Wise, *J. Am. Chem. Soc.* **100**, 6519 (1978).
[272] C. Dufraisse and R. Priou, *Bull. Soc. Chim. Fr.* [5] **5**, 502 (1938).
[273] E. de Barry Barnett, *J. Chem. Soc.*, 1326 (1935).
[274] M. P. Cava and J. McGrady, *J. Org. Chem.* **40**, 72 (1975).
[275] J. A. Moore, R. Muth, and R. Sorace, *J. Org. Chem.* **39**, 3799 (1974).

Footnotes continued from page 184

[276] G. Wittig, H. Härle, E. Knauss, and K. Niethammer, *Chem. Ber*, **93**, 951 (1960).
[277] T. Sasaki, K. Kanematsu, and K. Iizuka, *Heterocycles* **3**, 109 (1975).
[278] C. F. Allen and J. W. Gates, *J. Am. Chem. Soc.* **65**, 1502 (1943).
[279] A. Etienne and R. Heymes, *Bull. Soc. Chim. Fr.*, 1038 (1947).
[280] C. Dufraisse and P. Compagnon, *C.R. Hebd. Seances Acad. Sci., Ser. C* **207**, 585 (1938).
[281] E. Bergmann, *J. Chem. Soc.*, 1147 (1938).
[282] D. L. Fields and J. B. Miller, *J. Heterocycl. Chem.* **7**, 91 (1970).
[283] C. B. Quinn and J. R. Wiseman, *J. Am. Chem. Soc.* **95**, 6120 (1973).
[284] C. B. Quinn, J. R. Wiseman, and J. C. Calabrese, *J. Am. Chem. Soc.* **95**, 6121 (1973).
[285] W. J. le Noble, "Highlights of Organic Chemistry." Dekker, New York, 1974.
[286] K. L. Erickson and J. Wolinsky, *J. Am. Chem. Soc.* **87**, 1142 (1965).
[287] G. Wittig and R. Pohlke, *Chem. Ber.* **94**, 3276 (1961).
[288] G. Wittig, A. Krebs, and R. Pohlke, *Angew. Chem.* **72**, 324 (1960).
[289] G. Wittig and A. Krebs, *Chem. Ber.* **94**, 3260 (1961).
[290] G. Wittig and J. Heyn, *Justus Liebigs Ann. Chem.* **726**, 57 (1969).
[291] G. Wittig and U. Mayer, *Chem. Ber.* **96**, 329 (1963).
[292] D. E. Applequist, P. A. Gebauer, D. E. Gwynn, and L. H. O'Connor, *J. Am. Chem. Soc.* **94**, 4272 (1972).
[293] A. Krebs and H. Kimling, *Angew. Chem.* **83**, 540 (1971); *Angew. Chem., Int. Ed. Engl.* **10**, 509 (1971).
[294] A. Krebs and H. Kimling, *Tetrahedron Lett.*, 763 (1970).
[295] A. Krebs and H. Kimling, *Justus Liebigs Ann. Chem.*, 2074 (1974).
[296] G. Wittig and H.-L. Dorsch, *Justus Liebigs Ann. Chem.* **711**, 46 (1968).
[297] G. Wittig, H.-L. Dorsch, and J. Meske-Schüller, *Justus Liebigs Ann. Chem.* **711**, 55 (1968).
[298] P. E. Eaton and C. E. Stubbs, *J. Am. Chem. Soc.* **89**, 5722 (1967).
[299] E. Kloster-Jensen and J. Wirz, *Helv. Chim. Acta* **58**, 162 (1975).
[300] H. N. C. Wong and F. Sondheimer, *Angew. Chem.* **88**, 126 (1976); *Angew. Chem., Int. Ed. Engl.* **15**, 117 (1976).
[301] G. Schröder, H. Röttele, R. Merenyi, and J. F. M. Oth, *Chem. Ber.* **100**, 3527 (1967).
[302] R. Harrison, H. Heaney, and P. Lees, *Tetrahedron* **24**, 4589 (1968).
[303] G. Wittig, E. Knauss, and K. Niethammer, *Justus Liebigs Ann. Chem.* **630**, 10 (1960).
[304] G. Wittig, W. Stilz, and E. Knauss, *Angew. Chem.* **70**, 166 (1958).
[305] K. L. Shephard, *Tetrahedron Lett.*, 3371 (1975).
[306] G. F. Morrison and J. Hooz, *J. Org. Chem.* **35**, 1196 (1970).
[307] I. Fleming and T. Mah, *J. C. S., Perkin 1*, 1577 (1976).
[308] H. Heaney and J. M. Jablonski, *J. Chem. Soc. C*, 1895 (1968).
[309] G. A. Moser, F. E. Tibbets, and M. D. Rausch, *Organomet. Chem. Synth.* **1**, 99 (1971).
[310] G. Wittig and P. Fritze, *Angew. Chem.* **78**, 905 (1966); *Angew. Chem., Int. Ed. Engl.* **5**, 846 (1966); G. Wittig and P. Fritze, *Justus Liebigs Ann. Chem.* **711**, 82 (1968).
[311] G. Wittig and J. Meske-Schüler, *Justus Liebigs Ann. Chem.* **711**, 76 (1968).
[312] W. Tochtermann, D. Schäfer, and D. Pfaff, *Justus Liebigs Ann. Chem.* **764**, 1 (1972).
[313] H. Oda, Y. Ito, and Y. Kitahara, *Tetrahedron Lett.*, 2587 (1975).
[314] M. Kim and J. D. White, *J. Am. Chem. Soc.* **97**, 452 (1975).
[315] M. Kim and J. D. White, *J. Am. Chem. Soc.* **99**, 1172 (1977).
[316] G. Freslon and Y. Lepage, *C.R. Hebd. Seances Acad. Sci., Ser. C* **280**, 961 (1975).
[317] R. Weiss and F. Mayer, *Monatsh. Chem.* **71**, 6 (1939).
[318] R. Weiss and J. Koltes, *Monatsh. Chem.* **65**, 351 (1935).
[319] C. Dufraisse and Y. Lepage, *C.R. Hebd. Seances Acad. Sci., Ser. C* **258**, 1507 (1964).
[320] M. P. Stevens and F. Razmara, *Tetrahedron Lett.*, 1889 (1970).
[321] F. Razmara and M. P. Stevens, *J. Chem. Eng. Data* **17**, 511 (1972).

precedes the Diels–Alder reaction is generally not known.[322] With suitable olefinic compounds, two steroisomers (endo and exo) can be formed; in a number of instances both isomers have been isolated. Generally the endo adducts are less stable than the exo adducts; longer reaction times and higher temperatures tend to favor the formation of the latter. There has been a report, however, that an endo adduct survived temperatures up to 170°C.[277] Especially with higher substituted benzo[c]furans and/or olefins, it has frequently been observed, that the adduct formation is reversible even at room temperature, making purification difficult or impossible. Highly sterically hindered benzo[c]furans may fail to react. The fact that 1-phenyl-3-mesitylbenzo[c]furan readily gives a [4 + 2]-adduct with vinylene carbonate whereas 1,3-dimesitylbenzo[c]furan does not react at all was interpreted as evidence for a stepwise Diels–Alder reaction.[78]

1,3-Diarylbenzo[c]furans have often been used as trapping agents for unstable olefinic compounds (cycloolefins, cycloalkynes, arynes). Whereas almost all trapping experiments have been conducted in the presence of 1,3-diphenylbenzo[c]furan (**138**), there was a recent report where the use of 5,6-dimethyl-1,3-diphenylbenzo[c]furan was recommended[315]; the authors stated that this latter reagent is more stable than the former and has been found to give more easily characterized substances. The dehydration of the Diels–Alder adducts is generally carried out with inorganic acids (HCl, HBr, H_2SO_4) in ethanol or acetic acid; the reaction may fail.[107] More recently it has been found that the dehydration can be done conveniently with P_2S_5 in CS_2.[219,222] In this respect an adduct (**198**) prepared from 5,6-dimethyl-1,3-diphenylbenzo[c]furan and dibenzoylethylene deserves a remark.[104] After treatment with zinc in acetic acid a product was isolated as deep orange-yellow needles, which resembles in color and fluorescence other benzo[c]furans and which was formulated tentatively as **199**; with maleic anhydride and p-benzoquinone 1:1 adducts (elemental analysis) were obtained. The assigned structure (**199**) remains doubtful, however, since such a compound should be only slightly colored.

(198) (199)

[322] V. D. Kiselev and A. J. Konovalov, *J. Org. Chem. USSR* (*Engl. Transl.*) **10**, 4 (1974); V. D. Kiselev and J. G. Miller, *J. Am. Chem. Soc.* **97**, 4036 (1975).

Naturally, 1,3-diarylbenzo[c]furan–alkyne adducts cannot lose water on dehydration. As was found by Wittig and co-workers, in these cases a rearrangement takes place; cycloalkyne adducts of type **200** yield ketones (**201**).[237,287,291,296] The rearrangement may occur even during chromat-

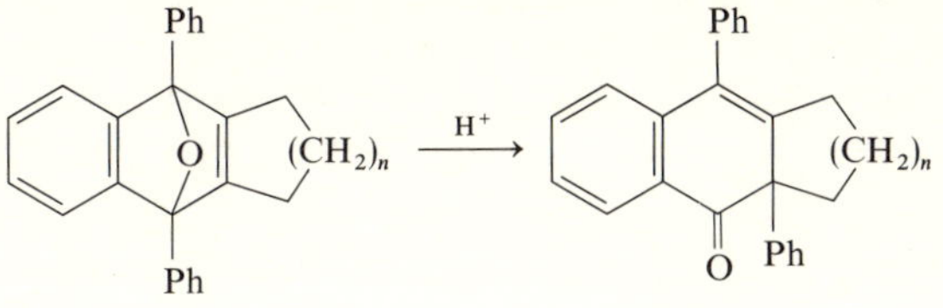

(**200**) $n = 1, 2, 3, 4$ (**201**)

ographic work-up on Al_2O_3.[291] The following self-explanatory mechanism (Scheme 12) has been suggested.[237]

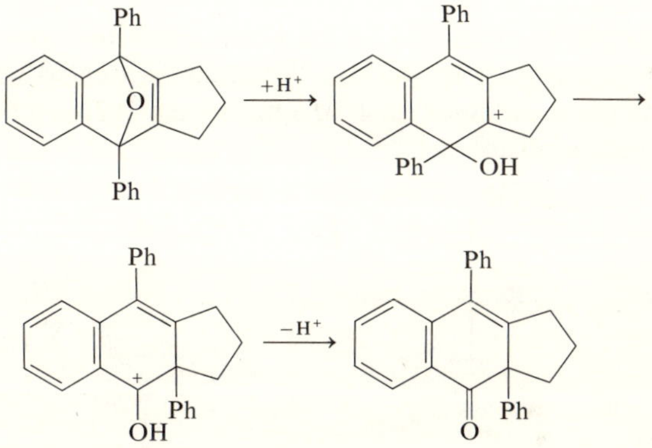

SCHEME 12

The question may arise as to whether the diene moiety of the benzene ring of benzo[c]furan may participate in a Diels–Alder reaction (Eq. 11). Such a reaction seems not to be known in the simple benzo[c]furan series, but

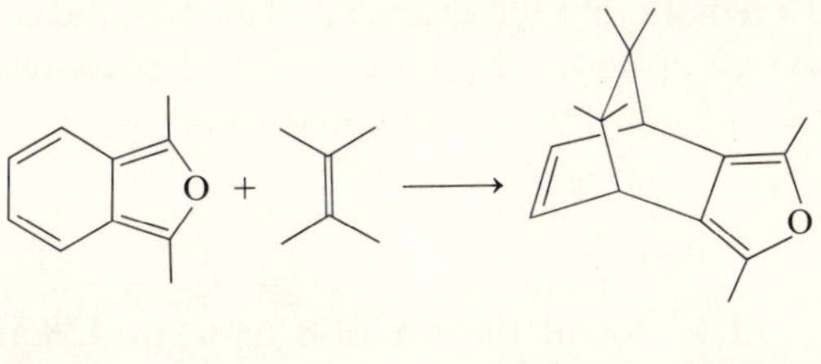

(11)

has been described for benz-annelated and hetero-substituted derivatives (Section VII). These observations are in accord with para localization energies (L_p) calculated for the attack of an olefinic compound across the 1,3 and

4,7 positions; whereas in the parent compound the 1,3 attack is strongly favored, progressive hetero substitution or linear benz-annelation reduces the difference ($|\Delta L_p|$) between 4,7- and 1,3-attack.

$L_p^{1,3}$ 2.863	2.494	2.784	2.595
$L_p^{4,7}$ 3.552	3.351	3.468	3.147
$|\Delta L_p|$ 0.869	0.857	0.684	0.552

SCHEME 13. Para localization energies (L_p) for 1,3- and 4,7-attack (parameters: $h_O = 2.0$, $h_N = 0.5$; $k_{C-O} = 0.8$, $k_{C=N} = 1.0$, $k_{C-N} = 0.8$, $k_{N=N} = 0.8$).

A further possibility exists in the participation of only one double bond; such a reaction seems to be known. When **202** is treated with *p*-chloranil in xylene, a dimer is obtained (**204**); the monomer (**203**) can be trapped with olefinic compounds (Table VII).[316]

(202)　　　　(203)　　　　(204)

$R^1 = CO_2Et$; $R^2 = Ph$

The Diels–Alder reaction between 1,3-diarylbenzo[*c*]furans and simple alkenes and alkynes has been much investigated kinetically. Glass and Smith[188] found a remarkable enhancement of dienophilicity by the trifluoromethane sulfonyl group in a series of substituted phenylacetylenes of type **205**; with **205** (R = SiMe$_3$) no reaction occurred.

$$Ph-C\equiv C-R$$

(205)　R = CHO, CN, CO$_2$Me, COCl, SO$_2$CF$_3$

Konovalov *et al.*[175] found the reaction between **138** and several aryl-substituted styrenes to be of the "diene-donor, dienophile-acceptor" type; they also investigated the kinetics and heats of the reaction between **138** and cyanoethylenes, maleic anhydride, and aryl substituted *N*-arylmaleimides[175,323] and the reaction of **138** and 5,6-dimethyl-1,3-di-*p*-tolylbenzo[*c*]-furan with aryl-substituted *N*-arylmaleimides.[175] The latter benzo[*c*]furan is more reactive and more selective than **138**.

Remarkable kinetic solvent effects reportedly occur in the reaction between **138** and acrylonitrile[324]; the observed large differences in the activation parameters between chloroform ($\Delta H^{\ddagger} = 5.4$ kcal mol^{-1}; $\Delta S^{\ddagger} = -50.3$ eu) and carbon tetrachloride ($\Delta H^{\ddagger} = 22.4$ kcal mol^{-1}; $\Delta S^{\ddagger} = +10.3$ eu) have been criticized.[325]

C. Higher Cycloaddition Reactions

As mentioned in Section III,A, benzo[*c*]furan (**4**) participates in thermal $[\pi_4 + \pi_6]$-cycloadditions. With tropone and substituted tropones, compounds of type **206** (exo; R = H, Cl, OMe) have been obtained,[55] and 6,6-dimethylfulvene yielded **207** (endo).[56] The extended Hückel method and

(206)   (207)

perturbation theory account for both the peri-selectivity and the stereochemistry.[326] Our own investigations[327] in this field differ somewhat: second-order perturbation treatment[328] using eigenvalues and coefficients obtained by the CNDO/2 method reveals that the $[\pi_4 + \pi_6]$-type is favored over the $[\pi_4 + \pi_2]$-type only when HOMO–LUMO interactions are involved. The reverse situation arises when interactions between further occupied and unoccupied orbitals are considered. The same result is obtained for the benzo[*c*]furan–tropone system.[327]

Also, **138** can be involved in a thermal $[\pi_4 + \pi_6]$-cycloaddition: it reacts with cycloheptatriene to give **208** (endo, 10%) and **209** (exo, mp 181–183°C, 20%).[246]

---

[323] V. D. Kiselev, A. N. Ustyagov, I. P. Breus, and A. I. Konovalov, *Dokl. Akad. Nauk SSSR* (*Engl. Transl.*) **234**, 320 (1977).
[324] J. Gillois and P. Rumpf, *Bull. Soc. Chim. Fr.*, 1823 (1959).
[325] A. Wassermann, "Diels-Alder Reactions." Elsevier, Amsterdam, 1965.
[326] H. Mametsuka, A. Mori, H. Takeshita, and H. Yamaguchi, *Heterocycles* **4**, 1867 (1976).
[327] W. Friedrichsen and I. Schwarz, unpublished results.
[328] M. J. S. Dewar, *J. Am. Chem. Soc.* **74**, 3341, 3345, 3350, 3353, 3357 (1952).

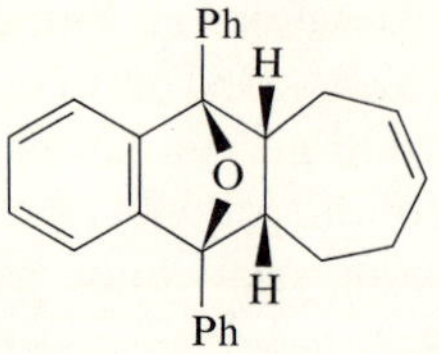

(208) 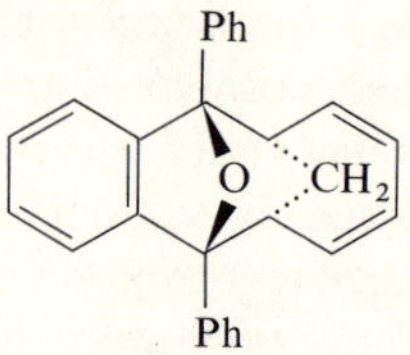 (209)

Unexpected reactions occur when benzo[c]furans are treated with o-quinones. Tedder and co-workers[148,329] reacted 5,6-dimethyl-1,3-diphenyl-benzo[c]furan (210) with o-benzoquinone (211, X = H) and o-chloranil (211, X = Cl) to give dioxoles (212). Subsequent investigations[109,330] have shown that $[\pi_4 + \pi_4]$-cycloadducts (213) are also formed; other benzo[c]furans

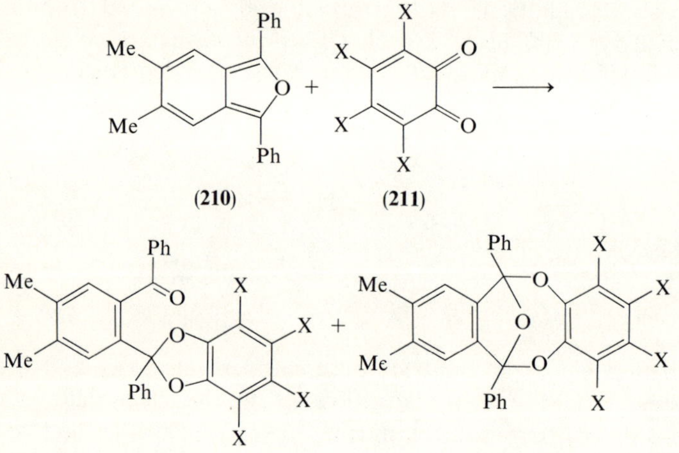

(210) (211)

(212) (213)

and o-quinones (alkyl-substituted quinones, α-naphthoquinone) react analogously. Unsymmetrical benzo[c]furans generally form both possible dioxoles. The reaction has been investigated mechanistically in some detail.[109] The formation of the dioxoles and the $[\pi_4 + \pi_4]$-adducts occurs indepen-

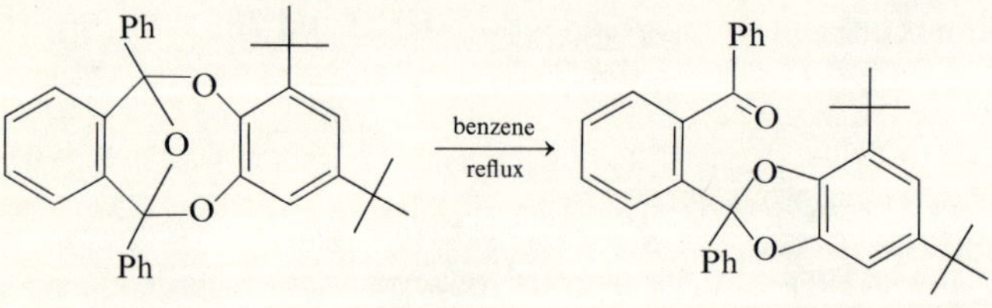

(214) —benzene reflux→ (215)

[329] W. M. Horspool, Q. Rev., Chem. Soc. 23, 204 (1969).
[330] W. Friedrichsen, Tetrahedron Lett., 4425 (1969).

dently; as shown for compound **214**, however, at higher temperatures intramolecular rearrangement to the dioxole (**215**) occurs.

The mechanisms of these unusual reactions are not known with certainty; the temperature and solvent dependence of the product distribution and the rate constants suggest that isopolar transition states rather than zwitterionic intermediates[148] are involved.[109]

An especially intriguing problem concerns the stereochemistry of the $[\pi_4 + \pi_4]$-adducts [e.g., **213** and **214**]. Dreiding models reveal two stable conformations, chair and boat (Fig. 1). Whereas interconversion appears

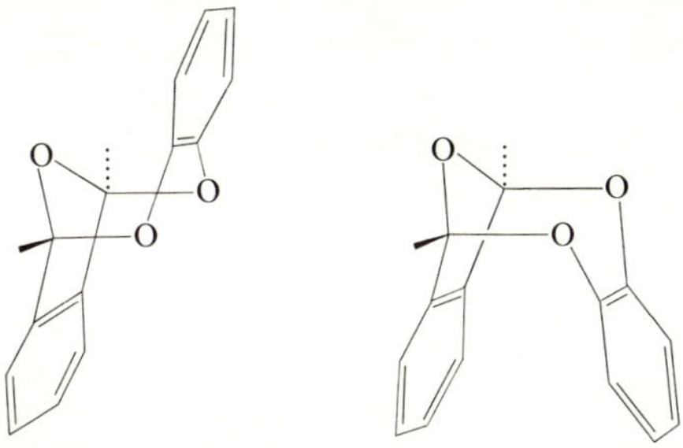

Fig. 1

to be rapid in the *o*-benzoquinone adducts, isomers result from the reaction of **138** with 4,5-dimethyl-*N*,*N*′-diphenylsulfonyl-*o*-benzoquinonediimine (**216**); both compounds (**217** and **218**) were isolated and their structures

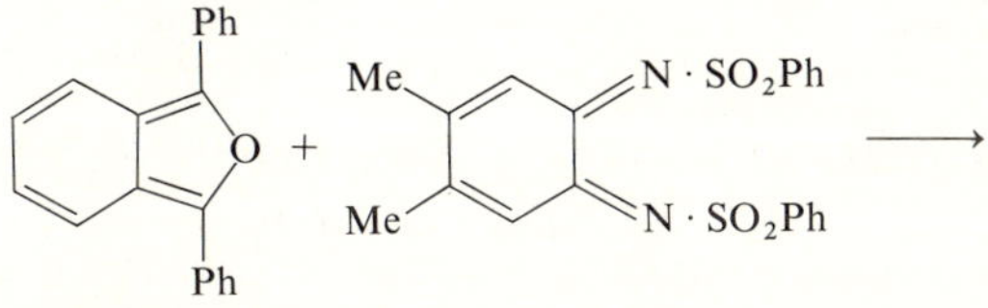

(138)   (216)

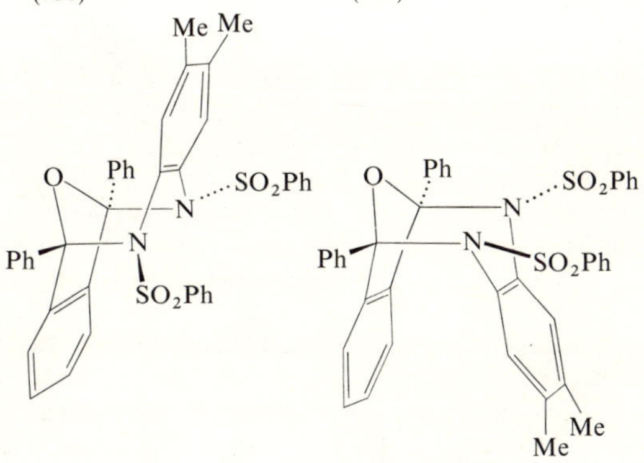

(217)   (218)

clarified through X-ray crystallography.[331] Further benzo[c]furans and o-benzoquinonediimines have been investigated; nitrogen analogs of the dioxoles have also been isolated.[332]

D. Reactions with Singlet Oxygen

1. *Preparative Aspects*

As was recognized by the first investigators, benzo[c]furans are sensitive to oxygen, especially in the presence of sunlight[12,333–336] (see also von Baeyer[16]). When **138** was subjected to these conditions, o-dibenzoylbenzene (**140**) was found to be the only product. Dufraisse and Ecary isolated a precursor of **140**, namely the photooxide **219**,[135,337], which seems to be formed irreversibly[338,339] from singlet oxygen ($^1\Delta_g$) and **138**. Compound **219** was described as a crystalline solid which exploded on warming to 18°C

(138) (219) (140)

but which was stable at −80°C for several hours. In solution at room temperature it was completely converted to **140**; the loss of an oxygen atom from the endo peroxide is not easily explained (see, e.g., Bergmann and McLean[334]). Recently **219** has been prepared in high yield by sensitized photooxidation of **138** with methylene blue absorbed on alumina at −50°C

---

[331] T. Debaerdemaeker, W. Friedrichsen, and M. Röhe, unpublished results.
[332] W. Friedrichsen and M. Röhe, unpublished results.
[333] A. Guyot and J. Catel, *Bull. Soc. Chim. Fr.* [3] **35**, 1135 (1906).
[334] W. Bergmann and M. J. McLean, *Chem. Rev.* **28**, 367 (1941).
[335] K. Gollnick and G. O. Schenck, in "1,4-Cycloaddition Reactions" (J. Hamer, ed.), Chapter 10. Academic Press, New York 1967.
[336] W. R. Adams in "Oxidation" (R. L. Augustine and D. J. Trecker, eds.), Vol. II, Chapter 2. Marcel Dekker, New York, 1974.
[337] C. Dufraisse and S. Ecary, *C.R. Hebd. Seances Acad. Sci., Ser. C* **223**, 735 (1946).
[338] C. Dufraisse and L. Enderlin, *C.R. Hebd. Seances Acad. Sci., Ser. C* **190**, 1229 (1930).
[339] G. Rio and M.-J. Scholl, *Chem. Commun.* 474 (1975).

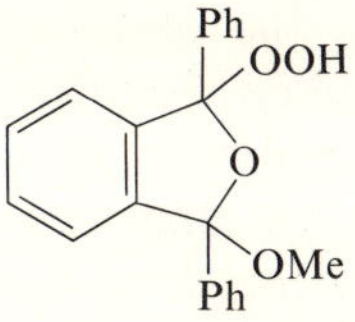

Sec. IV.D] BENZO[c]FURANS 195

(220) 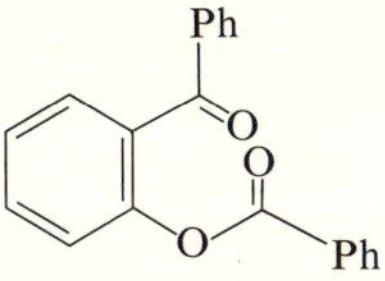 (221) 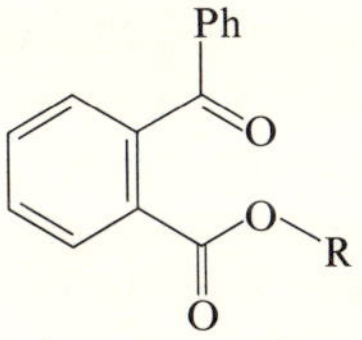 (222) R = H, Ph 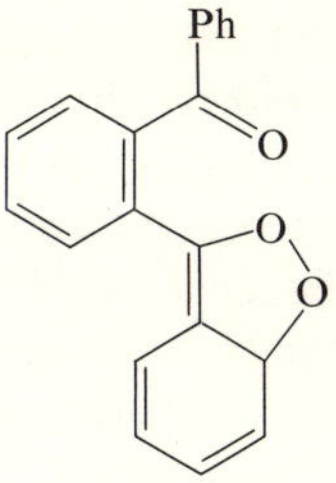 (223)

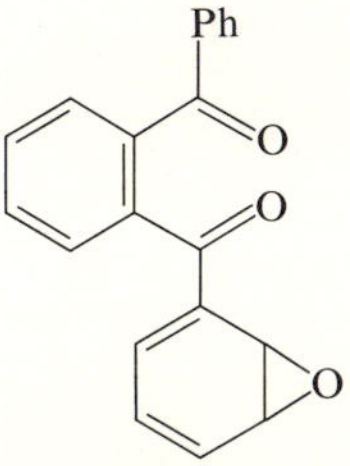

(224) 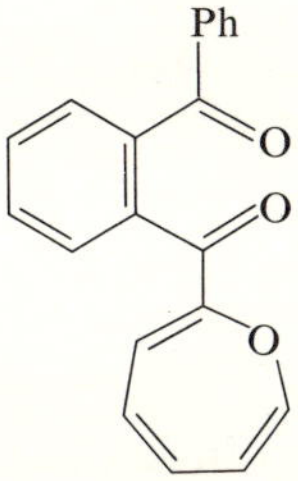 (225) 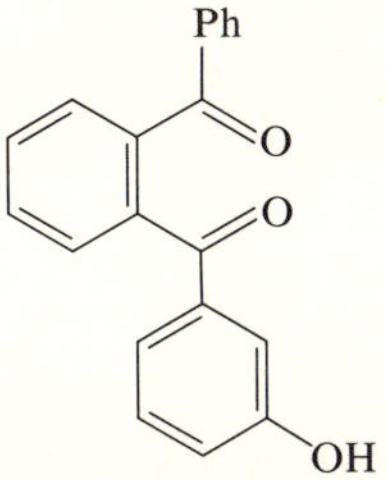 (226) 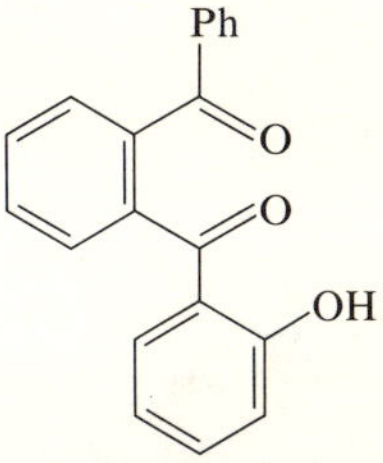 (227)

in ether[339]; its melting point is given as 112–114°C. On reduction with KI in acetic acid **140** is obtained; solvolysis with methanol yields **220**.

The photooxidation of **138** and the fate of **219** seem far more complicated than previously believed. Under various conditions, in addition to **140**, compounds **221**,[340] **222** (R = Ph), **224**, **225**, and **227** could be isolated. The formation of **224**, **225**, and **227** may occur by an arrangement of **219** to the

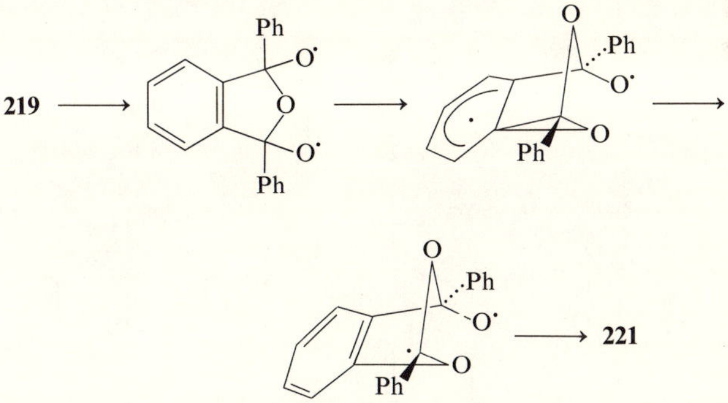

SCHEME 14

[340] F. Nahavandi, F. Razmara, and M. P. Stevens, *Tetrahedron Lett.*, 301 (1973).

hypothetical intermediate **223**, which in turn could give the observed products. In acid **224** isomerizes to **226** and **227**, whereas **225** on treatment with trifluoroacetic acid gives **227**, **222** (R = H), and, presumably, phenol. Ketoester **221** is believed to originate from a Wieland-type rearrangement (Scheme 14).[340] The course of the photooxidation evidently depends on the conditions employed.[147]

1,3,4,7-Tetraphenylbenzo[*c*]furan (**228**, R = H), and hexaphenylbenzo-[*c*]furan (**228**, R = Ph) on photooxidation yield **229** (see also Bergmann *et al.*[107]), **230**, and **231**.[341]

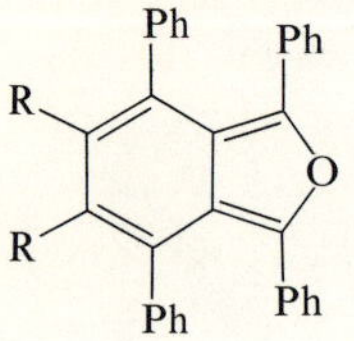

(228) 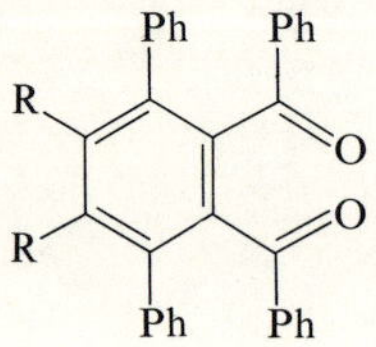   (229) 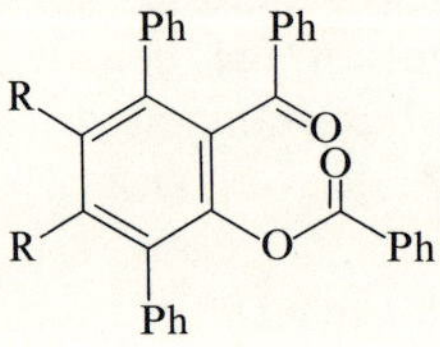   (230) 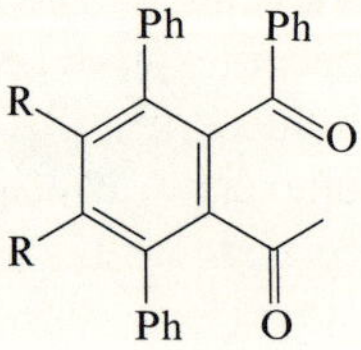   (231)

Only one other photooxide of a benzo[*c*]furan has been described.[129] Hydrocarbon **232** on treatment with oxygen in methanol gives **233** in 96% yield; further treatment with oxygen in benzene or tetrahydrofuran leads to **234**; probably the corresponding benzo[*c*]furan is involved. Whereas reduction with triphenylphosphine in benzene or KI in acetic acid gives **235**,[341] 2 hours reflux in benzene leads to **235**, **236**, and **237** in 7, 48, and 3% yield, respectively.

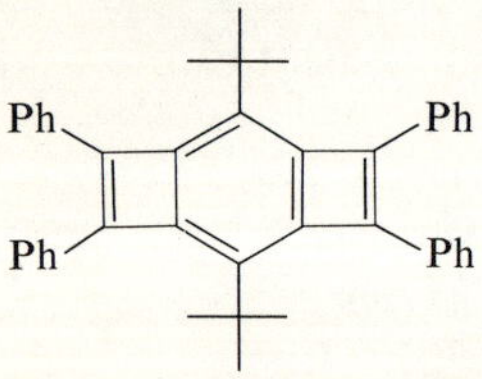

(232) 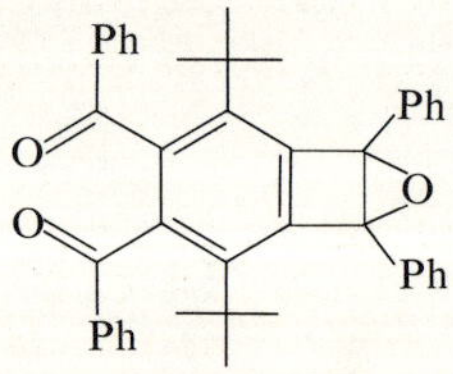   (233) 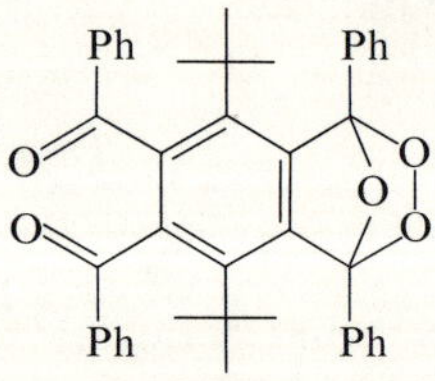   (234)

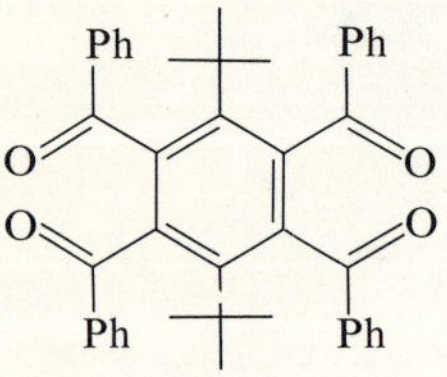

(235) 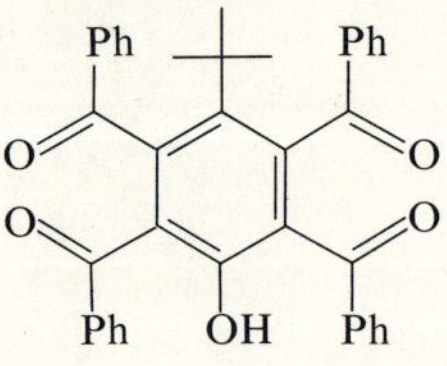   (236) 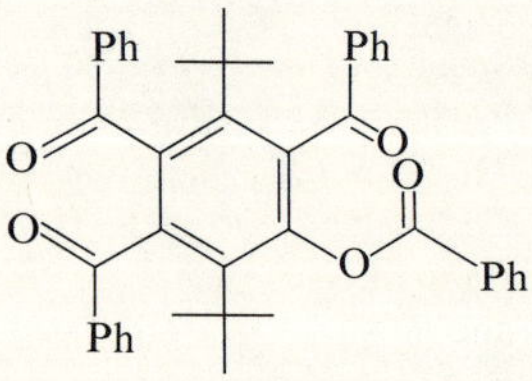   (237)

[341] F. Toda and Y. Takahara, *Bull. Chem. Soc. Jpn.* **49**, 2515 (1976).

2. Mechanistic Aspects

Detailed accounts[147,342-348] on the photooxygenation of 1,3-diphenylbenzo[c]furan (and other benzo[c]furans[344]) reveal a dual mechanistic process. In the high concentration region **138** reacts in its ground state with singlet oxygen ($^1\Delta_g$)[349] (generated either by dye sensitization [Scheme 15, step 1], self-sensitization [steps 2 and 3], or through direct excitation of $^3O_2$ by lasers [step 4]). As has been observed by Wilson,[342] the photooxidation cannot be completely inhibited by 2,3-dimethyl-2-butene; Olmsted and Akashah[344] have shown that the quantum yield of the uninhibited reaction becomes independent of the benzo[c]furan concentration when this is reduced to $\leqslant 10^{-6}$ M and suggested that this results from direct addition of unexcited $^3O_2$ to **138** in its singlet state (Scheme 15, step 5). Stevens et al. have confirmed this[345] and interpreted it in terms of a reencounter reaction of $^1O_2$ ($^1\Delta_g$) with **138** generated in the same triplet–triplet annihilation act.[345-348]

$$^3S + {}^3O_2 \rightarrow S + {}^1O_2 \quad (1)$$
$$^1A + {}^3O_2 \rightarrow {}^3A + {}^3O_2 \quad (2)$$
$$^3A + {}^3O_2 \rightarrow A + {}^1O_2 \quad (3)$$
$$A + {}^1O_2 \rightarrow AO_2 \quad (4)$$
$$^1A + {}^3O_2 \rightarrow AO_2 \quad (5)$$

SCHEME 15. Mechanistic scheme for the photoperoxygenation of **138** (A); S = sensitizer.

Compound **138** is a remarkably efficient quencher of methylene blue fluorescence; quenching occurs presumably by a redox process.[350] It is necessary to consider physical quenching as well as trapping in a kinetic treatment of $^1O_2$ reactions,[351-353] however, quenching of $^1O_2$ by **138** does

---

[342] T. Wilson, *J. Am. Chem. Soc.* **88**, 2898 (1966).
[343] E. Koch, *Tetrahedron* **24**, 6295 (1968).
[344] J. Olmsted and T. Akashah, *J. Am. Chem. Soc.* **95**, 6211 (1973).
[345] B. Stevens, J. A. Ors, and M. L. Pinsky, *Chem. Phys. Lett.* **27**, 157 (1974).
[346] B. Stevens, S. R. Perez, and J. A. Ors, *J. Am. Chem. Soc.* **96**, 6846 (1974).
[347] B. Stevens, *J. Photochem.* **3**, 393 (1974).
[348] B. Stevens and R. R. Williams, *Chem. Phys. Lett.* **36**, 100 (1975); B. Stevens and R. D. Small, *ibid.* **61**, 233 (1979).
[349] Reviews: A. M. Trozzolo, *Ann. N.Y. Acad. Sci.* **171**, 1 (1970); D. R. Kearns, *Chem. Rev.* **71**, 395 (1971); A. P. Schaap, "Singlet Molecular Oxygen. A Collection of Reprints with Comments." Dowden, Hutchinson, & Ross, Inc., Stroudsburg, Pennsylvania, 1976; B. Ranby and J. F. Rabek, "Singlet Oxygen." Wiley, New York, 1978; H. H. Wassermann and R. W. Murray, "Singlet Oxygen." Academic Press, New York, 1979.
[350] R. S. Davidson and K. R. Trethewey, *J. C. S., Perkin 2*, 169 (1977).
[351] S. R. Fahrenholtz, F. H. Dolliden, A. H. Trozzolo, and A. A. Lamola, *Photochem. Photobiol.* **20**, 505 (1974).
[352] C. S. Foote, T.-Y. Ching, and G. G. Geller, *Photochem. Photobiol.* **20**, 511 (1974).
[353] B. Stevens, R. D. Small, and S. R. Perez, *Photochem. Photobiol.* **20**, 515 (1974).

TABLE VIII
1,3-Diphenylbenzo[c]furan as Singlet Oxygen [$^1O_2(^1\Delta_g)$]
Acceptor in Preparative and Mechanistic Studies[a]

Source of $^1O_2$	Topic	References
Sens. (anthracene, phenanthrene, pyrene, 1,2-benzanthracene, benzophenone, benzophenone with 0.1 M naphthalene, MB[b])	Lifetime of $^1O_2$	363
Sens. (MB)	Lifetime of $^1O_2$, solvent effects, deuterium effects, reaction rate with **138**	354, 355, 357, 364, 365
He–Ne laser	$(^1\Delta_g)_2$ and $^1\Delta_g$ of $^1O_2$	359, 366
Sens. (MB, RB[c])	Lifetime of $^1O_2$ in various solvents	356
Sens. (MB, 2-acetonaphthone)	Lifetime and reactivity of $^1O_2$ in aqueous micellar systems	367
Sens. (rubrene)	Fate of oxygen in the quenching of excited singlet states	368
Sens. (pulse radiolysis of benzene; energy transfer to naphthalene, anthracene, benzophenone, benzil)	Determination of energy transfer efficiencies	369
Sens. (**138**, rubrene)	$^1O_2$ acceptor properties and reactivity	346
Nd–YAG laser	Production of $^1O_2$ by direct laser excitation	370
Sens. (MB)	Lifetime of $^1O_2$ in D_2O	371
Sens. (MB)	Lifetime of $^1O_2$; temperature dependence, solvent effects	372
Sens. (MB)	Quenching of $^1O_2$ by bilirubin	358
Nd–YAG laser	Reaction of $^1O_2$ with bilirubin and **138**	373
Sens. (MB)	Quenching of $^1O_2$ by α-tocopherol	352
Sens. (anthracene)	Quenching of $^1O_2$ by transition metal complexes and β-carotene	374

[a] For further details see also Ref. 349.
[b] Methylene blue.
[c] Rose bengal.
[d] 2,7-Diamino-10-ethyl-9-phenylphenanthridinium bromide.
[e] $^1O_2$ possibly involved.
[f] $^1O_2$ possibly not involved.

… Sec. IV.D] BENZO[c]FURANS 199

TABLE VIII (Continued)

Source of $^1O_2$	Topic	References
Sens. (MB)	Quenching of $^1O_2$ by nickel and zinc chelates	375
Sens. (RB, ethidium bromide[d])	Photosensitization abilities	376
Sens. (polystyrene)	Photooxidation of polymers	377
Sens. (MB, RB)	Reactivity and electrophilic character of $^1O_2$	378
Sens. (MB, RB, 138)	Mechanism of the photooxidation of 138	147, 342–348
Sens. (RB)	Solvent effects on the reaction rates of $^1O_2$ with 138	362
Sens. (MB)	Reactivity of 4 and 138 toward $^1O_2$	379
Sens. (MB, RB)	Quenching of $^1O_2$ by tertiary amines	380
Electrodeless discharge	Quenching of $^1O_2$ by tertiary amines	381
O_2^{-}/diacylperoxide	Generation of $^1O_2$; oxidation of 138 to 140 (>97%)	382
O_2^{-}/radical-cation of 1,4-dimethoxy-2,5-di-*tert*-butylbenzene[e]	Generation of $^1O_2$; oxidation of 138 to 140	383
Sens. (RB)	Quenching of $^1O_2$ by O_2^{-}	384
Microwave discharge	$^1O_2$ for preparative purposes; oxidation of 138 to 140 (91%)	385
Electrodeless discharge	$^1O_2$ for preparative purposes; oxidation of 138 to 140 (high yield)	386
Sens. (MB or RB on the surface of silica gel)	$^1O_2$ for preparative purposes; oxidation of 138 in various solvents	387, see also Ref. 388
Electrochemically generated	Generation of $^1O_2$; oxidation of 138 to 140	389
Diperoxychromium(VI) oxide etherate[f]	—	390, 391, 392, 393
Sens. (MB)	Sensitized photooxygenation as laboratory exercise	394

not seem to be a major decay pathway,[354–359] although another report stressed the importance of this step in the kinetic treatment[360] (for the photoperoxidation of other unsaturated organic molecules see also Stevens and Ors[361]).

1,3-Diphenylbenzo[c]furan is one of the fastest singlet oxygen acceptors known and has become a standard compound in preparative and kinetic studies in the $^1O_2$ ($^1\Delta_g$) field; its disappearance can easily be followed either by measuring the absorbance in the 415 nm region or by fluorescence emission (e.g., at 458 nm).[362] Table VIII summarizes work where **138** has been used either for preparative purposes or for mechanistic investigations.[147,342–349,352,354–359,362–394]

Cyclic peroxides may serve as a source of singlet oxygen. Wasserman *et al.*[395] reacted 9,10-diphenylanthracene peroxide (**238**, conveniently prepared as in Nilsson and Kearns[387]) with **138** to give **140**; rubrene peroxide proved to be considerably less efficient. Decomposition of anthracene peroxide alone takes another course.[132,396] When **138** is treated with phthaloyl peroxide (**239**), **140** is isolated in 59% yield; the reaction is accompanied by a weak chemiluminescence.[397,398] A bright yellow chemiluminescence has been observed when a solution of **240** in 1,2,4-trichlorobenzene is treated with dibenzoyl peroxide at about 210°C.[399] The generation of visible light from **138** under conditions where peroxides may present has been described.[400]

(**238**)   (**239**)   (**240**)

---

[354] P. B. Merkel and D. R. Kearns, *Chem. Phys. Lett.* **12**, 120 (1971).
[355] P. B. Merkel and D. R. Kearns, *J. Am. Chem. Soc.* **94**, 7244 (1972).
[356] R. H. Young, D. Brewer, and R. A. Keller, *J. Am. Chem. Soc.* **95**, 375 (1973).
[357] P. B. Merkel and D. R. Kearns, *J. Am. Chem. Soc.* **97**, 462 (1975).
[358] C. S. Foote and T.-Y. Ching, *J. Am. Chem. Soc.* **97**, 6209 (1975).
[359] D. F. Evans and J. N. Tucker, *J. C. S., Faraday 2*, 1661 (1976).
[360] I. B. C. Matheson, J. Lee, B. S. Yamanaski, and W. L. Wolbarsht, *J. Am. Chem. Soc.* **96**, 3343 (1974).
[361] B. Stevens and J. A. Ors, *J. Phys. Chem.* **80**, 2164 (1976).

Footnotes continued from page 200

[362] R. H. Young, K. Wehrly, and R. L. Martin, *J. Am. Chem. Soc.* **93**, 5774 (1971).
[363] D. R. Adams and F. Wilkinson, *J. C. S., Faraday 2* **68**, 586 (1972).
[364] P. B. Merkel and D. R. Kearns, *J. Am. Chem. Soc.* **94**, 1029 (1972).
[365] P. B. Merkel, R. Nilsson, and D. R. Kearns, *J. Am. Chem. Soc.* **94**, 1030 (1972).
[366] D. F. Evans, *Chem. Commun.*, 367 (1969).
[367] A. A. Gorman, G. Lovering, and M. A. J. Rodgers, *Photochem. Photobiol.* **23**, 399 (1976); A. A. Gorman and M. A. J. Rodgers, *Chem. Phys. Lett.* **55**, 52 (1978); Y. Usui, M. Tsukada, and H. Nakamura, *Bull. Chem. Soc. Jpn.* **51**, 379 (1978).
[368] P. B. Merkel and W. G. Herkstroeter, *Chem. Phys. Lett.* **53**, 350 (1978).
[369] A. A. Gorman, G. Lovering, and M. A. J. Rodgers, *J. Am. Chem. Soc.* **100**, 4527 (1978); *ibid.* **101**, 3050 (1979).
[370] I. B. C. Matheson and J. Lee, *Chem. Phys. Lett.* **7**, 475 (1970).
[371] I. B. C. Matheson, J. Lee, and A. D. King, *Chem. Phys. Lett.* **55**, 49 (1978).
[372] C. A. Long and D. R. Kearns, *J. Am. Chem. Soc.* **97**, 2018 (1975).
[373] I. B. C. Matheson, N. U. Curry, and J. Lee, *J. Am. Chem. Soc.* **96**, 3348 (1974).
[374] A. Farmilo and F. Wilkinson, *Photochem. Photobiol.* **18**, 447 (1973).
[375] J. Flood, K. E. Russell, and J. K. S. Wan, *Macromolecules* **6**, 669 (1973).
[376] J. Olmsted and D. R. Kearns, *Biochemistry* **16**, 3647 (1977).
[377] J. F. Rabek and B. Ranby, *J. Polym. Sci., Polym. Chem. Ed.* **12**, 273 (1974).
[378] R. H. Young, R. L. Martin, N. Chingh, C. Mallon, and R. H. Kayser, *Can. J. Chem.* **50**, 932 (1972); R. H. Young, R. L. Martin, K. Wehrly, and D. Feriozi, *Prepr. Div. Pet. Chem., Am. Chem. Soc.*, **16**, A89 (1971).
[379] R. H. Young and D. T. Feriozi, *Chem. Commun.*, 841 (1972).
[380] R. H. Young, D. Brewer, R. Kayser, R. Martin, D. Feriozi, and R. A. Keller, *Can. J. Chem.* **52**, 2889 (1974); R. H. Young and R. L. Martin, *J. Am. Chem. Soc.* **94**, 5183 (1972).
[381] C. Ouannes and T. Wilson, *J. Am. Chem. Soc.* **90**, 6527 (1968).
[382] W. C. Danen and R. L. Arudi, *J. Am. Chem. Soc.* **100**, 3944 (1978).
[383] A. Nishinaga, T. Shimizu, H. Tomita, and T. Matsuura, unpublished results [quoted after T. Matsuura, *Tetrahedron* **33**, 2881 (1977)].
[384] H. J. Guiraud and C. S. Foote, *J. Am. Chem. Soc.* **98**, 1984 (1976).
[385] J. R. Scheffer and M. D. Ouchi, *Tetrahedron Lett.*, 223 (1970).
[386] E. J. Corey and W. C. Taylor, *J. Am. Chem. Soc.* **86**, 3881 (1964).
[387] R. Nilsson and D. R. Kearns, *Photochem. Photobiol.* **19**, 181 (1974).
[388] K. A. Zaklika, A. L. Thayer, and A. P. Schaap, *J. Am. Chem. Soc.* **100**, 4916 (1978); C. W. Jefford, A. Exarchou, and P. A. Cadby, *Tetrahedron Lett.*, 2053 (1978).
[389] E. A. Mayeda and A. J. Bard, *J. Am. Chem. Soc.* **95**, 6223 (1973).
[390] H. W.-S. Chan, *Chem. Commun.*, 1550 (1970).
[391] H. W.-S. Chan, *J. Am. Chem. Soc.* **93**, 4632 (1971).
[392] J. E. Baldwin, J. C. Swallow, and H. W.-S. Chan, *Chem. Commun.*, 1407 (1971).
[393] H. W.-S. Chan, *Symp. Int. Oxyd. Lipides Catal. Met. (C.R.), 3rd*, 22 (1973).
[394] J. A. Bell and J. D. MacGillivray, *J. Chem. Educ.* **51**, 677 (1974).
[395] H. H. Wasserman and J. R. Scheffer, *J. Am. Chem. Soc.* **89**, 3073 (1967); H. H. Wasserman, J. R. Scheffer, and J. L. Cooper, *ibid.* **94**, 4991 (1972).
[396] J. Rigaudy, J. Baranne-Lafont, A. Defoin, and N. K. Cuong, *Tetrahedron* **34**, 73 (1978); A. Defoin, J. Baranne-Lafont, J. Rigaudy, and J. Guilhem, *ibid.* 83.
[397] K.-D. Gundermann, *Chem.-Ztg.* **99**, 279 (1975); K.-D. Gundermann and M. Steinfatt, *Angew. Chem.* **87**, 546 (1975); *Angew. Chem., Int. Ed. Engl.* **14**, 560 (1975).
[398] M. Steinfatt, Ger. Offen. 2,547,808 (1977).
[399] R. B. Kurtz, *Trans. N. Y. Acad. Sci.* [2] **16**, 399 (1954).
[400] M. M. Rauhut and A. M. Semsel (American Cyanamid Co.), U.S. Patent 3,329,621 (1967).

When 9,10-di-*p*-styrylanthracene is copolymerized with styrene and subsequently peroxidized an insoluble carrier for singlet oxygen is obtained; heating of the polymer in benzene in the presence of **138** yields **140** in up to 60% yield.[401]

(138)   (219)   (140)

Whether in all these cases the addition of $^1O_2$ to **138** with formation of the peroxide **219** takes place is open to question; it is strange that only **140** has been observed as reaction product despite the observation that decomposition of **219** may lead to other oxidation products as well (see above).

1,3-Diphenylbenzo[*c*]furan is quite often used in biochemical investigations as a probe for $^1O_2$ appearance. The oxidation of **138** by $^1O_2$ was significantly inhibited by superoxide dismutase[402,403] and catalase,[402] The system superoxide dismutase/H_2O_2 attacks **138**; H_2O_2 alone was ineffective.[404,405] The *in vitro* peroxidation of rat liver microsomal lipid, which was increased in the presence of paraquat (1,1′-dimethyl-4,4′-bipyridylium dichloride) could be inhibited by either superoxide dismutase or **138**.[406] In an investigation on the mechanism of liver microsomal lipid peroxidation, it was found that **138** inhibited the peroxidation promoted by xanthine and xanthine oxidase; the presence of $^1O_2$ formed from superoxide was suggested.[407,408] Buffered solutions of horseradish peroxidase and soybean lipoxidase in the presence of ethyl linoleate convert **138** to **140** (65% and 30% yield, respectively)[391,393,409]; whether $^1O_2$ is involved is unclear. The possible involvement of $^1O_2$ in prostaglandin biosynthesis and the oxygenation of

---

[401] I. Rosenthal and A. J. Acker, *Isr. J. Chem.* **12**, 897 (1974).
[402] A. Finazzi-Agro, C. Giovagnoli, P. DeSole, L. Calabrese, G. Rotilio, and B. Mondovi, *FEBS Lett.* **21**, 183 (1972); A. Finazzi-Agro, P. DeSole, G. Rotilio, and B. Mondovi, *Ital. J. Biochem.* **22**, 217 (1973).
[403] U. Weser, W. Paschen, and M. Younes, *Biochem. Biophys. Res. Commun.* **66**, 769 (1975).
[404] E. K. Hodgson and I. Fridovich, *Biochemistry* **14**, 5299 (1975).
[405] Review: A. M. Michelson, J. M. McCord, and I. Fridovich, eds., "Superoxide and Superoxide Dismutases." Academic Press, New York, 1977.
[406] J. S. Bus, S. D. Aust, and J. E. Gibson, *Biochem. Biophys. Res. Commun.* **58**, 749 (1974).
[407] T. C. Pedersen and S. D. Aust, *Biochem. Biophys. Res. Commun.* **52**, 1071 (1973).
[408] T. C. Pedersen and S. D. Aust, *Biochim. Biophys. Acta* **385**, 232 (1975).
[409] H.W.-S. Chan, *J. Am. Chem. Soc.* **93**, 2357 (1971).

138 to **140** by prostaglandin synthetase has been investigated.[410,411] The *in vitro* destruction of chlorophyll by bisulfite/oxygen could not be inhibited by **138**.[412] Singlet oxygen has been identified as a cytotoxic agent in photoinactivation of a murine tumor. **138** was used as $^1O_2$ trap; **140** was found to be formed nearly quantitatively (GLPC).[413]

The addition of O_2 (whether $^3\Sigma_g$ or $^1\Delta_g$) to **138** or **4** has apparently not been studied theoretically although selection rules for singlet oxygen reactions,[414] interpretation of substituent effects[415] and orbital correlations for homolytic fission and re-formation of the peroxide bond in 9,10-diphenylanthracene peroxide and rubrene peroxide[346,416] have been published.

Interesting "activations of oxygen" and peroxidation of organic substrates were reported by Barton *et al.*[417] When **138** was treated in dichloromethane either with triphenylmethyl hexachloroantimonate and oxygen in the presence of light (laboratory light was sufficient) or with tris(*p*-bromophenyl)ammoniumyl hexachloroantimonate and oxygen in the dark (in each case at $-78°C$) **140** was produced in 80 and 90% yield, respectively; both catalysts were effective in the oxygenation of ergosteryl acetate to the peroxide.

1,3-Diphenylbenzo[*c*]furan is reportedly attacked by oxygen without catalysts in the dark, even in the solid state.[394] When a solution of **138** in benzene (or toluene) is treated with oxygen, on precipitation with ether a white powder with mp 140°C is obtained in 82% yield; on heating to 150–170°C explosion occurs.[79] Iodometric titration in tetrahydrofuran showed a content of 94% peroxide; from this solution 37% *o*-dibenzoylbenzene (**140**) has been obtained. The molecular weight was determined to be 2830 (cryoscopic in benzene) and the structure formulated as **241** ($n = 8$–10).

(**241**)

---

[410] L. J. Marnett, P. Wlodawer, and B. Samuelson, *J. Biol. Chem.* **250**, 8510 (1975).
[411] A. Rahimtula and P. J. O'Brien, *Biochem. Biophys. Res. Commun.* **70**, 893 (1976).
[412] G. D. Peiser and S. F. Yang, *Plant Physiol.* **60**, 277 (1977).
[413] K. R. Weishaupt, C. J. Gomer, and T. J. Dougherty, *Cancer Res.* **36**, 2326 (1976).
[414] D. R. Kearns, *J. Am. Chem. Soc.* **91**, 6554 (1969).
[415] O. Chalvet, R. Daudel, C. Ponce, and J. Rigaudy, *Int. J. Quantum Chem.* **2**, 521 (1968).
[416] B. Stevens and R. D. Small, *J. Phys. Chem.* **81**, 1605 (1977).
[417] D. H. R. Barton, R. K. Haynes, G. Leclerc, P. D. Magnus, and I. D. Menzies, *J. C. S., Perkin 1*, 2055 (1975).

A slow, dark reaction of **138** with oxygen was also reported by other workers,[359,394] but it was recently claimed that **138** in solution does not absorb oxygen in the dark, autoxidation only took place in the presence of azobisisobutyronitrile.[147] A peroxide with mp 127.5–128.5°C and a molecular weight of 1350 ± 50 was obtained (**241**; $n = 4–5$); 35% *o*-dibenzoylbenzene could also be isolated. The peroxides decomposed in benzene (half-life ~ 72 h at 50°C) to give **140**.

E. Photochemical Reactions

As already described by Guyot and Catel,[12] 1,3-diphenylbenzo[*c*]furan dimerizes under the action of sunlight, resembling in this behavior anthracenes and other condensed hydrocarbons. The photoproduct was formulated as **242**; structures **243**[145] and **244**[80,145,418] can be rejected.[419] The photo-

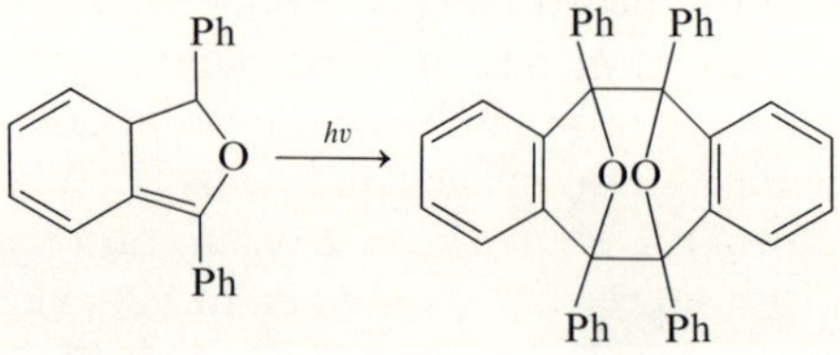

(138)   (242)

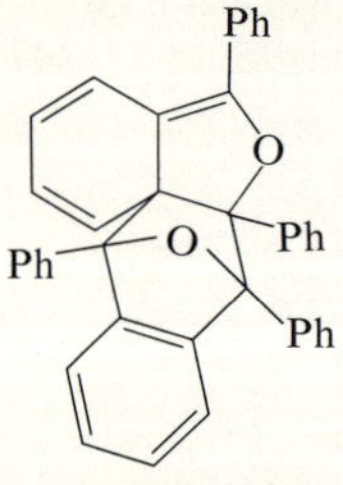

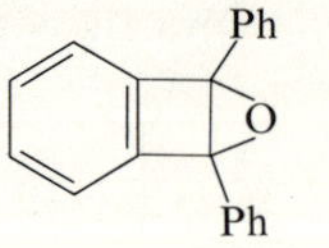

(243)   (244)

dimer is only slightly soluble in the majority of organic solvents. On heating to the melting point (151°C,[173] 190–200°C,[80] 210–220°C[145]), it is reconverted to **138**. Bromine with **242** gives *o*-dibenzoylbenzene (**140**) in 80% yield; reduction (sodium/*tert*-butanol/tetrahydrofuran) gives *o*-dibenzylbenzene (62%).[80] The stereochemistry of the photodimer (**245** or **246**) is unknown.

[418] A. Schönberg, A. Mustafa, and G. Aziz, *J. Am. Chem. Soc.* **76**, 4576 (1954).
[419] W. Friedrichsen, *Tetrahedron Lett.*, 1219 (1969).

(245)   (246)

Schönberg et al. reportedly obtained the photodimer in excellent yields when a solution of **138** in diethyl phthalate was heated to 270°C for 30 minutes and then rapidly chilled with ice water[420] (on slow cooling no dimer was obtained; see also Mustafa[177]). These observations could not be substantiated[80,145]; a product obtained in this way (mp 200°C) differed from the photodimer (comparison of IR spectra).[80] The structure of this product is not known. A dimer was also obtained when 1,3-bis(dimethylaminophenyl)benzo[c]furan was treated with acetic acid; with sulfuric acid it was reconverted to the monomer.[120]. There seem to be no reports concerning the photodimerization of other benzo[c]furans. As we have found, 1,3-bis(p-methylphenyl)- and 1,3-bis(p-methoxyphenyl)benzo[c]furan yield photodimers analogous to **138**, whereas 1-duryl-3-phenyl-, 1,3-dimesityl-, and 1,3,4,7-tetraphenylbenzo[c]furan are completely stable to irradiation (in absence of oxygen).[421]

More recently, photochemical reactions of **138** with cyclic and acyclic olefins have been described. When **138** is irradiated (Pyrex) with cycloheptatriene, [4 + 4]- and [4 + 6]-adducts (**247–250**, Scheme 16) are obtained[246,422]; in addition, photodimer **242** and o-dibenzoylbenzene (**140**) were isolated. The ratio of the [4 + 6]-adducts to the [4 + 4]-adducts [(**249** + **250**)/(**247** + **248**)] is increased in air-saturated benzene solutions compared with oxygen-free benzene solutions, and enhanced in heavy atom solvents (e.g., chloroform compared with cyclohexane); furthermore, this ratio is decreased in the presence of the triplet quencher azulene. These observations suggest that [4 + 6]- and [4 + 4]-adducts are formed by different mechanisms.

---

[420] A. Schönberg, A. Mustafa, M. Z. Barakat, N. Latif, R. Moubasher, and A. Mustafa, *J. Chem. Soc.*, 2126 (1948); A. Schönberg, A. Mustafa, and M. Z. Barakat, *Nature (London)* **160**, 401 (1947).
[421] W. Friedrichsen, unpublished results.
[422] T. Sasaki, K. Kanematsu, and K. Hayakawa, *Tetrahedron Lett.*, 343 (1974).

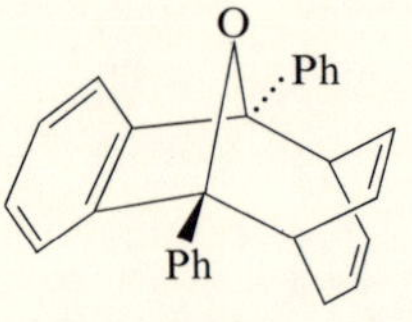

(247)

benzene (%): 7[a] (4.9)[b]
ethanol: 3.6 (4)[c]

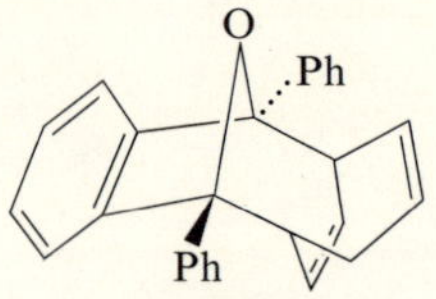

(248)

11.5[a] (8.0)[b]
10 (14)[c]

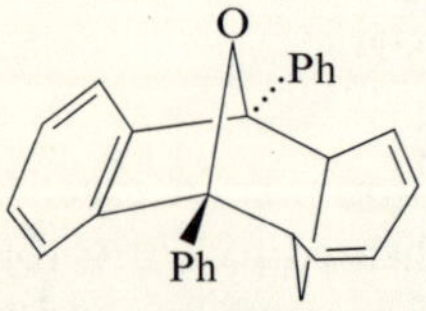

(249)

5.5[a] (4.1)[b]
11.5 (16)[c]

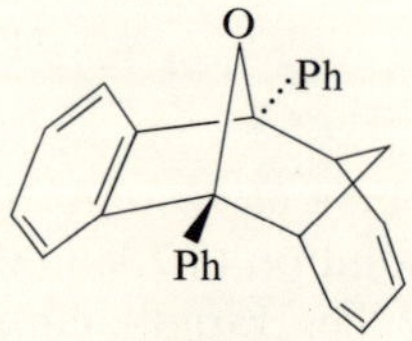

(250)

9.5[a] (6.6)[b]
14.5 (20)[c]

SCHEME 16. [a] Theoretical part; [b] experimental section, of Sasaki et al.[246] [c] The figures in parentheses for ethanol solution are from Sasaki et al.[422]

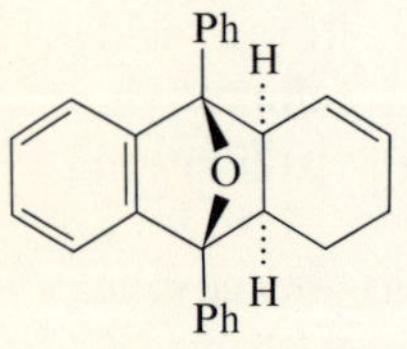

(251) (11%) 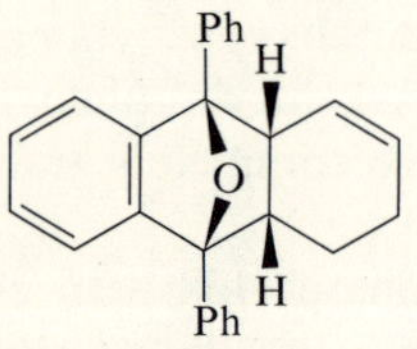   (252) (6%) 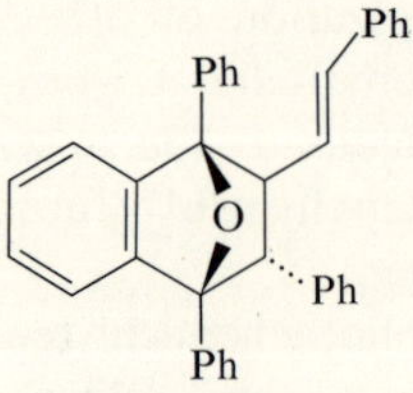   (253) (78%) 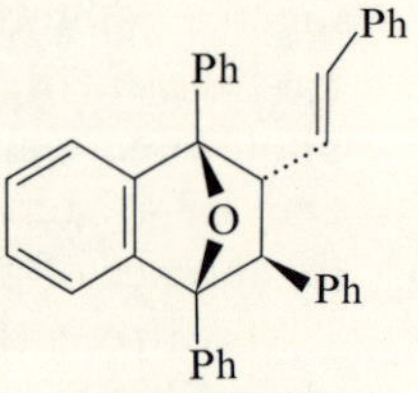   (254) (14%)

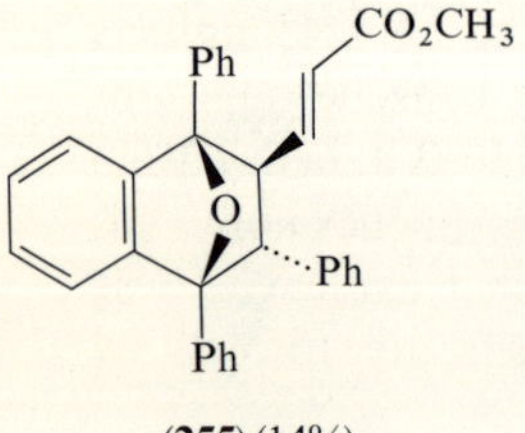

(255) (14%) 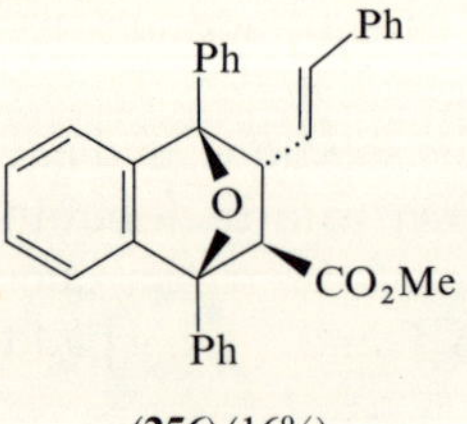   (256) (16%) 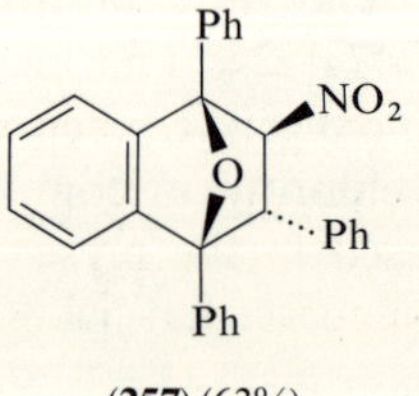   (257) (63%) 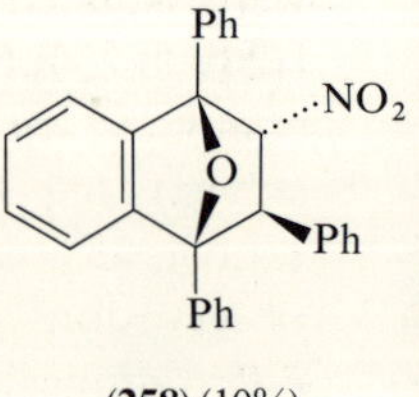   (258) (10%)

Irradiation of **138** ($\lambda > 370$ nm) together with 1,3-cyclohexadiene, 1,4-diphenylbutadiene,[423] 1-carbomethoxy-4-phenylbutadiene and ω-nitro-

---

[423] E. Teufel and G. Kaupp, "Organic Photochemical Syntheses," Vol. 3 (in press) (quoted after Kaupp and Teufel[173]).

styrene, leads to [4 + 2]-adducts exclusively (**251–258**); **242** is also isolated in varying amounts.[173] With methyl sorbate both [4 + 2]- and [4 + 4]-adducts (**259–262**) are obtained. 1,4-Dicarbomethoxybutadiene does not add to **138**; **242** is obtained in 90% yield. All photoadducts (**251–262**) are photolabile; on irradiation ($\lambda = 253.7$ nm) they are reconverted to **138** and the alkene. Significantly for preparative purposes, the product distributions (stereo-, regio-, and peri-selectivity) differ considerably from those obtained in the thermal Diels–Alder reactions (Section IV,B); mechanistic details are given in the original papers.

(**259**) (10%) (**260**) (6%)

(**261**) (22%) (**262**) (20%)

As described in Section IV,C, benzo[c]furans react with o-quinones to give [4 + 4]-adducts and/or dioxoles. The same products are obtained when higher o-quinones such as 9,10-phenanthrenequinone (**263**), benzo[h]-quinoline-5,6-dione (**264**) and 5,6-chrysenequinone (**265**) are irradiated with benzo[c]furans (Scheme 17)[418,421]; the dioxoles may react with a further molecule of excited **263** to give 2:1 adducts (**273**).

(**263**) (**264**) (**265**)

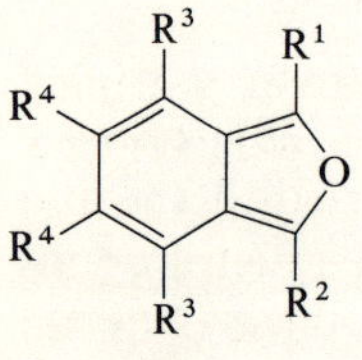

(266)a–h 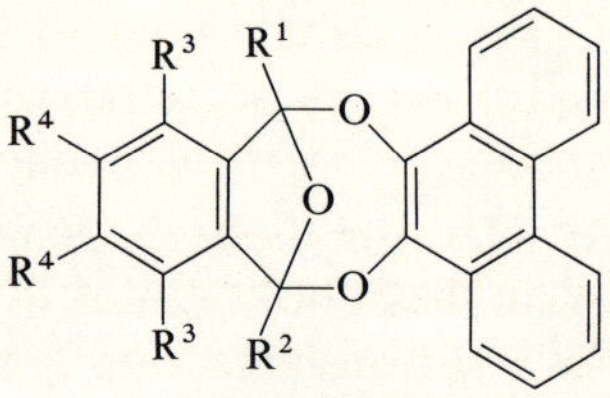 (267)a–f 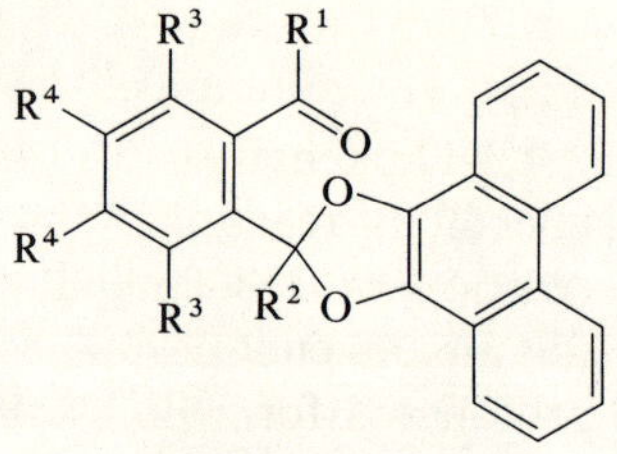 (268)a–c,d,[a]g,h

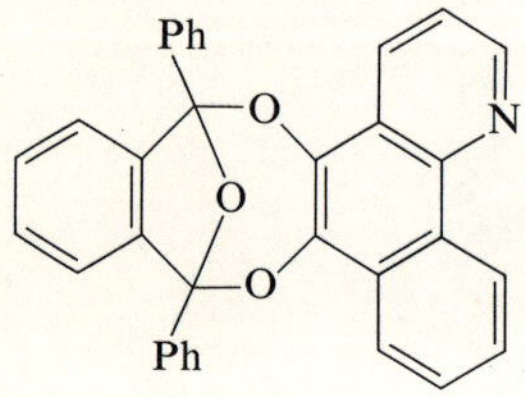

(269) 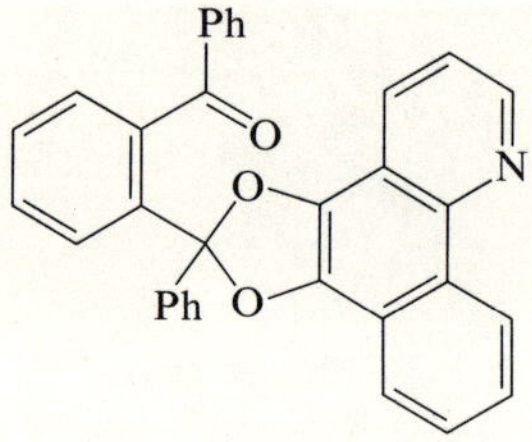 (270) 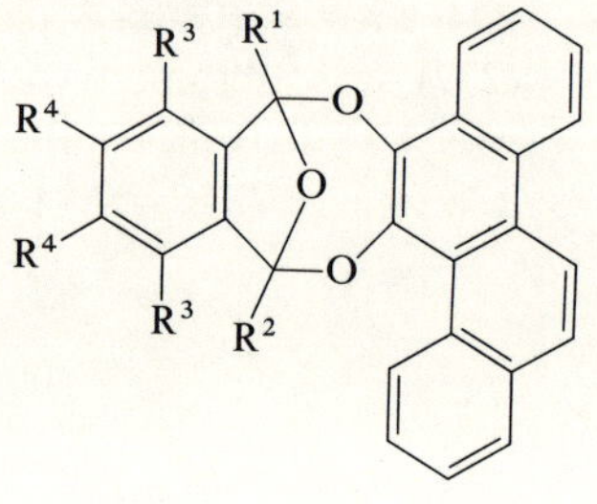 (271)e,f[b]

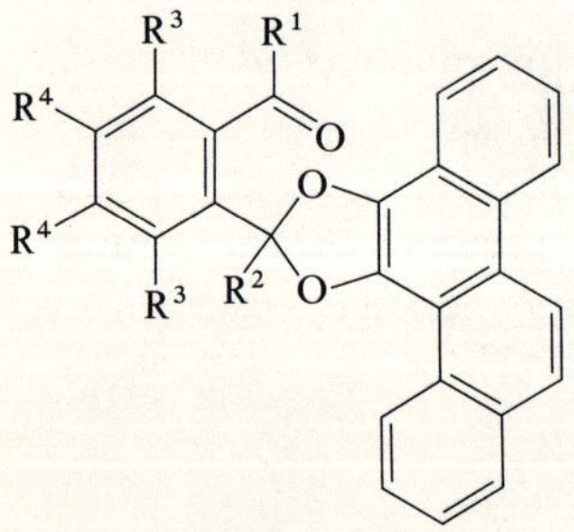

(272)a,e,g,[a]i 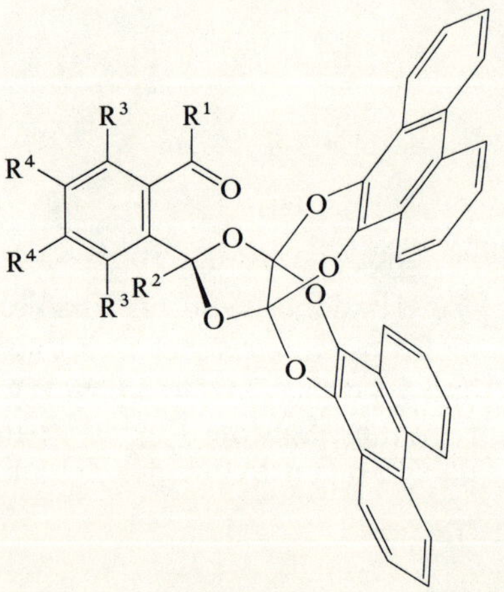 (273)a,b,c

a: $R^1 = R^2 = Ph$; $R^3 = R^4 = H$
b: $R^1 = R^2 = C_6H_4(p\text{-Me})$; $R^3 = R^4 = H$
c: $R^1 = R^2 = Ph$; $R^3 = H$; $R^4 = Me$
d: $R^1 = C_6H_4(p\text{-OMe})$; $R^2 = Ph$; $R^3 = R^4 = H$
e: $R^1 = R^2 = R^3 = Ph$; $R^4 = H$
f: $R^1 = PhCO$; $R^2 = Ph$; $R^3 = R^4 = H$
g: $R^1 = $ duryl; $R^2 = Ph$; $R^3 = R^4 = H$
h: $R^1 = $ mesityl; $R^2 = Ph$; $R^3 = R^4 = Ph$
i: $R^1 = R^2 = $ mesityl; $R^3 = R^4 = H$

SCHEME 17. [a] Both isomers isolated. [b] Positions of R^1 and R^2 possibly interchanged.

F. OTHER REACTIONS

1,3-Diphenylbenzo[c]furan (**138**) may act as an electron donor in triplet quenching reactions; absolute radical yields have been measured for the reaction of **138** (and a variety of other electron donors) with thionine (diaminophenothiazine) triplets (in methanolic buffer at pH = 8.6) and its protonated form (in methanolic solution containing phosphoric acid)[424]. Compound **138** has also been used as trapping agent for free radicals.[97,425,426] Phthalimidonitrene (**275**), generated by dehydrogenation of **274** with Pb(OAc)$_4$, reacts with **138** to give **276**; an analogous product (**277**) was obtained with 4,7-dihydro-1,3-diphenylbenzo[c]furan.[97] Thermolysis of oxadiazoline **278** yields a nitrene (**279**) which was trapped with **138** to give **281**; **280** was suggested as an intermediate.[425]

Few chemical reactions other than those described in Sections IV,B,C, and D have been reported. As pointed out in Section IV,D,1, the oxidation of **138** with $^1O_2$ yields o-dibenzoylbenzene (**140**). For preparative purposes this transformation is conveniently carried out with dichromate/acetic acid,[12,15,333] permanganate,[15] hexacyanoferrate(III),[427] peracids,[334,428] lead(IV) acetate[78] or sodium hypochlorite–hydrogen peroxide (as a source of $^1O_2$).[428] Aerial oxidation of **138** has often been used to destroy an excess of the reagent; chromatography on Al_2O_3 may result in oxidation of **138**

---

[424] U. Steiner, G. Winter, and H. E. A. Kramer, *Z. Naturforsch.*, Teil A **31**, 1019 (1976); *J. Phys. Chem.* **81**, 1104 (1977).
[425] G. Scherowsky and H. Franke, *Tetrahedron Lett.*, 1673 (1974).
[426] M. K. Huber, R. Martin, and A. S. Dreiding, *Helv. Chim. Acta* **60**, 1811 (1977).
[427] J. C. Emmett and W. Lwowski, *Tetrahedron* **22**, 1011 (1966).
[428] R. F. Boyer, C. G. Lindstrom, and B. Darby, *Tetrahedron Lett.*, 4111 (1975).

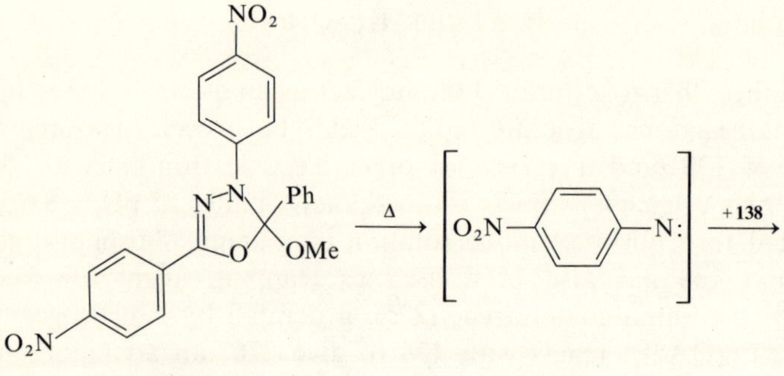

(278) (279)

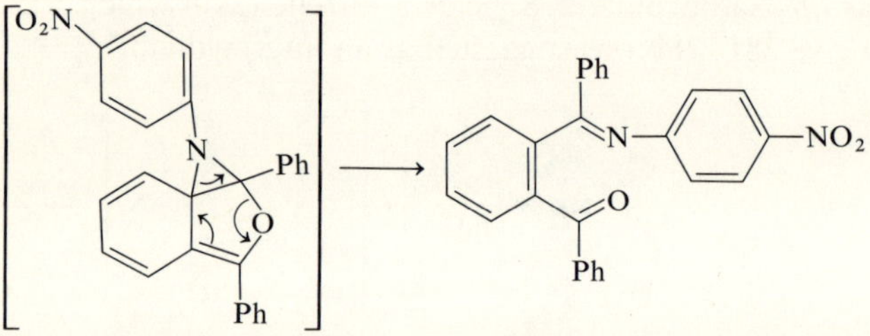

(280) (281)

to **140**.[236] *o*-Dibenzoylbenzene (**140**) has also been obtained by the action of *N,N'*-dibenzenesulfonyl-*p*-benzoquinonediimine (**282**) on **138** (Eq. 12).[429]

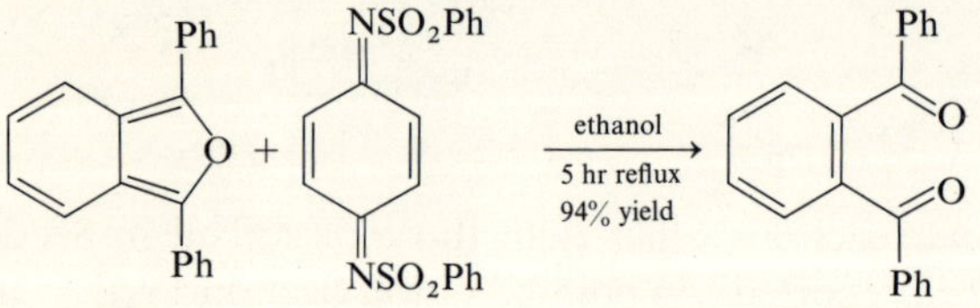

$$\text{(138)} + \text{(282)} \xrightarrow[\substack{5 \text{ hr reflux} \\ 94\% \text{ yield}}]{\text{ethanol}} \text{(140)} \qquad (12)$$

Reduction of 1,3-diarylbenzo[*c*]furans with sodium amalgam yields 1,3-dihydro-1,3-diarylbenzo[*c*]furans (Eq. 13).[12,158]

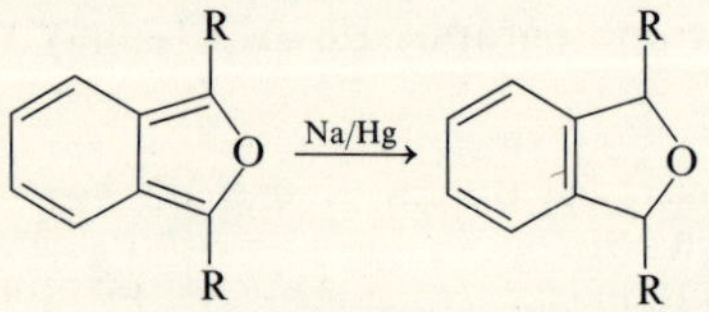

$$\xrightarrow{\text{Na/Hg}} \qquad (13)$$

R = Ph,[12] p-tolyl[158]

[429] R. Adams and C. R. Walter, *J. Am. Chem. Soc.* **73**, 1152 (1951).

Sec. IV.G] BENZO[c]FURANS

Electrophilic substitutions seem to have been reported only once, by Ecary; nitration of **138** with $NaNO_3/H_2SO_4$ gives 1-(3-nitrophenyl)-3-phenylbenzo[c]furan in 65% yield.[135,430]

Benzo[c]furans are suitable starting materials for the synthesis of benzo[c]thiophenes[2,316,431]; treatment of **138** with P_2S_5 yields 1,3-diphenylbenzo[c]thiophene[431] (for an improved procedure see Wittig et al.[303]).

Interestingly, **138** with ethenesulfonic acid, **283** (R = OH) [unlike acid derivatives (**283**; R = O-alkyl, NR_2) which yield crystalline Diels–Alder adducts (**284**) (Table VI)] gives an adduct which on treatment with hydrogen chloride or ethanol yields 1,4-diphenylnaphthalene (**285**) and a dimer of **138** (**286**; mp 236°C)[145,166] (Scheme 18). The mechanism for the generation of **286** is not known.

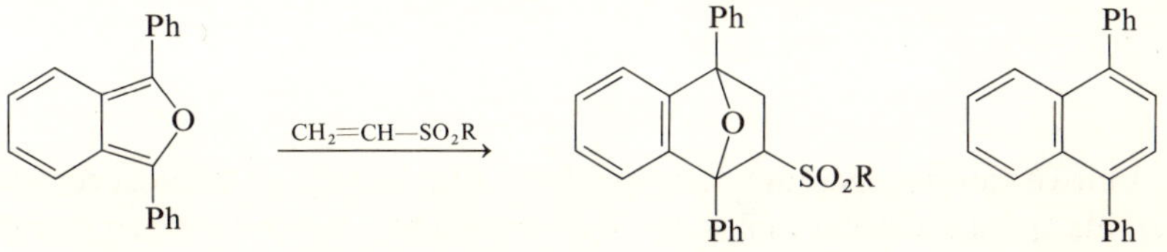

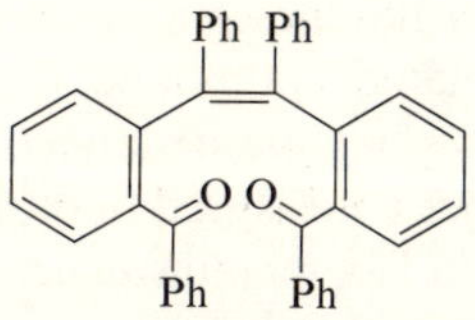

SCHEME 18

G. LUMINESCENCE, ELECTROLUMINESCENCE, AND LASING PROPERTIES

Both benzo[c]furans and 4,7-dihydrobenzo[c]furans exhibit strong blue or blue-green fluorescence in solution; both types of compounds have been

---

[430] S. Ecary, *C.R. Hebd. Seances Acad. Sci., Ser. C* **224**, 1828 (1947).
[431] C. Dufraisse and D. Daniel, *Bull. Soc. Chim. Fr.* [5] **4**, 2063 (1937).

used as oil additives.[102,103,432] The fluorescence of benzo[c]furans can also make the Edman method for determining the amino acid sequence of proteins more sensitive.[433] As known (Section IV,A,5; Scheme 9), 2,3-diphenylindenone (**188**) may be converted to **138** under alkaline conditions; in an analogous manner, treatment of the thiohydantoin derivatives (**287**) of amino acids with sodium ethylate produced intensely fluorescent derivatives of 1,3-diphenylbenzo[c]furans. For a number of benzo[c]furans,

(**287**)

quantitative fluorescence data are available (Table IX). Fluorescence rate constants are of the order of 3×10^8 sec^{-1}; evidently the first excited singlet state is of a (π,π^*)-type.[344] Triplet state energies could not be determined; no phosphorescence was found.[164] Even in the case of a heavy-atom-substituted benzo[c]furan (7-p-chlorophenyl-1,3,4-triphenylbenzo[c]furan) no intersystem crossing was observed; there was no delayed emission at 77°K. The quantum yield of luminescence (0.49) was nearly identical with the value of the unhalogenated compound (0.50). Attempts to sensitize phosphorescence with benzophenone irradiated at 254 and 313 nm were also unsuccessful.[164] Whether a measured phosphorescence ($E_T = 62.5$ kcal mol^{-1}; $\phi_p < 0.01$) was due to **138** remains uncertain.[413]

Certain aromatic hydrocarbons luminesce when raised to an excited electronic state by electrochemical energy. This phenomenon is called electroluminescence (ecl) and is shown by some benzo[c]furans. The ecl emission was examined in N,N'-dimethylformamide as solvent with tetra-n-butylammonium perchlorate as electrolyte.[155,165,434] The emission was identical with the normal fluorescence emission. Cyclic voltammograms were measured under the same conditions as used for the ecl studies; slowest scan rates at which rereduction of the cation or reoxidation of the anion

---

[432] French Patent 778660 (Wilmot and Cassidy, Inc.) (1934).
[433] C. Ivanov and I. Mancheva, *Tezisy Dokl.—Vses. Simp. Khim. Pept. Belkov*, 3rd, 542 (1974) [*CA* **85**, 59083 (1976)].
[434] D. L. Maricle and M. M. Rauhut (American Cyanamid Co.), (1974); U.S. Patent 3,816,795 U.S. Patent 3,654,525 (1974); Belgian Patent 666,750 (1966).

TABLE IX
FLUORESCENCE DATA OF BENZO[c]FURANS

R^1	R^2	R^3	R^4	λ_{max}	ϕ_f^a	References
Ph	H	H	H	486^b	0.96^c	101, 344
Ph	H	CH_3	H	486^b	0.94^c	101, 344
Ph	H	Ph	H	$520, 530^d$	$0.50^d, 0.96^c$	107, 165, 344
Ph	Ph	H	Ph	492^d	0.84^d	165
Ph	Ph	Ph	Ph	503^e	$0.64^e, 0.95^b$	165, 344
Ph	p-ClC$_6$H$_4$	H	Ph	533^d	0.49^d	164
Ph	9-Anthracenyl	H	Ph	530^d	0.23^d	165
Ph	p-OCH$_3$C$_6$H$_4$	H	p-OCH$_3$C$_6$H$_4$	515^d	0.33^d	165
p-OCH$_3$C$_6$H$_4$	Ph	H	Ph	568^d	0.53^d	165
p-OCH$_3$C$_6$H$_4$	p-OCH$_3$C$_6$H$_4$	H	p-OCH$_3$C$_6$H$_4$	547^d	0.49^d	165
p-C$_6$H$_5$C$_6$H$_4$	H	H	H	525^b	—	101
p-C$_6$H$_5$C$_6$H$_4$	H	CH_3	H	525^b	—	101
p-C$_6$H$_5$C$_6$H$_4$	Ph	H	Ph	540^d	0.57^d	165

[a] Quantum yield of fluorescence.
[b] In benzene.
[c] Relative to 9,10-diphenylanthracene in benzene.
[d] In dimethylformamide.
[e] In methylcyclohexane.

were detectable allowed an estimate of the lifetimes of the radical-ions (Table X).

As can be seen from Table X, bright ecl emission is observed only when both radical-anion and radical-cation are of moderate stability. The mechanism of the ecl emission has been studied in some detail.[164] A cation–anion annihilation delivers insufficient energy to reach the excited singlet state directly. Probably, ion-radical aggregates are involved and multiple electron transfer results in sufficient energy accumulation.

Ecl voltage and frequency brightness characteristics of 1,3-diphenyl and 1,3,4,7-tetraphenylbenzo[c]furan have also been studied.[435,436]

---

[435] E. V. Steblina and V. I. Steblin, *Vses. Inst. Nauk Tekh. Inf.* (*Moscow*), 785 (1974) [*CA* **86**, 179800 (1977)].

[436] E. V. Steblina and V. E. Steblin, *Zh. Prikl. Spektrosk.* **20**, 304 (1974).

TABLE X

Peak Potentials, Estimated Lifetimes of Radical-Ions, and Electrochemiluminescence (ecl) Emission Data for Some Benzo[c]furans
(in N,N'-Dimethylformamide)[165]

R^1	R^2	R^3	R^4	E^{ox} (V)	Lifetime (sec)	E^{red} (V)[a]	λ_{max} (ecl)[b]	Relative ecl intensity
Ph	H	H	H	+0.79	<0.1	−1.92	486	0.04
Ph	Ph	H	Ph	+0.98	1	−1.90	530	20
Ph	H	Ph	H	+0.85	<0.01	−1.79	492	0.2
Ph	Ph	Ph	Ph	+0.95	<0.01	−1.98	503[c]	—[d]
Ph	Ph	H	9-Anthracenyl	+0.93	1	−1.86	530	—[e]
p-PhC$_6$H$_4$	Ph	H	Ph	+0.94	5	−1.78	540	17
p-MeOC$_6$H$_4$	Ph	H	Ph	+0.76	0.6	−2.00	568	37
Ph	p-MeOC$_6$H$_4$	H	p-MeOC$_6$H$_4$	+0.91	4	−1.96	515	32
p-MeOC$_6$H$_4$	p-MeOC$_6$H$_4$	H	p-MeOC$_6$H$_4$	+0.74	0.5	−2.09	547	19

[a] Lifetime of the radical-anion > 20 sec.
[b] Identical with normal fluorescence.
[c] In methylcyclohexane.
[d] Not measured.
[e] Too dim for accurate measurement.

The ecl intensity of rubrene is enhanced about 100 to 200 times on addition of 1,3-diphenylbenzo[c]furan.[437] Probably the transfer of an electron from the anion of the benzo[c]furan to a rubrene cation with formation of an excited rubrene occurs; a mechanism has been proposed to explain the observation that the ecl of the mixture appears at lower voltages than those observed for the single compound. The shape of the luminescence pulse depends on the form of the alternating current applied.[438]

1,3-Diphenylbenzo[c]furan (**138**) has also been used as a dye for a flashlamp-pumped organic dye laser. The lasing wavelength range was from 484 to 518 nm[439] (480–515 nm)[440] with an extremely narrow bandwidth of 0.1 pm.[440] Dissolved oxygen was found to reduce the intensity virtually to zero.[439] Other chemical additives can alter the lasing performance; the pulse length and the lasing energy can be enhanced and reduced.[441,442] Organic photoconductive elements have been prepared by coating an electrically conducting support with a photoconductive layer of 1,3-diarylbenzo[c]furans or 1,3-diaryl-4,7-dihydrobenzo[c]furans.[151] The purity–property relationship in organic semiconductors containing benzo[c]furans has also been studied.[443]

V. Spectroscopic Properties

A. UV Spectra

Benzo[c]furan (**4**) exhibits a long-wave absorption band of medium intensity in the region of 340 nm.[19,20] Lack of solvent dependence together with mirror relationship to the fluorescence spectrum signifies a π–π^* band; a rotational analysis of the vapor phase spectrum led to an assignment as $^1B_2 \leftarrow {}^1A_1$.[34] 1,3-Diaryl-substituted benzo[c]furans show a strong absorption band in the region of 415 nm; in sterically hindered compounds, this

---

[437] Y. Angel and R. Signore, *C.R. Hebd. Seances Acad. Sci. Ser. B* **267**, 230 (1968).
[438] R. Signore, V. T. Han, and R. Zimmerman, *C.R. Hebd. Seances Acad. Sci., Ser. B* **268**, 571 (1969).
[439] J. B. Marling, D. W. Gregg, and S. J. Thomas, *IEEE J. Quantum Electron.* **6**, 570 (1970).
[440] G. Marowsky, *IEEE J. Quantum Electron.* **9**, 245 (1973).
[441] J. B. Marling, D. W. Gregg, and L. L. Wood, *Appl. Phys. Lett.* **17**, 527 (1970).
[442] J. B. Marling, L. L. Wood, and D. W. Gregg, *IEEE J. Quantum Electron.* **7**, 498 (1971).
[443] J. S. Driscoll, S. C. Kwan, and A. W. Berger, *U.S. C. F. S. T. I., AD Rep.* **AD-669352** (1968) [*CA* **70**, 24114 (1969)].

TABLE XI
UV DATA OF BENZO[c]FURANS

R^1	R^2	R^3	R^4	R^5	R^6	$\lambda_{max}(\log \varepsilon)$	References
H	H	H	H	H	H	303.0, 308.9, 316.0, 317.1, 322.3, 323.7, 326.0, 331.5[a]	34
						215 (4.17), 244 (3.4), 249 (3.37), 254 (3.35), 261 (3.12), 292(sh, 3.35), 299 (sh, 3.47), 305 (sh, 3.56), 313(3.7), 319 (3.7), 327 (3.87), 334 (3.66), 343 (3.79)[b]	20
CH_3	H	H	H	H	H	207, 236, 244, 251, 270, 282, 291, 309, 317, 324, 332, 340, 347, 357[c]	66
Ph	t-Bu	H	H	H	H	265 (3.813), 275 (3.806), 286 (3.763), 305 (3.58), 319 (3.69), 336 (3.724), 364 (4.0)[d]	71
Ph	Ph	H	H	H	H	261 (4.5), 270 (4.5), 310 (3.95), 415 (4.45)[e]	101, 445
						275 (4.5), 310 (3.95), 415 (4.45)[f]	165
Ph	Ph	H	Me	Me	H	249 (4.3), 258 (4.4), 269 (4.5), 277 (4.55), 310 (3.95), 415 (4.4)[e]	101
						258 (3.799), 276 (4.412), 410 (4.29)[g]	111
Ph	Du[h]	H	H	H	H	219 (4.428), 244 (4.092), 268 (4.066), 304 (3.689), 319.5 (3.657), 337.5 (3.664), 385 (4.124)[b]	109
Ph	Ph	OAc	H	H	OAc	395 (4.0)[i]	130
Ph	Ph	Ph	H	H	Ph	325 (3.66), 404 (4.08)[f]	165
						330 (3.9), 405 (4.25)[j]	156
						324 (3.4), 410 (4.04)[i]	107
Ph	Ph	H	Ph	Ph	H	333 (4.06), 428 (4.29), 450 (sh, 4.17)[f]	165
Ph	Ph	H	PhCO	PhCO	Ph	265 (sh, 4.53), 280 (4.47), 330 (4.12), 385 (4.05), 440 (4.02)[j]	157, 157a
Ph	Ph	Ph	Ph	Ph	Ph	273 (4.37), 285 (sh, 4.35), 399 (4.12)[k]	165

[a] Vapor phase; for further details see Robey and Ross.[34] [b] In cyclohexane. [c] In heptane. [d] In methanol. [e] In ether. [f] In N,N'-dimethylformamide. [g] In ethanol. [h] Du = duryl (2,3,5,6-tetramethylphenyl). [i] In chloroform. [j] In dichloromethane. [k] In methylcyclohexane.

band is shifted to shorter wavelengths and diminished in intensity.[444] In Table XI, UV data for benzo[c]furans are collected.[445]

---

[444] Relation between electronic absorption spectra and geometry of molecules: H. Suzuki, "Electronic Absorption Spectra and Geometry of Organic Molecules." Academic Press, New York, 1967.
[445] Qualitative study in the gas phase: B. Steyer and F. P. Schaefer, *Appl. Phys.* **7**, 113 (1975).

B. ¹H-NMR Spectra

The ¹H-NMR spectrum of the parent compound (**4**)[19,20,30,59] was analyzed[30] as an AA'BB'CC' system; the ratio $J_{4,5}/J_{5,6} = 0.73$ is as expected when compared both with hetero analogs and with benzene and 1,3-cyclohexadiene. Calculations on the ring current contributions to diamagnetic anisotropy and chemical shifts have been reported.[446,447] ¹H-NMR data for other benzo[c]furans have been published.[66,70,71,109,111,114,132]

C. Mass Spectra

As mentioned in Section III,B, benzo[c]furan (**4**) shows a strong parent peak; major peaks are also found at $m/e = 90 (M^+ - CO)$ and $m/e = 89$ $(90 - H; m^* = 88.2)$. The fragmention pattern of 1,3-diphenylbenzo[c]-furan (**138**) has been investigated in some detail[448]; in addition to the parent peak (M^+, $m/e = 270$) an M^{2+}-peak is observed. The main fragmentation paths are outlined in Scheme 19.[448]

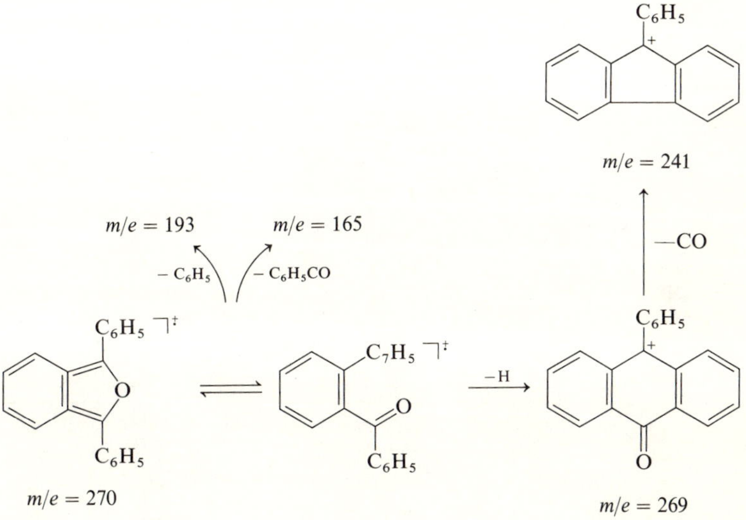

SCHEME 19

---

[446] P. J. Black, R. D. Brown, and M. L. Heffernan, *Aust. J. Chem.* **20**, 1305 (1967).
[447] E. Corradi, P. Lazzeretti, and F. Taddei, *Mol. Phys.* **26**, 41 (1973).
[448] J. P. Denhez, M. Ricard, and M. Corval, *Org. Mass. Spectrum.* **11**, 258 (1976).

D. Photoelectron Spectra

Photoelectron spectra have been reported for **2**, **4** and N-methylisoindole[30] and the ionization potentials (IP) assigned in the light of nonempirical calculations using Koopmans' theorem. Linear correlations of the type $IP_{obs} = a \cdot IP_{calc} + b$ were obtained in all three cases. As was noted, extended Hückel, PPP, and other semiempirical calculations also led to satisfactory correlations of the first three IPs, but the scatter was generally larger. The first IP of **4** lies in the order of 7.9 eV (Fig. 1 of Palmer and Kennedy[30]); a value of 7.91 eV has been reported by other authors.[379] In comparison, the first IP of 1,3-diphenylbenzo[c]furan is 7.09 eV.[379]

VI. Cyclobuta[c]furans and Cyclobutabenzo[c]furans

3-Oxabicyclo[3.2.0]hepta-1,4,6-triene (**289**), a planar 8π-electron analog of **4**, has been prepared by flow pyrolysis of **288** (both cis and trans) in approximately 10% yield (>95% purity)[449]; **289** is an extremely sensitive compound, polymerizing instantaneously on exposure to oxygen. In solution, where it is stable for several days, it slowly dimerizes to give the known [450] compound **291**; the pentacyclic intermediate **290** is possibly involved. In Diels–Alder reactions, **289** behaves like an olefin; with cyclopentadiene it reacts immediately to give **292**. Hydrogenation occurs at the same site.[451]

Whether both the reactivity and the ¹H-NMR spectrum (CCl₄: AA′BB′ system centered at 6.11 and 6.33 ppm; $J_1 = 0.4$ Hz, $J_2 = 0.3$ Hz) can be

---

[449] K. P. C. Vollhardt and R. G. Bergman, J. Am. Chem. Soc. **94**, 8950 (1972); C. Müller, A. Schweig, W. Thiel, W. Grahn, R. G. Bergman, and K. P. C. Vollhardt, ibid. **101**, 5579 (1979).
[450] J. A. Elix, M. V. Sargent, and F. Sondheimer, J. Am. Chem. Soc. **89**, 5080 (1967).
[451] R. G. Bergmann and K. P. C. Vollhardt, Chem. Commun., 214 (1973).

accounted for by considering **289** as "schizoaromatic" is open to question. Attempts to prepare **293** and **294** have failed (quoted as footnote 14 in Vollhardt and Bergman[449]).

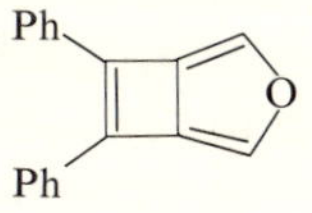

(293) 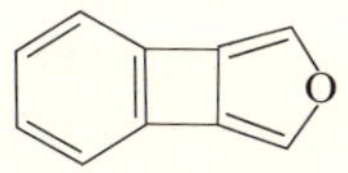   (294)

Another benz-analog of **289**, namely **296**, has been prepared by ring closure of bisacetylene **295** with subsequent dehalogenation in 10% yield. **296** forms orange-red stable crystals with indefinite melting point; a solution in benzene is photochemically unstable and is destroyed in a few minutes.[452]

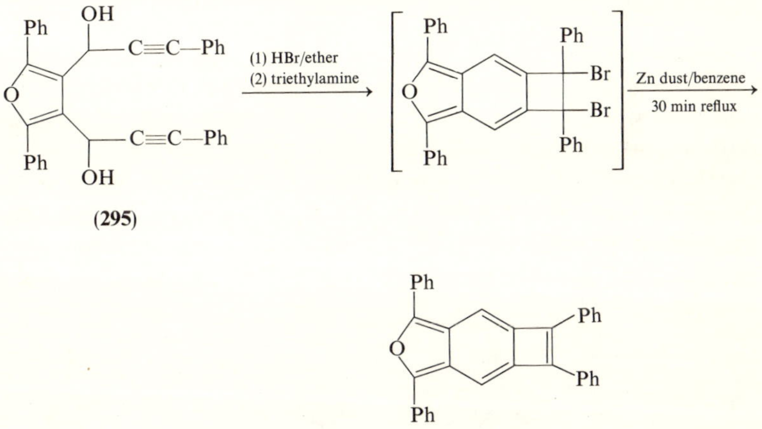

(295)

(296)

VII. Benz-Annelated and Hetero-Substituted Benzo[c]furans and Larger Ring[c]-Fused Furans

Linear benz-annelation enhances the reactivity and decreases the stability of benzo[c]furan, as with other o-quinonoid heterocycles.[453] 1,3-Diphenylnaphtho[2,3-c]furan (**299**) has been obtained from **297** via lactol **298** as deep red glistening plates with mp 148–154°C in 89% yield.[218] Although dried crystals of **299** could be stored for several months in the absence of light and air, solutions in organic solvents decolorized slowly on standing

---

[452] H. Firouzabadi and N. Maleki, *Tetrahedron Lett.*, 3153 (1978).
[453] W. Friedrichsen and P. Kaschner, *Justus Liebigs Ann. Chem.*, 1959 (1977).

and recrystallization was never achieved. The compound reacts very rapidly with such dienophiles as N-phenylmaleimide, maleic anhydride, tetracyanoethylene, ω-nitrostyrene, α-naphthoquinone, p-benzoquinone, acenaphthylene, bis(trifluoromethyl)acetylene, and dimethyl acetylenedicarboxylate to give the corresponding Diels–Alder adducts[218]; with sulfone **40** (R = Me) and phenylvinyl sulfoxide, longer reaction times are needed.[46,176] In the case of acenaphthylene both stereoisomers could be isolated.

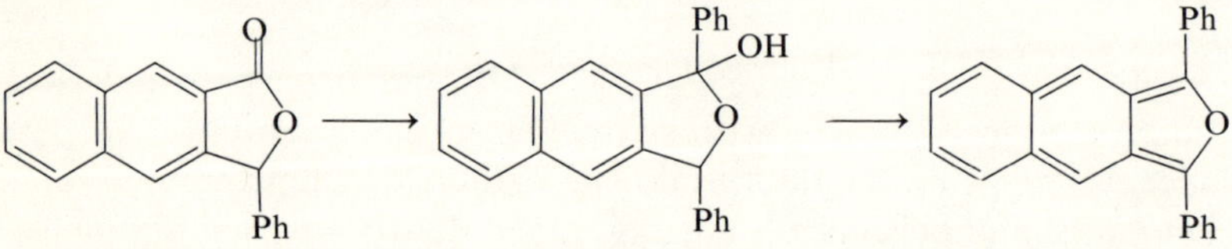

(297)   (298)   (299)

Compound **299** has also been used as a trapping agent. o-Quinones react with **299** in the same manner as benzo[c]furans (Section IV,C)[109]; it is noteworthy that 9,10-phenanthrenequinone gives a [4 + 4]-adduct and a dioxole under thermal conditions (unlike benzo[c]furans).[109]

6,7-Dibenzoyl-1,3-diphenylnaphtho[2,3-c]furan (**301**, violet crystals, mp 287°C) is available from **300** on dehydrogenation with p-chloranil and is evidently more stable than **299**. It reacts with 1,4-naphthoquinone both at positions 1,3 and 4,9 to give the Diels–Alder adducts **302** and **303**[157a] (stereochemistry unknown).

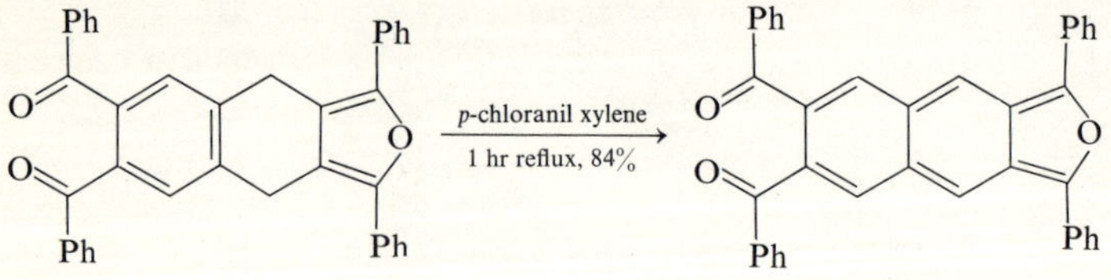

(300)   (301)

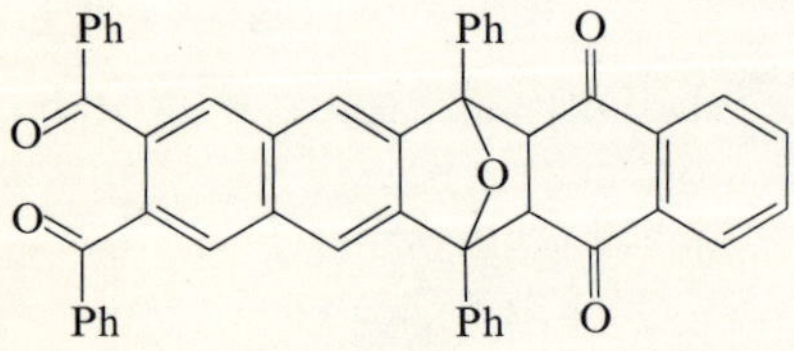

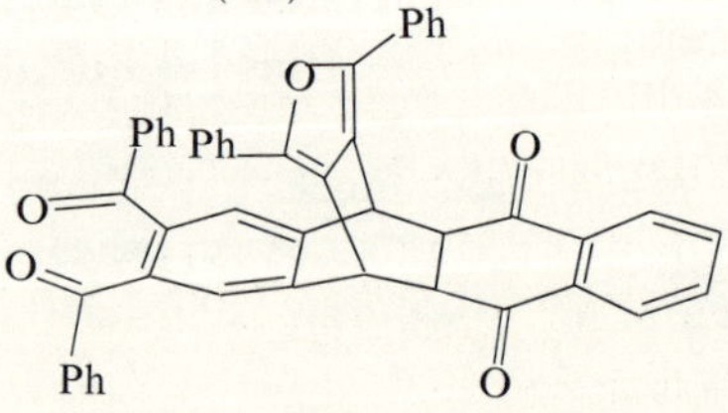

(302)   (303)

Angular benz-annelated benzo[c]furans (naphtho[1,2-c]furans) are also known. Compounds **307** have been synthesized in the usual manner (Scheme 20).[114]

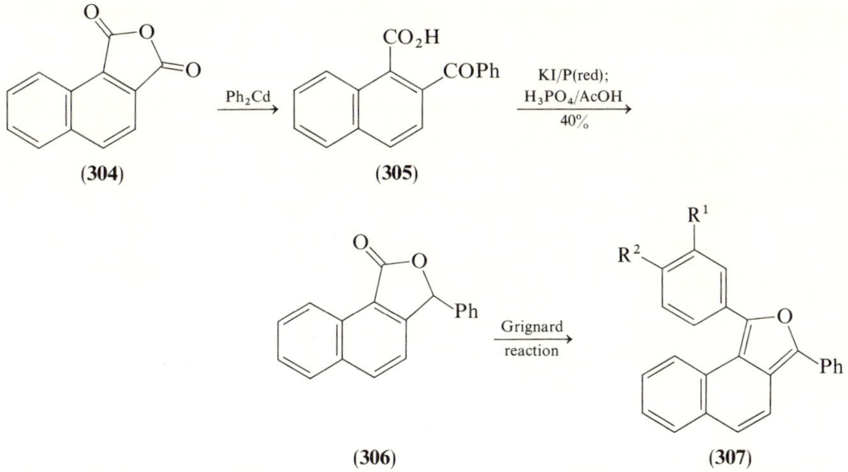

Sec. VII] BENZO[c]FURANS 221

(304) → Ph₂Cd → (305) → KI/P(red); H₃PO₄/AcOH 40% →

(306) → Grignard reaction → (307)

a: $R^1 = R^2 = H$ (mp 89–91°; 77%)
b: $R^1 = H$; $R^2 = Me$ (mp 128–130°; 60%)
c: $R^1 = Me$; $R^2 = H$ (oily; 60%)

SCHEME 20

Phenanthro[9,10-c]furans (311) were obtained either by reduction of 9,10-diaroylphenanthrenes (309), which are in turn available through methylene blue sensitized photooxygenation of the corresponding phencyclones (308),[454,455] or by photooxygenation of triarylphenanthrenocyclopentadienols (310); in this latter reaction hemiketals (312 and/or 313) are involved.[455] Partially hydrogenated derivatives of this type of phenanthrofuran have been described.[456]

Recently, an interesting synthesis of 311a was described starting from a diacetylene compound (314)[457,458]; reaction with Rh(Ph₃P)₃Cl in benzene yields a complex (315)[458] (Eq. 14) which on treatment with oxygen gives 311a in 10% yield (besides 51% of diketone 309a). The furan 311a under forcing conditions reacts with N-phenylmaleimide, maleic anhydride, and 1,4-dihydro-1,4-oxidonaphthalene (19), to give endo Diels–Alder adducts; in the last case two stereoisomers (endo–exo, exo–exo) were obtained.[459]

---

[454] W. Dilthey, S. Henkels, and M. Leonhard, *J. Prakt. Chem.* [N. S.] **151**, 97 (1938).
[455] J. J. Basselier, J. P. LeRoux, F. Caumartin, and J. C. Cherton, *Bull. Soc. Chim. Fr.*, 2950 (1974).
[456] F. Bergmann and H. E. Eschinazi, *J. Am. Chem. Soc.* **65**, 1405 (1943); F. Bergmann, H. E. Eschinazi, and M. Neeman, *J. Org. Chem.* **8**, 179 (1943).
[457] E. Müller, R. Thomas, M. Sauerbier, E. Lange, and D. Streichfuss, *Tetrahedron Lett.*, 521 (1971).
[458] Review: E. Müller, *Synthesis*, 761 (1974).
[459] T. Sasaki, K. Kanematsu, K. Iizuka, and N. Izumichi, *Tetrahedron* **32**, 2879 (1976).

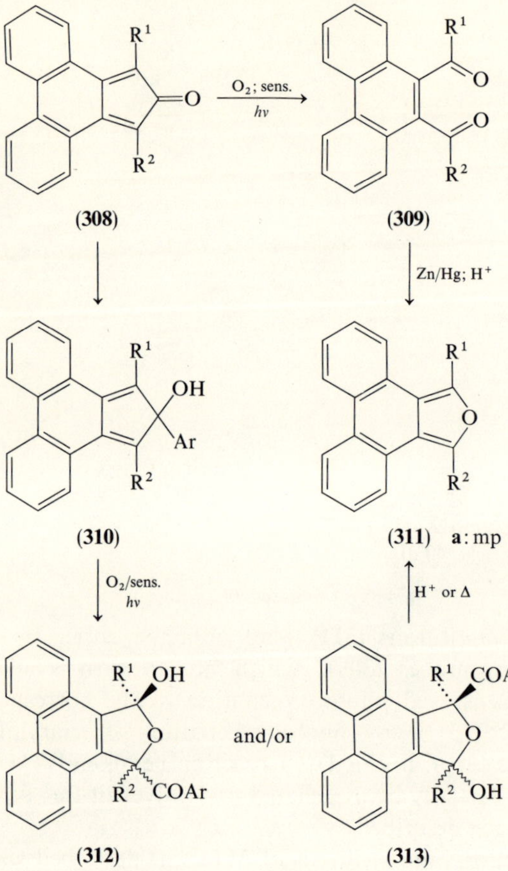

(308) → (309) O₂; sens. / hν

(309) → (311) Zn/Hg; H⁺

(308) → (310)

(310) → (312) and/or (313) O₂/sens. hν

(313) → (311) H⁺ or Δ

(311) a: mp 184°[454]; 170°[455]

a: R¹ = R² = Ph; b: R¹ = Ph, R² = p-MeOC$_6$H$_4$;
c: R¹ = Ph, R² = p-ClC$_6$H$_4$

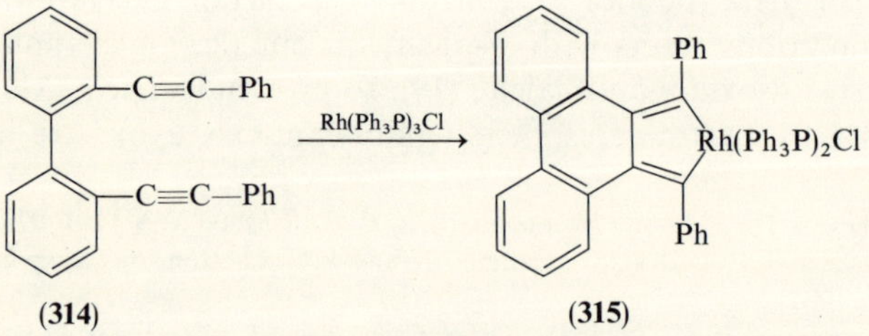

(314) → (315) Rh(Ph$_3$P)$_3$Cl   (14)

Acenaphtho[1,2-c]furan **317** has been obtained in high yield through sensitized photooxygenation of cyclopentadienol **316** and subsequent treatment with acid; hemiketals of the type described above are involved.[455] The rhodium complex pathway to **317** was unsuccessful; oxidation of **319**,

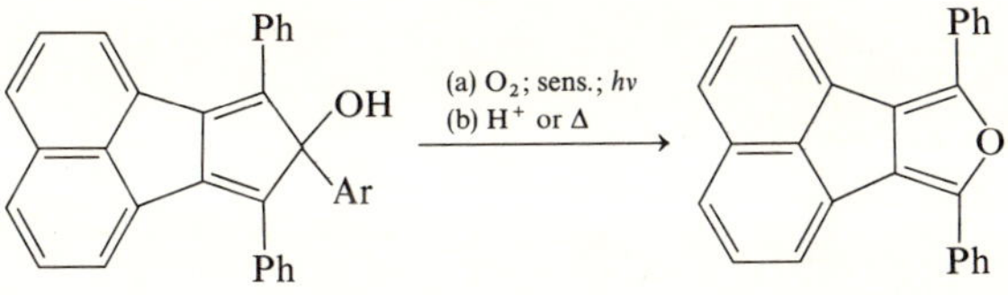

(316) → (317)

(a) O$_2$; sens.; hν
(b) H$^+$ or Δ

which is in turn available from **318**, yields diketone **320** in 44% yield.[457]

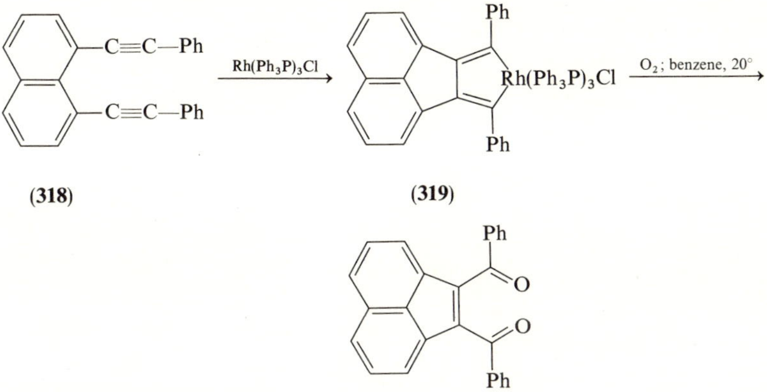

(318) —Rh(Ph$_3$P)$_3$Cl→ (319) —O$_2$; benzene, 20°→ (320)

Benz-annelated derivatives of **317** (fluorantheno[8,9-c]furans) have also been described (**324**). Grignard reaction of acenaphthenequinone (**321**) with methylmagnesium halide yields a mixture of diols (**322a,b**)[460,461]; *trans*-diol **322a** on treatment with *trans*-di-*p*-toluylethylene in acetic anhydride in the presence of hydroquinone gives a dihydrofluoranthenofuran (**323**), which is dehydrogenated with *p*-chloranil to give **324** (R^1 = H, R^2 = Me) as orange-red prisms with mp 289–290°C. Phenomenologically, this compound resembles 1,3-diarylbenzo[c]furans; in benzene solution it displays a magnificent yellowish-orange fluorescence. In addition **324** is a potent diene; with maleic anhydride it reacts almost instantaneously to give a Diels-Alder adduct (mp 220–264°C, depending on the rate of heating) which could be dehydrated to the corresponding aromatized product.[462] Another route to fluoranthenofurans starts with the (dimeric) acecyclone **325**. Diels-Alder

---

[460] N. Maxim, *Bull. Soc. Chim. Fr.* **45**, 1137 (1929).
[461] Improved procedure: R. Criegee, L. Kraft, and B. Rank, *Justus Liebigs Ann. Chem.* **507**, 176, (1933).
[462] N. Campbell and R. S. Gow, *J. Chem. Soc.*, 1555 (1949).

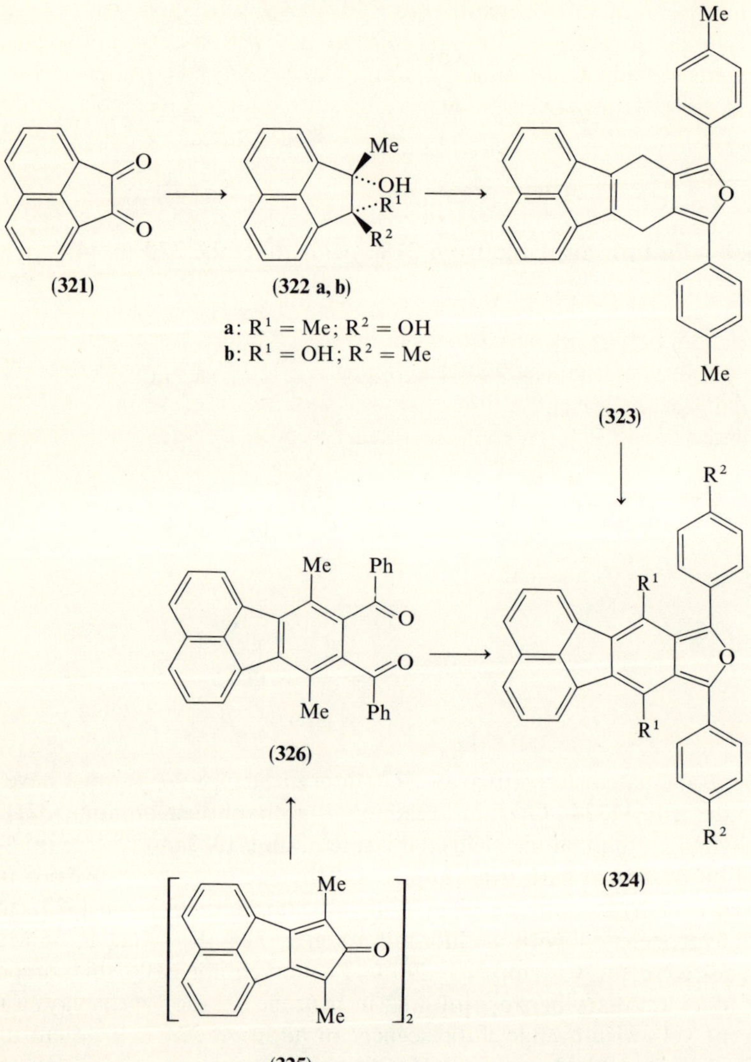

(321) → (322 a, b) →

a: R^1 = Me; R^2 = OH
b: R^1 = OH; R^2 = Me

(323)

↓

(326) → (324)

↑

(325)

reaction with dibenzoylacetylene yields diketone **326** which on reduction gives **324** (R^1 = Me, R^2 = H).[109]

Recently a number of heteroatom-substituted benzo[c]furans have been described. Flash vacuum thermolysis of **327** yields furo[3,4-c]pyridine (**328**) as white crystals, which are collected in a liquid-nitrogen-cooled trap. The compound melts below room temperature, polymerizing to a viscous mass;

the half-life is 50 hr (5% solution in CCl_4 at 4°C). In Diels–Alder reactions **328** shows a reactivity comparable to **4**; with maleic anhydride, *N*-phenylmaleimide, and 1,4-naphthoquinone, it reacts instantaneously to give the corresponding Diels–Alder adducts in quantitative yield (endo/exo mixtures).[463]

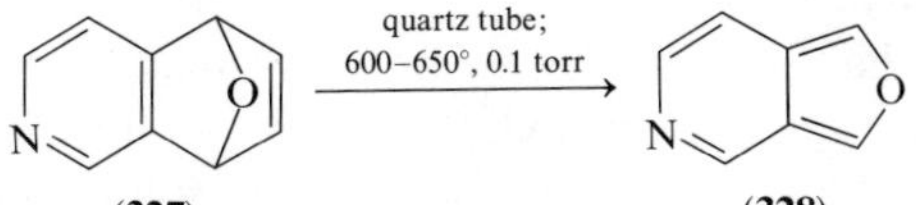

quartz tube;
600–650°, 0.1 torr

(327)   (328)

Introduction of nitrogen atoms into both positions 5 and 6 in **4** enhances the stability of the system tremendously. Furo[3,4-*d*]pyridazine (**330**) is available from **329** (Eq. 15) as yellowish crystals (mp 161°C) which can be sublimed at 145°C/0.1 torr.[464] A further number of furo[3,4-*d*]pyridazines

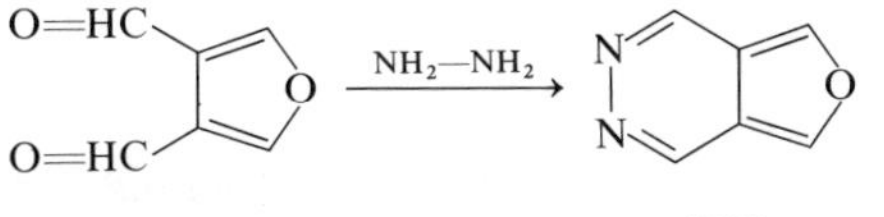

(329)   (330)   (15)

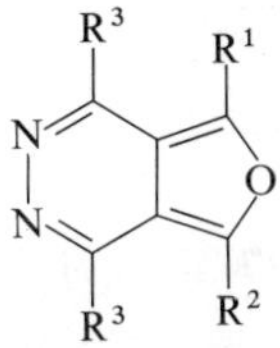

(**331a**,[465–468] **b**,[468] **c**[467,469])

a: $R^1 = R^2 = R^3$ = Me   a: mp 144° (monohydrate)
b: $R^1 = R^3$ = Me; $R^2 = CH_2OH$ b: mp 164–166°
c: $R^1 = R^2 = R^3$ = Ph   c: mp 255°

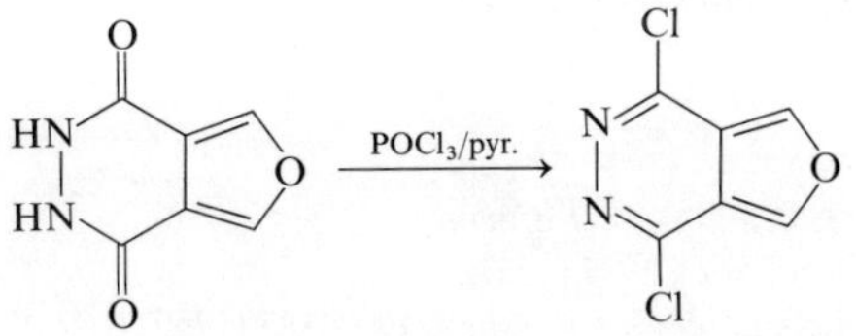

(332) $POCl_3$/pyr.   (333)   (16)

---

[463] U. E. Wiersum, C. D. Eldred, P. Vrijhof, and H. C. van der Plaas, *Tetrahedron Lett.*, 1741 (1977).

[464] M. Robba and M.C. Zaluski, *C. R. Hebd. Seances Acad. Sci., Ser. C* **263**, 301 (1966).

(331a–c) have been obtained on the same way.[465–469] The action of $POCl_3$ on hydrazide 332 in pyridine yields 4,7-dichlorofuro[3,4-d]pyridazine (333) as colorless crystals with mp 132°C (Eq. 16).

Furo[3,4-d]pyridazines have also been used in Diels–Alder reactions (331a with maleic anhydride, acrylic acid, 1,4-naphthoquinone, dibenzoylethylene, 1,4-benzoquinone, benzo[c]furandione[470]; 331c with maleic anhydride); 331a has been shown to be more reactive than 331c.[467] 1,3-Diphenylfuro[3,4-b]quinoxaline (335) has been obtained[471] from phthalide 334 (Eq. 17) as a green crystalline, quite stable solid (mp 244–246°C). In DMSO (deep blue solution), 335 reacts instantaneously with such dienophiles

$$(334) \xrightarrow[\text{(b) acetic acid}]{\text{(a) PhMgBr}} (335) \quad (17)$$

as maleic anhydride, maleimide, N-phenylmaleimide, dimethyl acetylenedicarboxylate and 1,4-naphthoquinone, to give crystalline adducts, which, when heated to or above their melting points, decompose giving a blue color presumably due to reversal of the Diels–Alder reactions. Compound 335 seems to be the only known member in the furo[3,4-b]quinoxaline series.[472]

Furo[3,4-g]phthalazines (337, R = H,[152] Ph[157,157a]) have been obtained from the dicarbonyl compounds 336 (R = H,Ph). The phthalazine 337 (R = Ph) crystallizes as fine violet needles which fluoresce red in solution; with maleic anhydride a Diels–Alder adduct is obtained in 99% yield.

Attempts to prepare 338 (X = O) by the action of hydrazine on 301 were unsuccessful; whether a transient green color is due to 338 is unclear. The corresponding sulfur compound (338, X = S) is known (green crystals).[157,157a]

---

[465] W. Bradley and L. J. Watkinson, *J. Chem. Soc.*, 319 (1956).
[466] W. L. Mosby, *J. Chem. Soc.*, 3997 (1957).
[467] L. Lomme and Y. Lepage, *Bull. Soc. Chim. Fr.*, 4183 (1969).
[468] G. A. Adembri, F. de Sio, R. Nesi, and M. Scotton, *J. Chem. Soc. C*, 1536 (1970).
[469] H. Keller and H. von Halban, *Helv. Chim. Acta* **27**, 1253 (1944).
[470] Y. Lepage and D. Villesot, *C. R. Hebd. Seances Acad. Sci., Ser. C* **274**, 1466 (1972).
[471] M. J. Haddadin, A. Yavrouian, and C. H. Issidorides, *Tetrahedron Lett.*, 1409 (1970).
[472] Routes in the synthesis of furo[3,4-b]quinoxalines have been explored: M. M. Roland, Ph.D. Dissertation, Utah State University, Salt Lake City (1975) [*Diss. Abstr. B* **36**, 244 (1975)].

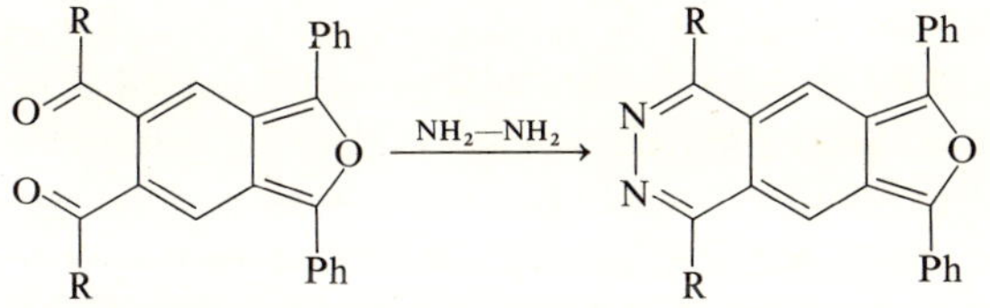

Sec. VII] BENZO[c]FURANS 227

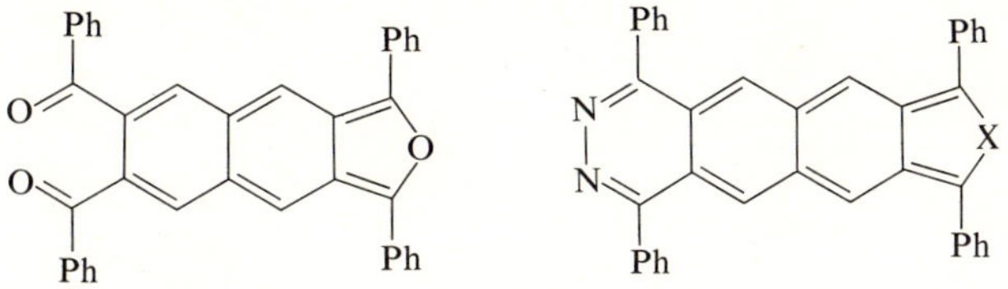

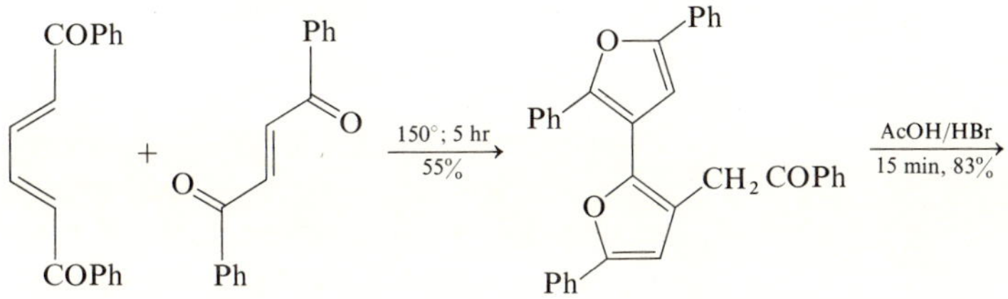

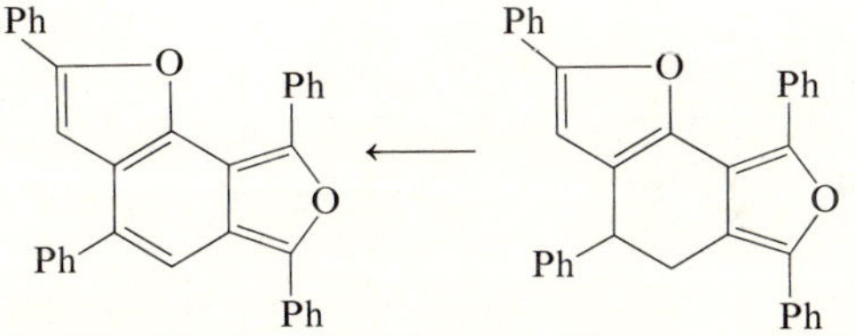

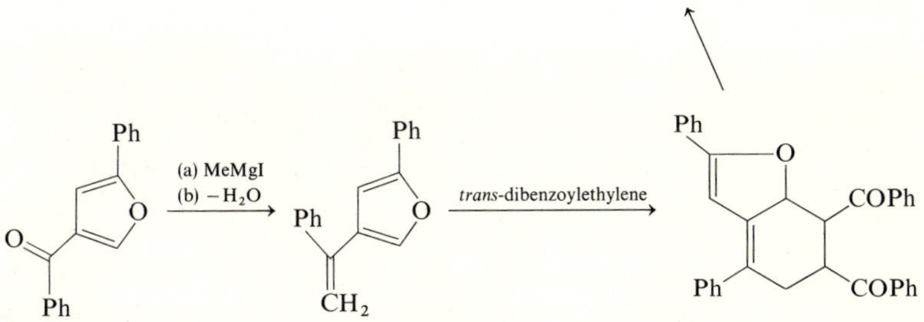

SCHEME 21

Two furo-annelated benzo[c]furans, a benzo[2,1-b:3,4-c']difuran (**341**) and a benzo[1,2-b:3,4-c']difuran (**344**) have been described. Acid-catalyzed ring closure of bisfuran **340**, which is available from *trans-trans*-1,4-dibenzoylbutadiene[473](**339**) and *trans*-dibenzoylethylene, yields **341** in 83% yield. An independent synthesis which starts from 4-benzoyl-2-phenylfuran (**343**) is outlined in Scheme 21; the isomeric compound has been obtained similarly (Scheme 22).

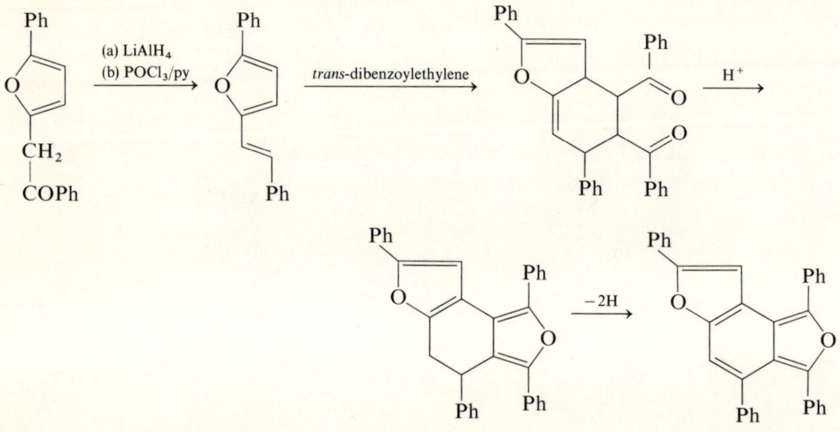

SCHEME 22

Diels–Alder reactions of **341** with maleic anhydride and *N*-phenylmaleimide and of **344** with maleic anhydride occur quite easily at 20°C; on dehydration with acetic acid the aromatized products are obtained in high yields.[153]

Fusion of thiophene to a furan nucleus gives rise (*inter alia*) to the two positional isomers[474] **345** (thieno[2,3-c]furan) and **346** (thieno[3,4-c]furan).

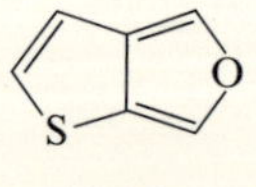

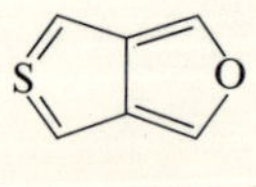

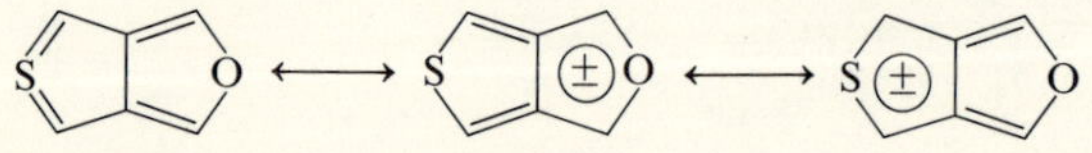

---

[473] P. S. Bailey and J. H. Ross, *J. Am. Chem. Soc.* **71**, 2370 (1949).
[474] N. Trinajstić, *Rec. Chem. Prog.* **32**, 85 (1971).

Topological resonance energy (TRE) values suggest that **346** is considerably less stable than **345**.[475] CNDO/2 calculations with inclusion of d-orbital participation for **346** show considerable π-bonding between carbon and sulfur of which over half is attributable to pπ–dπ overlap; the degree of charge separation indicates that canonical resonance forms of type **b** (carbonyl ylide type) are more important than those of type **c** (thiocarbonyl ylide type).[476]

Whereas compounds of type **345** are not known, the transient existence of tetraphenylthieno[3,4-c]furan (**348**) has been demonstrated. Sulfoxide **347** refluxed in acetic anhydride under nitrogen gave a pale violet color which was attributed to **348**; in the presence of dimethyl acetylenedicarboxylate a Diels–Alder adduct (**349**) was formed in 70% yield.[476–479]. The dehydration of **347** can also be effected by base; treatment with hydroxide ion in benzene/water with a phase-transfer catalyst affords a deep-blue

air-sensitive solution ($\lambda_{max} = 577$ nm) the color of which is attributed to the thienofuran **348**.[480] Attempts to prepare **351** from **350** in the way indicated were unsuccessful.[481]

---

[475] M. Milun and N. Trinajstić, *Croat. Chem. Acta* **49**, 107 (1977).
[476] M. P. Cava, M. A. Sprecker, and W. R. Hall, *J. Am. Chem. Soc.* **96**, 1817 (1974).
[477] M. P. Cava and M. A. Sprecker, *J. Am. Chem. Soc.* **94**, 6214 (1972).
[478] Review on nonclassical condensed thiophenes: M. P. Cava and M. V. Lakshikantham, *Acc. Chem. Res.* **8**, 139 (1975).
[479] Reviews on heteropentalenes: K. T. Potts, *in* "The Chemistry of Heterocyclic Compounds" (A. Weissberger and E. C. Taylor, eds.), Vol. 30, p. 317. Wiley, New York, 1977; C. A. Ramsden, *Tetrahedron* **33**, 3203 (1977).
[480] C. J. Horner, L. E. Saris, M. V. Lakshmikantham, and M. P. Cava, *Tetrahedron Lett.*, 2581 (1976).
[481] L. Grehn and H. Lindberg, *Chem. Scr.* **11**, 199 (1978).

Higher $4n$ and $4n + 2$ π-analogs of benzo[c]furan have recently been described. Cycloocta[1,2-c]furan (**353**) is obtained either[482] by reduction of 1,2-dicarbomethoxycyclooctatetraene (**352**) with $LiAlH_4/AlCl_3$ and subsequent dehydration, or by oxidation of diol **354** with manganese dioxide[483,484] or nickel peroxide,[483] as a bright yellow-orange mobile liquid.

It reacts rapidly with oxygen forming a white polymeric solid; with trinitrofluorenone and trinitrobenzene, very weak π-complexes were formed, which appear to be stable only in the solid phase. Tetracyanoethylene does not add at the furan moiety but reacts with the cyclooctatetraene part of the molecule to give **355**. The stability of **353** is such that prolonged heating with tributylphosphine has no effect.

The 18 π-electron system furo[3,4-c]octalene (**356**) is available by a Wittig reaction starting from cyclooctatetraene-1,2-dicarboxaldehyde; it is obtained as an unstable pale yellow liquid, which is rapidly oxidized in air. The electronic spectrum resembles that of **353**, showing little extended conjugation. Compound **356** underwent Diels–Alder addition at the terminal cyclooctatetraene ring; thus treatment with dimethyl fumarate leads to monoadduct **357**.[450,484]

---

[482] E. LeGoff and R. B. LaCount, *Tetrahedron Lett.*, 2787 (1965).
[483] R. Breslow, W. Horspool, H. Sugiyama, and W. Vitale, *J. Am. Chem. Soc.* **88**, 3677 (1966).
[484] J. A. Elix, M. V. Sargent, and F. Sondheimer, *J. Am. Chem. Soc.* **92**, 973 (1970).

(356)   (357)

Cycloocta[1,2-c:5,6-c']difuran (**291**), a 16 π-electron analog of benzo[c]-furan, has been obtained by a Wittig reaction as a colorless crystalline solid (mp 131–133°C), rapidly becoming orange in air. In the absence of air the compound is stable. Thus, treatment of **291** with an excess of dimethyl fumarate in boiling benzene for 96 hr gave 76% of adduct **358** (based on unrecovered **291**; 56% of **291** was unreacted).[484]

A further route to the cycloocta[1,2-c:5,6-c']difuran system makes use of a Perkin-type reaction. Condensation of furan-3,4-dicarboxaldehyde with furan-3,4-diacetic acid in acetic anhydride and triethylamine at room temperature, followed by esterification with methanol and sulfuric acid, gave **359** (mp 134–135°C, 33% yield).[484]

(291)   (358)

(359)   (360)

A benz-analog of **353**, namely **360**, has been obtained in a Wittig reaction with either o-phthalaldehyde or furan-3,4-dialdehyde; as expected, **360** was found to be more stable than **353**. Diels–Alder reaction with dimethyl fumarate resulted in cycloaddition to the furan ring.[484]

Oxidative cyclization of **361** and **363** yields 18 π-electron analogs of benzo[c]furan (**362**, **364**); in solution these possibly exist as several conformers, as do the Diels–Alder adducts obtained with dimethyl acetylenedicarboxylate.[485]

^{1}H-Chemical shifts of cyclododeceno[c]furan (an annulated [12]-annulene) and cyclotetradeceno[c]furan (an annulated [14]annulene) have been calculated.[486] Remarkably stable compounds result when a carbonyl

[486] H. Vogler and G. Ege, *Tetrahedron* **32**, 1789 (1976).
[485] P. J. Beeby, R. T. Weavers, and F. Sondheimer, *Angew. Chem.* **86**, 163 (1974); *Angew. Chem., Int. Ed. Engl.* **13**, 138 (1974); R. T. Weavers and F. Sondheimer, *Angew. Chem.* **86**, 165 (1974); *Angew. Chem., Int. Ed. Engl.* **13**, 139 (1974).

group is inserted into the benzene ring of the benz[c]furan system: furo-[4,5-c]tropone (365)[487] is available by condensation of furan-3,4-dicarbox-

aldehyde with acetone. In concentrated sulfuric acid, the stable solution contains the carbonyl oxygen protonated species; on dilution with water 365 is recovered in 86% yield. In D_2SO_4/D_2O the protons at C-1 and C-3 are exchanged successively. Compound 365 does not show benzo[c]furan-like behavior in cycloaddition reactions; with tetracyanoethylene no reaction occurs. Probably, a less electron-deficient olefin would add to 365.

Further derivatives of 365 have been described.[487,487a] A theoretical study predicts furotropones to be nonaromatic, with polyenoid structures.[488]

[487] M. J. Cook and E. J. Forbes, *Tetrahedron* 24, 4501 (1968).
[487a] M. El Borai, R. Guilard, and P. Fournari, *Bull. Soc. Chim. Fr.*, 1383 (1974), G. Seitz and R. A. Olsen, *Chem.-Ztg.* 100, 142 (1976); B. Serpaud and Y. Lepage, *ibid.* 539; M. El Borai, R. Guilard, P. Fournari, Y. Dusausoy, and J. Protos, *ibid.* 75 (1977); T. Asao, N. Morita, and K. Kato, *Heterocycles* 11, 287 (1979); K. Kato, M. Oda, S. Kuroda, N. Morita, and T. Asao, *Chem. Lett.* 43 (1979).
[488] N. Zambelli and N. Trinajstić, *Z. Naturforsch., Teil B* 26, 1007 (1971).

Dibenzofuro[4,5-c]tropone (8H-dibenzo[a,e]furo[3,4-c]cyclohepten-8-one†; **368**, X = CO) and the corresponding thiepin-5,5-dioxide (8H-dibenzo-[b,f]furo[3,4-d]thiepin-5,5-dioxide; **368**, X = SO$_2$) were first prepared (79% and 66%, respectively) by Tochtermann and co-workers, using the now well-established retro Diels–Alder route.[489] In refluxing benzene, the intermediate (**367**, X = CO) can be isolated.[53] Further routes to **368** (X = CO) by retro Diels–Alder reaction exist.[53] The 1-methyl derivative of **368** (X = CO) has been trapped as a maleic anhydride adduct.[489]

The cycloaddition reactions of **368** (X = CO) with N-phenylmaleimide, p-benzoquinone, dimethyl acetylenedicarboxylate, and tetracyanoethylene afforded [4 + 2]-cycloadducts with endo-stereochemistry[53]; ethyl acrylate gave two isomers (probably endo and exo).[489] 1,4-Dihydro-1,4-oxidonaphthalene (**19**) yielded the endo–exo isomer.[53] The Diels–Alder reaction of **368** (X = SO$_2$) with ethyl acrylate gave a mixture of isomers, from which the higher melting product could be isolated in pure form.[489]

(366)   (367)

(368)   (369)

X = CO, SO$_2$

R^1 = ethyl, butyl
R^2 = ethyl, propyl, amyl
R^1 = H, R^2 = Me: mp 147–148°C
R^1 = R^2 = Me: mp 80–86°C

---

[489] W. Tochtermann, C. Franke, and D. Schäfer, *Chem. Ber.* **101**, 3122 (1968); M. Nogradi, *Acta Chim., Acad. Sci. Hung.* **96**, 393 (1978).

† Nomenclature according to Tochtermann *et al.*[489] Strictly speaking, compounds of type **368** (X = CO) and **369** should be named from furan as the "base system" (8H-dibenzo[3,4:6,7]-cyclohepta[1,2-c]furan-8-one).

Cycloalkylidene derivatives of **368** [4(8H-dibenzo[a,e]furo[3,4-c]cyclohepten-8-ylidene)piperidines, **369**] have been obtained by a route similar to that described for **368**.[490]

Preparation the linear tropone-annelated benzo[c]furans **371** (X = O) by direct condensation of dialdehyde **370** (X = O) with acetone, 1-phenylacetone, and 1,3-diphenylacetone, in alkaline medium was—in contrast to the corresponding benzo[c]thiophene (X = S)—unsuccessful. Interestingly, this condensation has been brought about with the Diels–Alder adduct **372**; subsequent heat yields **371** (X = O), a rare case where N-phenylmaleimide is used as a protective group. Compounds **371** react with N-phenylmaleimide even at room temperature to give adducts **373**.[152]

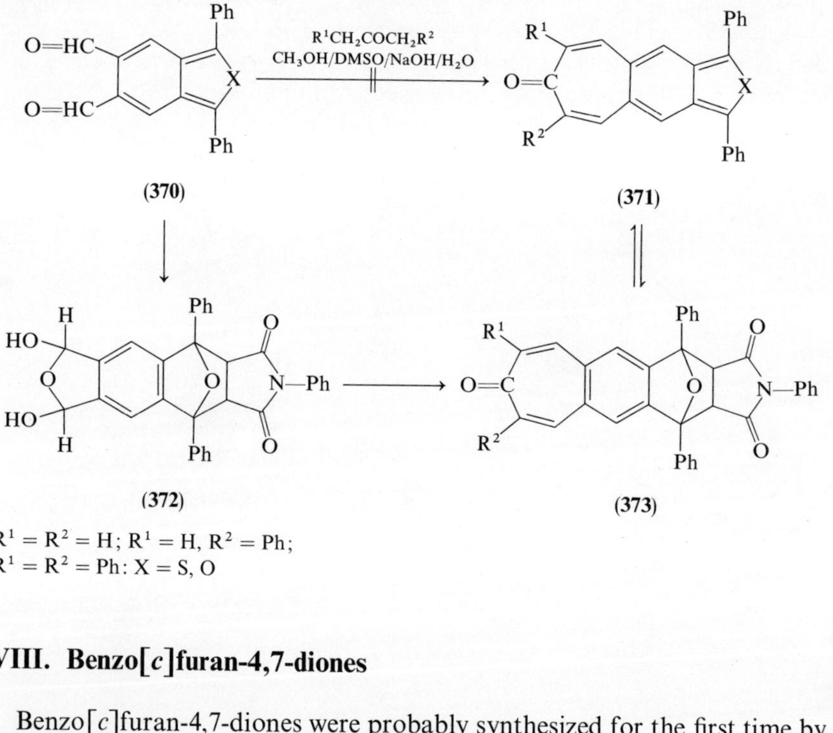

(370) (371)

(372) (373)

$R^1 = R^2 = H$; $R^1 = H$, $R^2 = Ph$;
$R^1 = R^2 = Ph$: X = S, O

VIII. Benzo[c]furan-4,7-diones

Benzo[c]furan-4,7-diones were probably synthesized for the first time by Pummerer and co-workers.[491,492] Condensation of benzoin and hydro-

---

[490] D. C. Remy (Merck and Co., Inc.), U.S. Patent 3,974,285 (1976). D. C. Remy, A. W. Raab, K. E. Rittle, E. L. Engelhardt, A. Scriabine, and V. J. Lotti, *J. Med. Chem.* **20**, 836 (1977). D. C. Remy (Merck and Co., Inc.), U.S. Patent 4,044,143 (1977).

[491] R. Pummerer, E. Buchta, E. Deimler, and E. Singer, *Ber. Dtsch. Chem. Ges.* **75**, 1976 (1942).

[492] R. Pummerer and G. Marondel, *Chem. Ber.* **89**, 1454 (1950).

quinone with sulfuric acid[493] (results obtained with boric acid anhydride[494] could not be duplicated[492]) yields a mixture of two bisfurans (**374, 375**; for a convenient separation see ref. 130); oxidation of the angular isomer (**375**) gives 1,2-dibenzoyl-3,6-dibenzoyloxybenzene (**376**) which on treatment with sulfuric acid furnished 1,3-diphenylbenzo[c]furan-4,7-dione (**377**). As Eugster and co-workers found,[495] photolysis of the light-sensitive[496] 2-(α-furyl)-3-acetyl-1,4-benzoquinones (**378**) also yields benzo[c]furan-4,7-diones (**379**); this reaction has been extended to benz-annelated derivatives (naphtho[2,3-c]furan-4,9-diones).

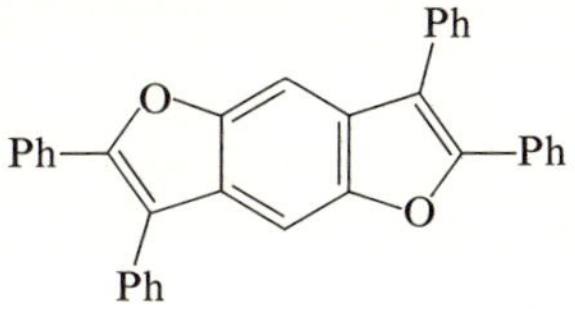

(**374**) 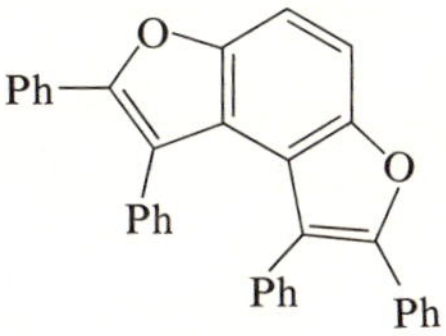 (**375**)

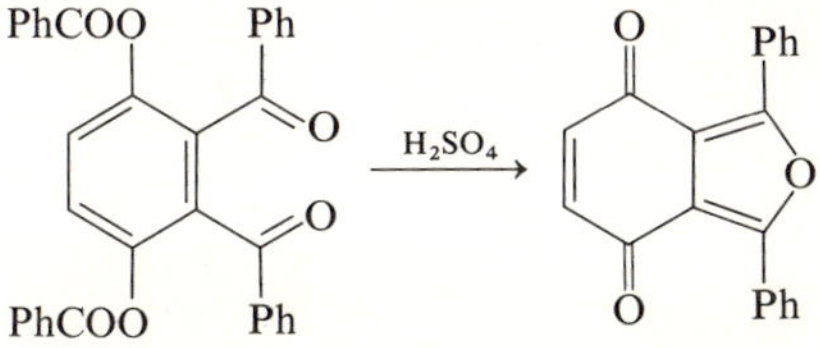

(**376**) (**377**)

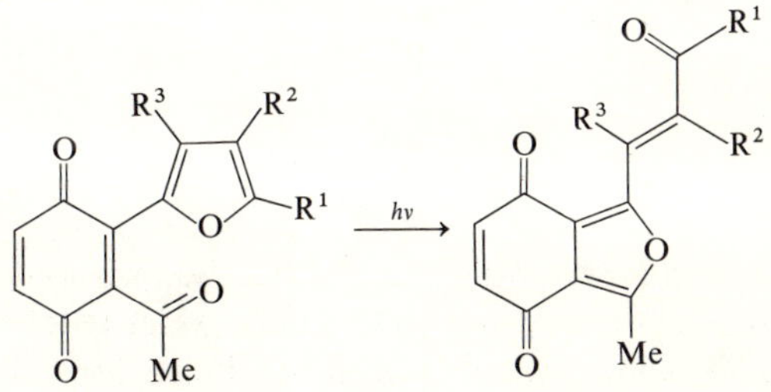

(**378**) (**379**)

R^1, R^2, R^3; see Ref. 496

---

[493] O. Dischendorfer, *Monatsh. Chem.* **66**, 201 (1935).
[494] O. Dischendorfer and W. Limontschew, *Monatsh. Chem.* **80**, 58 (1949).
[495] G. Weisgerber and C. H. Eugster, *Helv. Chim. Acta* **49**, 1806 (1966).
[496] C. H. Eugster and P. Bosshard, *Helv. Chim. Acta* **46**, 815 (1963).

The already mentioned Rh(I)-catalyzed cyclization (Section VII) has also been applied in the benzo[c]furandione series; thus **380** on treatment with $(PPh_3)_3Rh(I)Cl$ and subsequent oxidation with *m*-chloroperbenzoic acid or nitrosobenzene gives **381**.[497]

(380) → (a) $(PPh_3)_3Rh(I)Cl$ (b) *m*-chloroperbenzoic acid → (381)

The parent compound (**5**), which was first prepared by Eugster and co-workers,[498] is readily available by the tetrazine route.[500] Benzo[c]furan-4,7-diones can also be prepared by dehydrogenation of the saturated ketone as was shown by the synthesis of **383** from **382**.[501]

In contrast to benzo[c]furan itself, quinonoid derivatives of benzo[c]-

[OMe compound] →AgO/HNO₃→ [diketone] →3,6-dipyridyltetrazine→ (5)

[BrCH₂CO furan] → (382) →DDQ→ (383)

furan are stable compounds, which can be handled without special precautions. Benzo[c]furandiones do not show Diels–Alder reactivity at the diene site.[495] In Table XII some benzo[c]furan-4,7-diones are collected.

[497] J. Hambrecht, H. Straub, and E. Müller, *Tetrahedron Lett.*, 1789 (1976); J. Hambrecht and E. Müller, *Justus Liebigs Ann. Chem.*, 387 (1977).
[498] I. Wyrsch-Walraf, Diploma Thesis, University of Zürich (1968) (quoted in Fumagalli and Eugster[499]); A. A. Hofmann, I. Wyrsch-Walraf, P. X. Iten, and C. H. Eugster, *Helv. Chim. Acta* **62**, 2211 (1979).
[499] S. E. Fumagalli and C. H. Eugster, *Helv. Chim. Acta* **54**, 959 (1971).
[500] R. G. F. Giles and G. H. P. Roos, *Chem. Commun.* 260 (1975); G. M. L. Cragg, R. G. F. Giles, and G. H. P. Roos, *J. C. S., Perkin 1*, 1339 (1975).
[501] E. Ghera, Y. Gaoni, and D. H. Perry, *Chem. Commun.*, 1034 (1974).

TABLE XII
BENZO[c]FURAN-4,7-DIONES

R^1	R^2	R^3	R^4	mp (°C)	UV $\lambda(\log \varepsilon)$	References
H	H	H	H	140–142	235 (4.15), 340 (3.42)[a]	499
					240 (4.10), 340 (3.43)	500
Me	Me	H	H	176	—	501
Me	C_3H_3O[b]	H	H	130	226 (4.48), ∼ 300 (4.07), 308 (4.12), 319 (4.02), 389 (4.0)[c]	495
Ph	Ph	H	H	177 (corr.)	230 (4.56), 334 (4.24), 430 (3.85)[d]	492
Ph	Ph	Br	H	168	—	491
Ph	Ph	Br	Br	229	—	491
Ph	Ph	Me	Me	236	—	491
Ph	Ph	Ph	Ph	264	—	491

[a] Ethanol.

[b] (structure: CH=CH–CHO group)

[c] Diethyl ether. Either on photolysis or on standing in concentrated solution dimers are obtained (dimerization possibly at the acyclic double bond).[495]

[d] Cyclohexane.

On catalytic hydrogenation, 5,6-dihydrobenzo[c]furandiones are formed [130]; bromine adds also to the 5,6-double bond.[491] Benzo[c]furan-4,7-diones are unstable in alcoholic solutions in the presence of light, although no products were isolated[495]; photolysis of the parent compound in benzene gives 5-hydroxy-1,4-naphthoquinone (80%).[500] Treatment of (**377**) with acid anhydrides in the presence of sulfuric acid leads to 3,6-diacyloxy-1,2-dibenzoylbenzenes.[492,502]

Benz-annelated benzo[c]furandiones—naphtho[2,3-c]furan-4,9-diones—(Table XIII) were prepared in moderate to good yields by a Friedel–Crafts

[502] W. Limontschew, *Chem. Ber.* **86**, 1362 (1953).

TABLE XIII
Naphtho[2,3-c]furan-4,9-diones

R^1	R^2	R^3	R^4	R^5	R^6	References and notes
H	H	Ph	Ph	Ph	Ph	458, 509
Me	Me	H	H	H	H	503
Me	Me		Various Me$_n$			503
Me	CH=CH—COR^a	H	H	H	H	495, b, c
Ph	Ph	H	H	H	H	503, 507, 508
Ph	Ph		Various Me$_n$			503
Ph	Ph	OR^d	H	H	OR'^{d}	504, b

^{a} R = H and Me.
^{b} UV spectrum given.
^{c} Either on photolysis or on standing in concentrated solution, dimers are obtained (dimerization possibly at the acyclic double bond).
^{d} R, R' = H, Me, or Ac.

reaction between furan-3,4-dicarboxylic acid dichlorides and aromatic compounds (Eq. 18)[503,504]; 1,3-diphenylnaphtho[2,3-c]furandione (**385**) was also obtained by a Friedel–Crafts condensation between phthalic anhydride and 2,5-diphenylfuran[505] and by Müller's procedure[506–508] of Rh(I)-catalyzed ring closure of bisacetylene ketone **386**. This latter reaction sequence has also proved to be of value in the synthesis of other benzo[c]-furandiones (**387** to **390**[458,507,509–511]).

---

[503] D. V. Nightingale and B. Sukornick, *J. Org. Chem.* **24**, 497 (1959).
[504] D. Villesot and Y. Lepage, *C. R. Hebd. Seances Acad. Sci., Ser. C* **274**, 85 (1972).
[505] Y. Lepage and C. Champredon, unpublished work (quoted after Villesot and Lepage[504]).
[506] E. Müller, *Quinone Symp., University of Aberdeen, 1971*, Abstr. Pap. (Synthesis of quinones using transition metal complexes).
[507] E. Müller, C. Beissner, H. Jäkle, E. Langer, H. Muhm, G. Odenigbo, M. Sauerbier, A. Segnitz, D. Streichfuss, and R. Thomas, *Justus Liebigs Ann. Chem.* **754**, 64 (1971).
[508] E. Müller and G. Odenigbo, *Justus Liebigs Ann. Chem.*, 1435 (1975).
[509] E. Müller and W. Winter, *Justus Liebigs Ann. Chem.* **761**, 14 (1972).
[510] E. Müller and W. Winter, *Chem. Ber.* **105**, 2523 (1972).
[511] E. Müller and W. Winter, *Justus Liebigs Ann. Chem.*, 605 (1975).

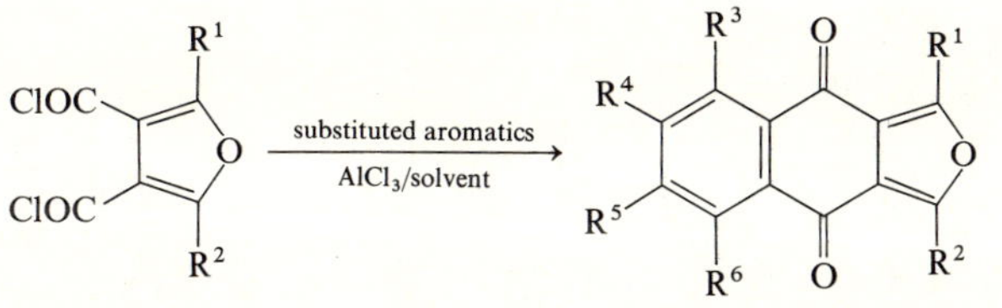

Sec. VIII] BENZO[c]FURANS 239

(18)

(384)

R^1 to R^6; see Table XIII

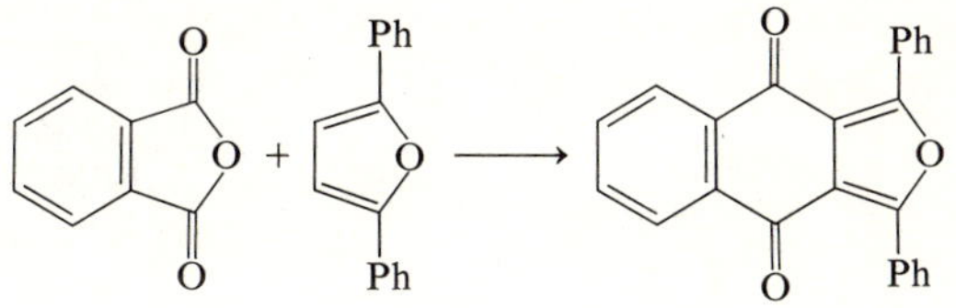

(385)

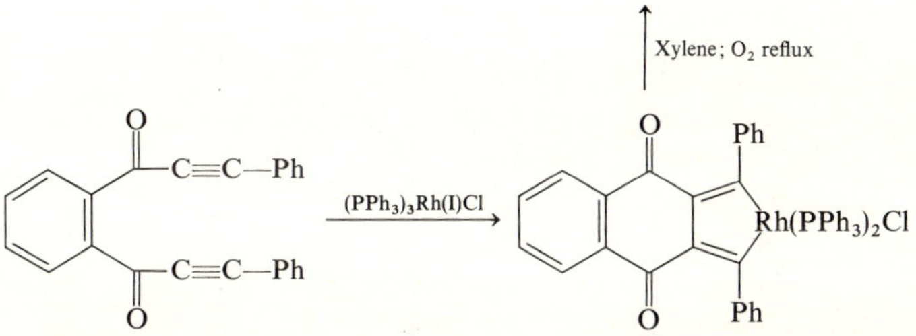

Xylene; O_2 reflux

(386)

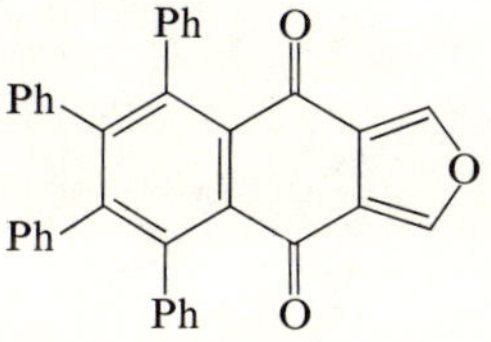

(387)[458,509]

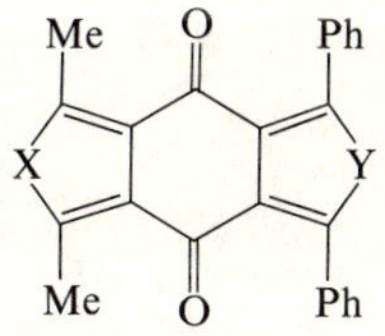

(388)[510,511]

X = NMe, S; Y = O

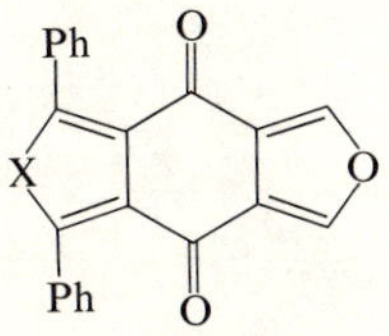

(389)[458,509]

X = O, S, Se

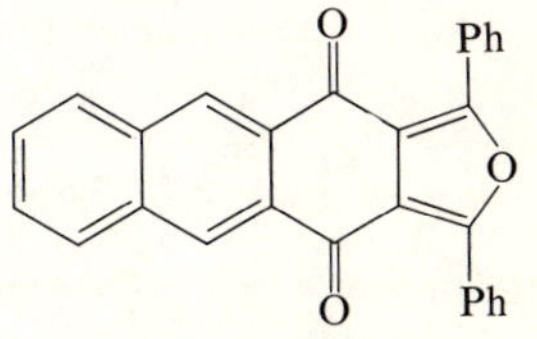

(390)[130,510]

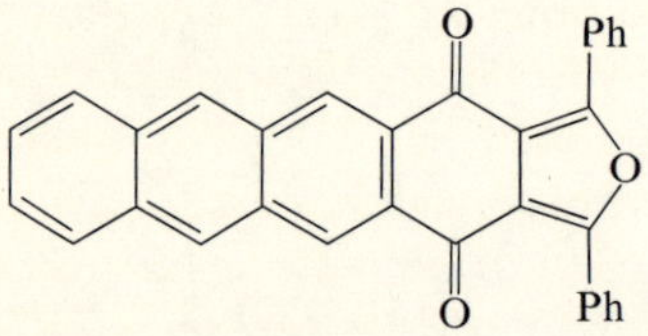

(391)[130]

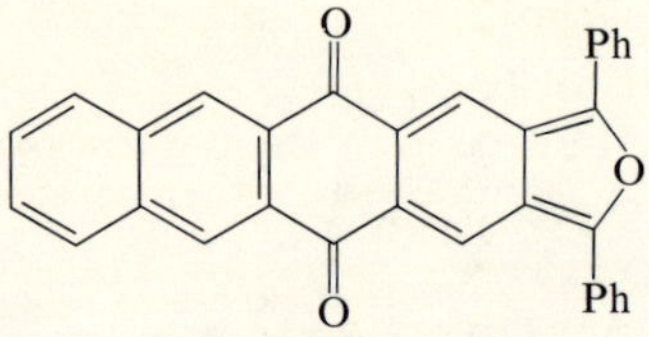

(392)[130]

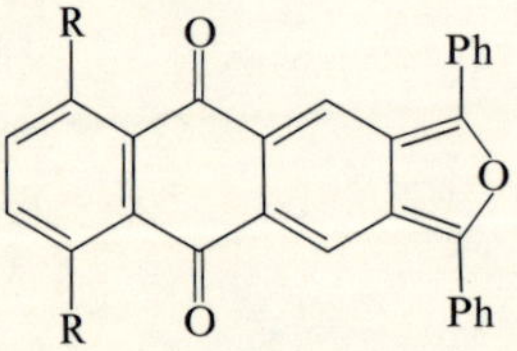

(393)[130,157,512]

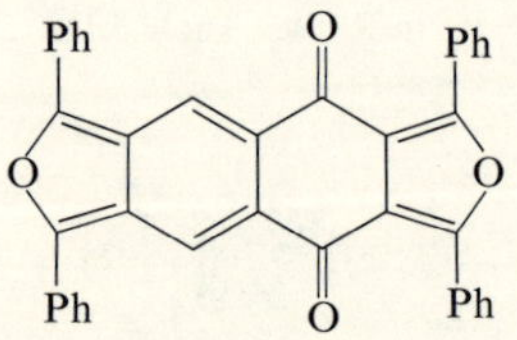

(394)[130]

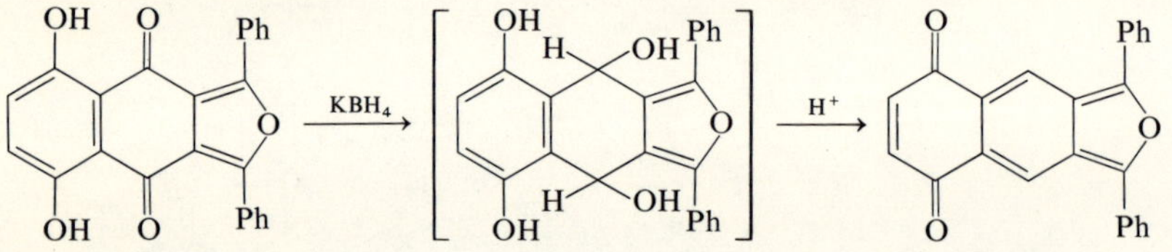

(395) —KBH₄→ [not isolated] —H⁺→ (396)

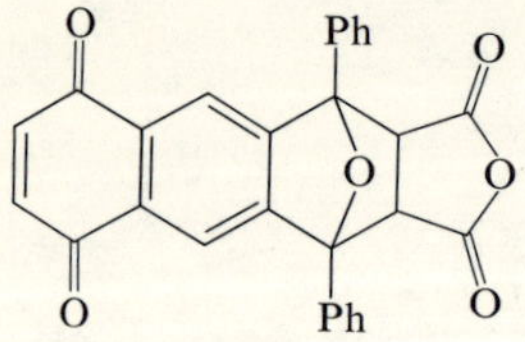

(397)

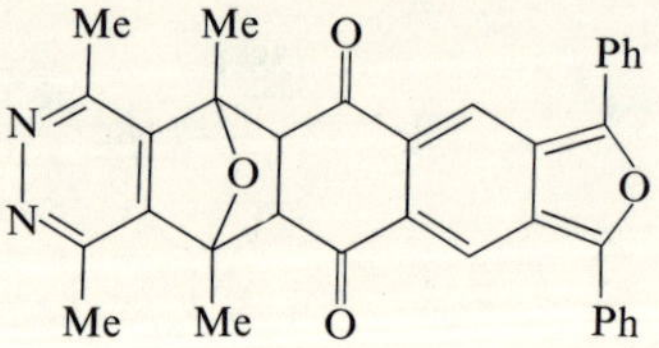

(398)

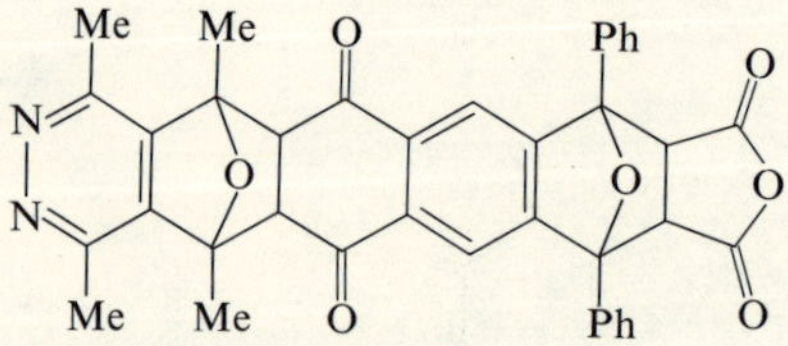

(399)

Naphtho[2,3-c]furandione **390** was also obtained by condensation of o-phthalaldehyde and 1,3-diphenyl-5,6-dihydrobenzo[c]furandione in dimethyl sulfoxide in the presence of sodium methylate[130]; in an analogous manner **391, 392, 393**, (R=OH,[157,512] R=Ph[130]) and **394** were prepared.[130]

The halfwave reduction potential (in acetonitrile with tetraethylammonium perchlorate as supporting electrolyte) is more negative for **387** than for **385** (difference: 70 mV) and becomes more negative when going from Y = Se over S to O (in **388**: X = N—Ph, Y = O,S,Se; in **389**: X = O,S,Se).[513]

An interesting translocation of carbonyl groups takes place when **395** is reduced with KBH_4 and subsequently treated with acid (for an analogous sequence see also Villesot and Lepage[152]). Compound **396** exhibits properties characteristic of both a normal benzo[c]furan (addition of maleic anhydride to give **397**) and an olefin (reaction with **331a** to **398**; catalytic reduction to a dihydro compound); **398** may react further with maleic anhydride to give **399**.[470]

[512] Method: H. Waldmann and H. Mathiowetz, *Ber. Dtsch. Chem. Ges.* **64**, 1713 (1931).
[513] W. Dilger, Ph.D. Dissertation, University of Tübingen (1973).

Cumulative Index of Titles

A

Acetylenecarboxylic acids and esters, reactions with N-heterocyclic compounds, **1**, 125
Acetylenecarboxylic esters, reactions with nitrogen-containing heterocycles, **23**, 263
Acetylenic esters, synthesis of heterocycles through nucleophilic additions to, **19**, 297
Acid-catalyzed polymerization of pyrroles and indoles, **2**, 287
t-Amino effect, **14**, 211
Aminochromes, **5**, 205
Anils, olefin synthesis with, **23**, 171
Annulenes, N-bridged, cyclazines and, **22**, 321
Anthracen-1,4-imines, **16**, 87
Anthranils, **8**, 277
Applications of NMR spectroscopy to indole and its derivatives, **15**, 277
Applications of the Hammett equation to heterocyclic compounds, **3**, 209; **20**, 1
Aromatic azapentalenes, **22**, 183
Aromatic quinolizines, **5**, 291
Aromaticity of heterocycles, **17**, 255
Aza analogs of pyrimidine and purine bases, **1**, 189
7-Azabicyclo[2.2.1]hepta-2,5-dienes, **16**, 87
Azapentalenes, aromatic, chemistry of, **22**, 183
Azines, reactivity with nucleophiles, **4**, 145
Azines, theoretical studies of, physicochemical properties of reactivity of, **5**, 69
Azinoazines, reactivity with nucleophiles, **4**, 145
1-Azirines, synthesis and reactions of, **13**, 45

B

Base-catalyzed hydrogen exchange, **16**, 1
1-, 2-, and 3-Benzazepines, **17**, 45
Benzisothiazoles, **14**, 43
Benzisoxazoles, **8**, 277
Benzoazines, reactivity with nucleophiles, **4**, 145
1,5-Benzodiazepines, **17**, 27
Benzo[b]furan and derivatives, recent advances in chemistry of, Part I, occurrence and synthesis, **18**, 337
Benzo[c]furans, **26**, 135
Benzo[c]cinnolines, **24**, 151
Benzofuroxans, **10**, 1
2H-Benzopyrans (chrom-3-enes), **18**, 159
Benzo[b]thiophene chemistry, recent advances in, **11**, 177
Benzo[c]thiophenes, **14**, 331
1,2,3-(Benzo)triazines, **19**, 215
Biological pyrimidines, tautomerism and electronic structure of, **18**, 199

C

Carbenes, reactions with heterocyclic compounds, **3**, 57
Carbolines, **3**, 79
Cationic polar cycloaddition, **16**, 289 (**19**, xi)
Chemistry
  of aromatic azapentalenes, **22**, 183
  of benzo[b]furan, Part I, occurrence and synthesis, **18**, 337
  of benzo[b]thiophenes, **11**, 178
  of chrom-3-enes, **18**, 159
  of diazepines, **8**, 21
  of dibenzothiophenes, **16**, 181
  of 1,2-dioxetanes, **21**, 437
  of furans, **7**, 377
  of isatin, **18**, 1
  of isoxazolidines, **21**, 207
  of lactim ethers, **12**, 185
  of mononuclear isothiazoles, **14**, 1
  of 4-oxy- and 4-keto-1,2,3,4-tetrahydroisoquinolines, **15**, 99
  of phenanthridines, **13**, 315
  of phenothiazines, **9**, 321
  of 1-pyrindines, **15**, 197
  of tetrazoles, **21**, 323
  of 1,3,4-thiadiazoles, **9**, 165
  of thienothiophenes, **19**, 123
  of thiophenes, **1**, 1

Chrom-3-ene chemistry, advances in, **18,** 159

Claisen rearrangements, in nitrogen heterocyclic systems, **8,** 143

Complex metal hydrides, reduction of nitrogen heterocycles with, **6,** 45

Covalent hydration
  in heteroaromatic compounds, **4,** 1, 43
  in nitrogen heterocycles, **20,** 117

Current views on some physicochemical aspects of purines, **24,** 215

Cyclazines, and related N-bridged annulenes, **22,** 321

Cyclic enamines and imines, **6,** 147

Cyclic hydroxamic acids, **10,** 199

Cyclic peroxides, **8,** 165

Cycloaddition, cationic polar, **16,** 289 (**19,** xi)

(2 + 2)-Cycloaddition and (2 + 2)-cycloreversion reactions of heterocyclic compounds, **21,** 253

D

Developments in the chemistry
  of furans (1952–1963), **7,** 377
  of Reissert compounds (1968–1978), **24,** 187

2,4-Dialkoxypyrimidines, Hilbert–Johnson reaction of, **8,** 115

Diazepines, chemistry of, **8,** 21

1,4-Diazepines, 2,3-dihydro-, **17,** 1

Diazirines, diaziridines, **2,** 83; **24,** 63

Diazo compounds, heterocyclic, **8,** 1

Diazomethane, reactions with heterocyclic compounds, **2,** 245

Dibenzothiophenes, chemistry of, **16,** 181

2,3-Dihydro-1,4-diazepines, **17,** 1

1,2-Dihydroisoquinolines, **14,** 279

1,2-Dioxetanes, chemistry of, **21,** 437

Diquinolylmethane and its analogs, **7,** 153

1,2- and 1,3-Dithiolium ions, **7,** 39

E

Electrolysis of N-heterocyclic compounds, **12,** 213

Electronic aspects of purine tautomerism, **13,** 77

Electronic structure of biological pyrimidines, tautomerism and, **18,** 199

Electronic structure of heterocyclic sulfur compounds, **5,** 1

Electrophilic substitutions of five-membered rings, **13,** 235

F

Ferrocenes, heterocyclic, **13,** 1

Five-membered rings, electrophilic substitutions of, **13,** 235

Free radical substitutions of heteroaromatic compounds, **2,** 131

Furans, development of the chemistry of (1952–1963), **7,** 377

G

Grignard reagents, indole, **10,** 43

H

Halogenation of heterocyclic compounds, **7,** 1

Hammett equation, applications to heterocyclic compounds, **3,** 209; **20,** 1

Hetarynes, **4,** 121

Heteroannulenes, medium-large and large π-excessive, **23,** 55

Heteroaromatic compounds
  free-radical substitutions of, **2,** 131
  homolytic substitution of, **16,** 123
  nitrogen, covalent hydration in, **4,** 1, 43
  prototropic tautomerism of, **1,** 311, 339; **2,** 1, 27; Suppl. 1
  quaternization of, **22,** 71

Heteroaromatic N-imines, **17,** 213

Heteroaromatic nitro compounds, ring synthesis of, **25,** 113

Heteroaromatic radicals, Part I, general properties; radicals with Group V ring heteroatoms, **25,** 205

Heteroaromatic substitution, nucleophilic, **3,** 285
Heterocycles
   aromaticity of, **17,** 255
   nomenclature of, **20,** 175
   photochemistry of, **11,** 1
   by ring closure of ortho-substituted t-anilines, **14,** 211
   synthesis of, through nucleophilic additions to acetylenic esters, **19,** 279
   thioureas in synthesis of, **18,** 99
Heterocyclic betaine derivatives of alternant hydrocarbons, **26,** 1
Heterocyclic chemistry, literature of, **7,** 225
Heterocyclic chemistry, literature of, Part II, **25,** 303
Heterocyclic compounds
   application of Hammett equation to, **3,** 209; **20,** 1
   (2 + 2)-cycloaddition and (2 + 2)-cycloreversion reactions of, **21,** 253
   halogenation of, **7,** 1
   isotopic hydrogen labeling of, **15,** 137
   mass spectrometry of, **7,** 301
   quaternization of, **3,** 1; **22,** 71
   reactions of, with carbenes, **3,** 57
   reactions of diazomethane with, **2,** 245
N-Heterocyclic compounds
   electrolysis of, **12,** 213
   reaction of acetylenecarboxylic acids and esters with, **1,** 125; **23,** 263
Heterocyclic diazo compounds, **8,** 1
Heterocyclic ferrocenes, **13,** 1
Heterocyclic oligomers, **15,** 1
Heterocyclic pseudo bases, **1,** 167
Heterocyclic pseudobases, **25,** 1
Heterocyclic sulphur compounds, electronic structure of, **5,** 1
Heterocyclic synthesis, from nitrilium salts under acidic conditions, **6,** 95
Hilbert–Johnson reaction of 2,4-dialkoxypyrimidines, **8,** 115
Homolytic substitution of heteroaromatic compounds, **16,** 123
Hydrogen exchange
   base-catalyzed, **16,** 1
   one-step (labeling) methods, **15,** 137
Hydroxamic acids, cyclic, **10,** 199

I

Imidazole chemistry, advances in, **12,** 103
N-Imines, heteroaromatic, **17,** 213
Indole Grignard reagents, **10,** 43
Indole(s)
   acid-catalyzed polymerization, **2,** 287
   and derivatives, application of NMR spectroscopy to, **15,** 277
Indolizine chemistry, advances in, **23,** 103
Indolones, isatogens and, **22,** 123
Indoxazenes, **8,** 277
Isatin, chemistry of, **18,** 1
Isatogens and indolones, **22,** 123
Isoindoles, **10,** 113
Isoquinolines
   1,2-dihydro-, **14,** 279
   4-oxy- and 4-keto-1,2,3,4-tetrahydro-, **15,** 99
Isothiazoles, **4,** 107
   recent advances in the chemistry of monocyclic, **14,** 1
Isotopic hydrogen labeling of heterocyclic compounds, one-step methods, **15,** 137
Isoxazole chemistry, recent developments in, **2,** 365
Isoxazole chemistry since 1963, **25,** 147
Isoxazolidines, chemistry of, **21,** 207

L

Lactim ethers, chemistry of, **12,** 185
Literature of heterocyclic chemistry, **7,** 225
Literature of heterocyclic chemistry, Part II, **25,** 303

M

Mass spectrometry of heterocyclic compounds, **7,** 301
Medium-large and large π-excessive heteroannulenes, **23,** 55
Meso-ionic compounds, **19,** 1
Metal catalysts, action on pyridines, **2,** 179
Monoazaindoles, **9,** 27

Monocyclic pyrroles, oxidation of, **15**, 67
Monocyclic sulfur-containing pyrones, **8**, 219
Mononuclear isothiazoles, recent advances in chemistry of, **14**, 1

N

Naphthalen-1,4-imines, **16**, 87
Naphthyridines, **11**, 124
Nitriles and nitrilium salts, heterocyclic syntheses involving, **6**, 95
Nitrogen-bridged six-membered ring systems, **16**, 87
Nitrogen heterocycles
  covalent hydration in, **20**, 117
  reactions of acetylenecarboxylic esters with, **23**, 263
  reduction of, with complex metal hydrides, **6**, 45
Nitrogen heterocyclic systems, Claisen rearrangements in, **8**, 143
Nomenclature of heterocycles, **20**, 175
Nuclear magnetic resonance spectroscopy, application to indoles, **15**, 277
Nucleophiles, reactivity of azine derivatives with, **4**, 145
Nucleophilic additions to acetylenic esters, synthesis of heterocycles through, **19**, 299
Nucleophilic heteroaromatic substitution, **3**, 285

O

Olefin synthesis with anils, **23**, 171
Oligomers, heterocyclic, **15**, 1
1,2,4-Oxadiazoles, **20**, 65
1,3,4-Oxadiazole chemistry, recent advances in, **7**, 183
1,3-Oxazine derivatives, **2**, 311; **23**, 1
Oxaziridines, **2**, 83; **24**, 63
Oxazole chemistry, advances in, **17**, 99
Oxazolone chemistry
  new developments in, **21**, 175
  recent advances in, **4**, 75
Oxidation of monocyclic pyrroles, **15**, 67
3-Oxo-2,3-dihydrobenz[d]isothiazole-1,1-dioxide (Saccharin) and derivatives, **15**, 233
4-Oxy- and 4-keto-1,2,3,4-tetrahydroisoquinolines, chemistry of, **15**, 99

P

Pentazoles, **3**, 373
Peroxides, cyclic, **8**, 165 (see also 1,2-Dioxetanes)
Phenanthridine chemistry, recent developments in, **13**, 315
Phenanthrolines, **22**, 1
Phenothiazines, chemistry of, **9**, 321
Phenoxazines, **8**, 83
Photochemistry of heterocycles, **11**, 1
Physicochemical aspects of purines, **6**, 1; **24**, 215
Physicochemical properties
  of azines, **5**, 69
  of pyrroles, **11**, 383
π-Excessive heteroannulenes, medium-large and large, **23**, 55
3-Piperideines, **12**, 43
Polymerization of pyrroles and indoles, acid-catalyzed, **2**, 287
Prototropic tautomerism of heteroaromatic compounds, **1**, 311, 339; **2**, 1, 27; Suppl. 1
Pseudo bases, heterocyclic, **1**, 167
Pseudobases, heterocyclic, **25**, 1
Purine bases, aza analogs of, **1**, 189
Purines
  physicochemical aspects of, **6**, 1; **24**, 215
  tautomerism, electronic aspects of, **13**, 77
Pyrazine chemistry, recent advances in, **14**, 99
Pyrazole chemistry, progress in, **6**, 347
Pyridazines, **9**, 211; **24**, 363
Pyridine(s)
  action of metal catalysts on, **2**, 179
  effect of substituents on substitution in, **6**, 229
  1,2,3,6-tetrahydro-, **12**, 43
Pyridoindoles (the carbolines), **3**, 79
Pyridopyrimidines, **10**, 149
Pyrimidine bases, aza analogs of, **1**, 189
Pyrimidines
  2,4-dialkoxy-, Hilbert–Johnson reaction of, **8**, 115
  tautomerism and electronic structure of biological, **18**, 199
1-Pyrindines, chemistry of, **15**, 197
Pyrones, monocyclic sulfur-containing, **8**, 219
Pyrroles
  acid-catalyzed polymerization of, **2**, 287
  oxidation of monocyclic, **15**, 67
  physicochemical properties of, **11**, 383

Pyrrolizidine chemistry, **5,** 315; **24,** 247
Pyrrolodiazines, with a bridgehead nitrogen, **21,** 1
Pyrrolopyridines, **9,** 27
Pyrylium salts, syntheses, **10,** 241

Q

Quaternization
   of heteroaromatic compounds, **22,** 71
   of heterocyclic compounds, **3,** 1
Quinazolines, **1,** 253; **24,** 1
Quinolizines, aromatic, **5,** 291
Quinoxaline chemistry
   developments 1963–1975, **22,** 367
   recent advances in, **2,** 203
Quinuclidine chemistry, **11,** 473

R

Reduction of nitrogen heterocycles with complex metal hydrides, **6,** 45
Reissert compounds, **9,** 1; **24,** 187
Ring closure of ortho-substituted *t*-anilines, for heterocycles, **14,** 211
Ring synthesis of heteroaromatic nitro compounds, **25,** 113

S

Saccharin and derivatives, **15,** 233
Selenazole chemistry, present state of, **2,** 343
Selenium–nitrogen heterocycles, **24,** 109
Selenophene chemistry, advances in, **12,** 1
Six-membered ring systems, nitrogen bridged, **16,** 87
Substitution(s),
   electrophilic, of five-membered rings, **13,** 235
   homolytic, of heteroaromatic compounds, **16,** 123
   nucleophilic heteroaromatic, **3,** 285
   in pyridines, effect of substituents, **6,** 229
Sulfur compounds, electronic structure of heterocyclic, **5,** 1
Synthesis and reactions of 1-azirines, **13,** 45
Synthesis of heterocycles through nucleophilic additions to acetylenic esters, **19,** 279

T

Tautomerism
   electronic aspects of purine, **13,** 77
   and electronic structure of biological pyrimidines, **18,** 199
   prototropic, of heteroaromatic compounds, **1,** 311, 339; **2,** 1, 27; Suppl. 1
Tellurophene and related compounds, **21,** 119
1,2,3,4-Tetrahydroisoquinolines, 4-oxy- and 4-keto-, **15,** 99
1,2,3,6-Tetrahydropyridines, **12,** 43
Theoretical studies of physicochemical properties and reactivity of azines, **5,** 69
Tetrazole chemistry, recent advances in, **21,** 323
1,2,4-Thiadiazoles, **5,** 119
1,2,5-Thiadiazoles, chemistry of, **9,** 107
1,3,4-Thiadiazoles, recent advances in the chemistry of, **9,** 165
Thiathiophthenes (1,6,6aS^{IV}-Trithiapentalenes), **13,** 161
1,2,3,4-Thiatriazoles, **3,** 263; **20,** 145
1,4-Thiazines and their dihydro-derivatives, **24,** 293
4-Thiazolidinones, **25,** 83
Thienopyridines, **21,** 65
Thienothiophenes and related systems, chemistry of, **19,** 123
Thiochromanones and related compounds, **18,** 59
Thiocoumarins, **26,** 115
Thiophenes, chemistry of, recent advances in, **1,** 1
Thiopyrones (monocyclic sulfur-containing pyrones), **8,** 219
Thioureas in synthesis of heterocycles, **18,** 99
Three-membered rings with two heteroatoms, **2,** 83; **24,** 63
1,3,5-, 1,3,6-, 1,3,7-, and 1,3,8-Triazanaphthalenes, **10,** 149
1,2,3-Triazines, **19,** 215
1,2,3-Triazoles, **16,** 33
1,6,6aS^{IV}-Trithiapentalenes, **13,** 161

Date Due

SEP 11 1980